Wind and Solar Power Systems

Wind and Solar Power Systems

Design, Analysis, and Operation

Mukund R. Patel and Omid Beik

CRC Press
Taylor & Francis Group
Boca Raton London New York

CRC Press is an imprint of the
Taylor & Francis Group, an **informa** business

Third edition published 2021
by CRC Press
6000 Broken Sound Parkway NW, Suite 300, Boca Raton, FL 33487-2742

and by CRC Press
2 Park Square, Milton Park, Abingdon, Oxon, OX14 4RN

© 2021 Mukund R. Patel and Omid Beik

First edition published by CRC Press 1999

CRC Press is an imprint of Taylor & Francis Group, LLC

The right of Mukund R. Patel and Omid Beik to be identified as authors of this work has been asserted by them in accordance with sections 77 and 78 of the Copyright, Designs and Patents Act 1988.

Reasonable efforts have been made to publish reliable data and information, but the author and publisher cannot assume responsibility for the validity of all materials or the consequences of their use. The authors and publishers have attempted to trace the copyright holders of all material reproduced in this publication and apologize to copyright holders if permission to publish in this form has not been obtained. If any copyright material has not been acknowledged please write and let us know so we may rectify in any future reprint.

Except as permitted under U.S. Copyright Law, no part of this book may be reprinted, reproduced, transmitted, or utilized in any form by any electronic, mechanical, or other means, now known or hereafter invented, including photocopying, microfilming, and recording, or in any information storage or retrieval system, without written permission from the publishers.

For permission to photocopy or use material electronically from this work, access www.copyright.com or contact the Copyright Clearance Center, Inc. (CCC), 222 Rosewood Drive, Danvers, MA 01923, 978-750-8400. For works that are not available on CCC please contact mpkbookspermissions@tandf.co.uk

Trademark notice: Product or corporate names may be trademarks or registered trademarks and are used only for identification and explanation without intent to infringe.

ISBN: 978-0-367-47693-9 (hbk)
ISBN: 978-0-367-71620-2 (pbk)
ISBN: 978-1-003-04295-2 (ebk)

Typeset in Times
by SPi Global, India

Dedications

*Dedicated to my parents Shakariba and Ranchhodbhai who practiced ingenuity,
and to my grandchildren Rayna, Dhruv, Naiya, Sevina, Viveka and Mira
for keeping me young
Mukund Patel*

*Dedicated to my family
Omid Beik*

Contents

Preface xiii
Acknowledgements xv
Author Biographies xvii
List of Abbreviations and Conversion of Units xix
Glossary xxi

PART A   Wind Power Systems

Chapter 1   Introduction 3

       1.1   Industry Overview 3
       1.2   History of Renewable Energy Development 4
       1.3   Utility Perspective 7
       Further Reading 8

Chapter 2   Wind Power 9

       2.1   Wind Power in the World 9
       2.2   U.S. Wind Power Development 12
       References 13

Chapter 3   Wind Speed and Energy 15

       3.1   Speed and Power Relation 15
       3.2   Power Extracted from the Wind 17
       3.3   Rotor-Swept Area 19
       3.4   Air Density 20
       3.5   Wind Speed Distribution 21
       3.6   Wind Speed Prediction 34
       References 34

Chapter 4   Wind Power Systems 37

       4.1   System Components 37
       4.2   Turbine Rating 45
       4.3   Power vs. Speed and TSR 47
       4.4   Maximum Energy Capture 49
       4.5   Maximum Power Operation 50
       4.6   System-Design Trade-offs 52
       4.7   System Control Requirements 55

Chapter 5 Electrical Generators ... 63

- 4.8 Environmental Aspects ... 57
- 4.9 Potential Catastrophes ... 60
- 4.10 System-Design Trends ... 61
- References ... 62

Chapter 5 Electrical Generators ... 63
- 5.1 Turbine Conversion Systems ... 63
- 5.2 Synchronous Generator ... 66
- 5.3 Induction Generator ... 70
- 5.4 Doubly Fed Induction Generator ... 82
- 5.5 Direct-Driven Generator ... 83
- 5.6 Unconventional Generators ... 85
- 5.7 Multiphase Generators ... 87
- References ... 89

Chapter 6 Generator Drives ... 91
- 6.1 Speed Control Regions ... 92
- 6.2 Generator Drives ... 95
- 6.3 Drive Selection ... 98
- 6.4 Cutout Speed Selection ... 99
- References ... 100

Chapter 7 Offshore Wind Farms ... 101
- 7.1 Environmental Impact ... 102
- 7.2 Ocean Water Composition ... 103
- 7.3 Wave Energy and Power ... 105
- 7.4 Ocean Structure Design ... 106
- 7.5 Corrosion ... 108
- 7.6 Foundation ... 108
- 7.7 Materials ... 111
- 7.8 Maintenance ... 112
- References ... 113

Chapter 8 AC Wind Systems ... 115
- 8.1 Overview ... 115
- 8.2 Wind Turbine and Wind Farm Components ... 118
- 8.3 System Analyses ... 120
- 8.4 Challenges ... 123
- References ... 123

Contents ix

Chapter 9 DC Wind Systems 125
    9.1   Making a Case for All-DC Wind System 125
    9.2   Overview.. 125
    9.3   All-DC System Components 128
    9.4   System Analyses 133
    9.5   Variable Voltage Collector Grid 136
    References 139

PART B Photovoltaic Power Systems

Chapter 10 Photovoltaic Power 143
    10.1   Building-Integrated PV System 145
    10.2   PV Cell Technologies 147
    References 153

Chapter 11 Photovoltaic Power Systems 155
    11.1   PV Cell 155
    11.2   Module and Array 156
    11.3   Equivalent Electrical Circuit 157
    11.4   Open-Circuit Voltage and Short-Circuit Current 159
    11.5   I-V and P-V Curves 160
    11.6   Array Design 162
    11.7   Peak-Power Operation 171
    11.8   System Components of Stand-Alone System 172
    References 174

Chapter 12 Solar Power Conversion Systems 175
    12.1   Overview... 175
    12.2   Solar Power Electronics Systems 175
    12.3   Challenges 182
    12.4   Trend and Future.. 183
    References 184

PART C System Integration

Chapter 13 Energy Storage 189

    13.1 Battery 189
    13.2 Types of Battery 191
    13.3 Equivalent Electrical Circuit 193
    13.4 Performance Characteristics 195
    13.5 More on Lead-Acid Battery 204
    13.6 Battery Design 206
    13.7 Battery Charging 207
    13.8 Charge Regulators 208
    13.9 Battery Management 209
    13.10 Flywheel 212
    13.11 Superconducting Magnet 217
    13.12 Compressed Air 220
    13.13 Technologies Compared 222
    13.14 More on Lithium-Ion Battery 223
    References 226

Chapter 14 Power Electronics 229

    14.1 Basic Switching Devices 229
    14.2 AC-DC Rectifier 232
    14.3 AC-DC Inverter 233
    14.4 IGBT/MOSFET-Based Converters 235
    14.5 Control Schemes 237
    14.6 Multilevel Converters 239
    14.7 HVDC Converters 241
    14.8 Matrix Converters 243
    14.9 Cycloconverter 244
    14.10 Grid Interface Controls 244
    14.11 Battery Charge/Discharge Converters 246
    14.12 Power Shunts 249
    References 251

Chapter 15 Stand-Alone Systems 253

    15.1 PV Stand-Alone 253
    15.2 Electric Vehicle 254
    15.3 Wind Stand-Alone 256
    15.4 Hybrid Systems 257
    15.5 System Sizing 267
    15.6 Wind Farm Sizing 271
    References 273

Contents

Chapter 16 Grid-Connected Systems 275
    16.1   Interface Requirements 276
    16.2   Synchronizing with the Grid 278
    16.3   Operating Limit 282
    16.4   Energy Storage and Load Scheduling 286
    16.5   Utility Resource Planning Tools 286
    16.6   Wind Farm–Grid Integration 287
    16.7   Grid Stability Issues 288
    16.8   Distributed Power Generation 290
    References 292

Chapter 17 Electrical Performance 295
    17.1   Voltage Current and Power Relations 295
    17.2   Component Design for Maximum Efficiency 296
    17.3   Electrical System Model 298
    17.4   Static Bus Impedance and Voltage Regulation 299
    17.5   Dynamic Bus Impedance and Ripples 300
    17.6   Harmonics 301
    17.7   Quality of Power 303
    17.8   Renewable Capacity Limit 312
    17.9   Lightning Protection 316
    References 318

Chapter 18 Plant Economy 319
    18.1   Energy Delivery Factor 319
    18.2   Initial Capital Cost 320
    18.3   Availability and Maintenance 320
    18.4   Energy Cost Estimates 323
    18.5   Sensitivity Analysis 323
    18.6   Profitability Index 326
    18.7   Project Finance 327
    References 327

Chapter 19 The Future 329
    19.1   World Electricity to 2050 329
    19.2   Future of Wind Power 331
    19.3   PV Future 334
    19.4   Declining Production Cost 335
    19.5   Market Penetration 337
    References 340

PART D Ancillary Power Technologies

Chapter 20 Solar Thermal System 343

    20.1   Energy Collection 344
    20.2   Solar-II Power Plant 345
    20.3   Synchronous Generator 347
    20.4   Commercial Power Plants 354
    20.5   Recent Trends 355
    References 355

Chapter 21 Ancillary Power Systems 357

    21.1   Heat-Induced Wind Power.. 357
    21.2   Marine Current Power 357
    21.3   Ocean Wave Power 361
    21.4   Jet-Assisted Wind Turbine 362
    21.5   Bladeless Wind Turbine 363
    21.6   Solar Thermal Microturbine 363
    21.7   Thermophotovoltaic System 365
    References 366

Index 369

Preface

The 3rd edition of this book is an expanded, revised, and updated version of the 2nd edition with new chapters such as AC wind systems, HVDC and all-DC wind systems, multiphase and DC wind turbine conversion systems, and solar power electronics for on-grid and off-grid systems. The new edition is the result of teaching the course to inquisitive students and short courses to professional engineers that enhanced the contents in many ways. The book is designed and tested to serve as textbook for a semester course for university seniors in electrical and mechanical engineering fields. The practicing engineers will get detailed treatment of this rapidly growing segment of the power industry. The government policy makers would benefit by overview of the material covered in the book. The book is divided into four parts in 21 chapters.

Part A covers the wind power technologies. It includes the engineering fundamentals, the probability distributions of the wind speed, the annual energy potential of a site, and the wind power system operation and the control requirements. Since most wind plants use induction generators for converting the turbine power into electrical power, the theory of the induction machine performance and operation is reviewed. The electrical generator speed control for capturing the maximum energy under wind fluctuations over the year is presented. The rapidly developing offshore wind farms with their engineering and operational aspects are covered in detail. Included in Part A are also new chapters on AC wind systems and DC wind systems.

Part B covers the solar photovoltaic technologies and the current developments around the world. It starts with the energy conversion characteristics of the photovoltaic cell, and then the array design, the effect of the environment variables, the sun-tracking methods for the maximum power generation, the controls, and the emerging trends are discussed. A new chapter on modern power electronics needed for solar power conversion for on-grid and off-grid systems is included in Part B.

Part C starts with large-scale energy storage technologies often required to augment non-dispatchable energy sources, such a wind and PV, to improve the availability of power to the users. It covers characteristics of various batteries, their design methods using the energy balance analysis, factors influencing their operation, and the battery management methods. The energy density and the life and operating cost per kWh delivered are presented for various batteries, such as lead-acid, nickel-cadmium, nickel-metal-hydride, and lithium-ion. The energy storage by the flywheel, compressed air, the superconducting coil, and their advantages over the batteries are reviewed. The basic theory and operation of the power electronic converters and inverters used in the wind and solar power systems are then presented. The grid-connected renewable power systems are covered with voltage and frequency control methods needed for synchronizing the generator with the grid. The theory and the operating characteristics of the interconnecting transmission line, the voltage regulation, the maximum power transfer capability, and the static and dynamic stability are covered. About two billion people in the world not yet connected to the utility grid are the largest potential market of stand-alone power systems using wind and photovoltaic systems in hybrid with diesel generators or fuel cells which are also covered.

Part C continues with the overall electrical system performance, the method of designing system components to operate at their maximum possible efficiency, the static and dynamic bus performance, the harmonics, and the increasingly important quality of power issues applicable to the renewable power systems. The total plant economy and the costing of energy delivered to the paying customers are presented. It also shows the importance of a sensitivity analysis to raise confidence level of the investors. The past and present trends of the wind and PV power, the declining price model based on the learning curve, and the Fisher-Pry substitution model for predicting the future market growth of the wind and PV power based on historical data on similar technologies are presented.

Part D covers the ancillary power system derived from the sun, the ultimate source of energy on the earth. It starts with the utility-scale solar thermal power plant using concentrating heliostats and molten salt steam turbine. It then covers the solar-induced wind power, the marine current power, and the ocean wave power.

List of acronyms and conversion of units are given at the Starting of the book.

Acknowledgements

Any book of this nature on emerging technologies such as wind and photovoltaic power systems cannot possibly be written without help from many sources. We have been extremely fortunate to receive support from many organizations and individuals. They not only encouraged us to write on this timely subject but also provided valuable suggestions and comments during the development of the book.

The 3rd edition of this book is the extension of the successful 1st and 2nd editions. Therefore, our gratitude remains to Dr. Nazmi Shehadeh, Head of the EE Department at the University of Minnesota, Duluth, who gave Dr. Patel an opportunity to develop and teach this course as a Visiting Professor. Dr. Elliott Bayly of the World Power Technologies in Duluth shared with us his long experience in pioneering wind power systems in the USA. For the 2nd edition, we remain grateful to Prof. Jose Femenia, Head of Marine Engineering at the U.S. Merchant Marine Academy, Kings Point, New York, for supporting our research and publications in this field. For the 3rd edition, as was for the 1st and 2nd editions, several institutions worldwide provided current data and reports on these rather rapidly developing technologies. They are the American Wind Energy Association, the Canadian Wind Energy Association, the American Solar Energy Association, the European Wind Energy Association, the National Renewable Energy Laboratory, the Riso National Laboratory in Denmark, the Tata Energy Research Institute in India, the California Energy Commission, and many corporations engaged in the wind and solar power technologies.

We wholeheartedly acknowledge the valuable support that we have received from all during the course of preparing the book.

Mukund R. Patel. Ph.D., P.E.
Yardley, Pennsylvania

Omid Beik, Ph.D., SMIEEE
Toronto, Canada

Author Biographies

Mukund R. Patel, PhD, PE, has served as a Professor at the U.S. Merchant Marine Academy in Kings Point, NY; Principal Engineer at General Electric Space Division in Valley Forge, PA; Fellow Engineer at Westinghouse Research Center in Pittsburgh, PA; Senior Staff Engineer at Lockheed Martin in Princeton, NJ; Development Manager at Bharat Bijlee (Siemens) in Mumbai, India; and 3M McKnight Distinguished Visiting Professor at the University of Minnesota, Duluth. He has over 50 years of internationally recognized experience in research, development, design, and education of the state-of-the-art electrical power equipment and systems.

Dr. Patel obtained his Ph.D. degree in Electrical Power from the Rensselaer Polytechnic Institute, Troy, NY; M.S. in Engineering Management from the University of Pittsburgh; M.E. in Electrical Machine Design with Distinction from Gujarat University and B.E.E. with Distinction from Sardar University, India. He is a Fellow of the Institution of Mechanical Engineers (UK), Associate Fellow of the American Institute of Aeronautics and Astronautics, Senior Life Member of IEEE, Registered Professional Engineer in Pennsylvania, Chartered Mechanical Engineer in the UK, and a member of Eta Kappa Nu, Tau Beta Pi, Sigma Xi, and Omega Rho.

Dr. Patel is an Associate Editor of Solar Energy, the journal of the International Solar Energy Society. He has presented and published over 50 research papers at national and international conferences and journals, holds several patents, has earned NASA recognition for exceptional contribution to the power system design for the UAR Satellite, and was nominated by NASA for an IR-100 award. He has authored five text books published by CRC Press, and major chapters on electrical power in handbooks such as the International Handbook on Space Technologies published by Praxis and Springer, Energy Storage in Technology, Humans, and Society Towards a Sustainable World, and Solar Power in Marine Engineering Handbook published by SNAME.

Omid Beik, Ph.D., SMIEEE (S'14–M'16–SM'20) received the B.Sc. degree (Hons.) with highest distinction in electrical engineering from Yazd University, Yazd, Iran, in 2007, the M.Sc. degree with highest distinction in electrical engineering from Shahid Beheshti University, Abbaspour School of Engineering, Tehran, Iran, in 2009, and the Ph.D. degree in electrical engineering from McMaster University, Hamilton, ON, Canada, in 2016. He was a Postgraduate Researcher with the Power Conversion Group, University of Manchester, U.K. (2011–2012) and a Postdoctoral

Research Fellow at McMaster University, Hamilton, ON, Canada (2016–2017). Dr. Beik is author of the book '*DC Wind Generation Systems: Design, Analysis, and Multiphase Turbine Technology*', and has authored/co-authored a diverse portfolio of peer-reviewed journal, conference and magazine papers, and patent applications. Dr. Beik is currently an Adjunct Faculty member at the Department of Electrical and Computer Engineering at McMaster University, he serves as an Associate Editor for the IEEE Transactions on Transportation Electrification, IEEE Transactions on Energy Conversion, and IEEE Electrification eNewesletter. He was the chair of industry relations for IEEE Hamilton Section, and has served as session chair, panel moderator and organizer for IEEE conferences. Dr. Beik has received several awards including three International Excellence Awards from McMaster University. He is a currently a Senior Member of the IEEE.

List of Abbreviations and Conversion of Units

AC	Alternating Current
ASES	American Solar Energy Society
AWEA	American Wind Energy Association
Ah	Ampere-hour of the battery capacity
BIPV	Building-integrated photovoltaics
BWEA	British Wind Energy Association
Cp	Rotor energy conversion efficiency
C/D	Charge/discharge of the battery
CTP	Constant power
DC	Direct current
DOD	Depth of discharge of the battery
DOE	Department of Energy
DWIA	Danish Wind Industry Association
ECU	European currency unit
EDF	Energy delivery factor
EPRI	Electric Power Research Institute
EWEA	European Wind Energy Association
GW	Gigawatts (10^9 watts)
GWh	Gigawatthours
HV	High voltage
HVDC	High-voltage direct current (transmission)
IEA	International Energy Agency
IEC	International Electrotechnical Commission
ISES	International Solar Energy Society
kW	Kilowatts
kWh	Kilowatthours
LV	Low voltage
LVRT	Low voltage ride-through
MVDC	Medium voltage DC
MVA	Mega volt amperes
MPPT	Maximum power point tracking
MW	Megawatts
MWa	Megawatts accumulated
MWe	Megawatts electric
MWh	Megawatthours
NEC	National Electrical Code®
NOAA	National Oceanic and Atmospheric Administration
NREL	National Renewable Energy Laboratory
NWTC	National Wind Technology Center

PM	Permanent Magnet
PV	Photovoltaic
PWM	Pulse width modulation
QF	Qualifying facility
RPM	Revolutions per minute
SiC	Silicon Carbide
SOC	State of charge of the battery
SRC	Specific rated capacity (kW/m^2)
THD	Total harmonic distortion in quality of power
THM	Top head mass (nacelle + rotor)
TPV	Thermophotovoltaic
TSR	Tip speed ratio (of rotor)
TWh	Trillion (10^{12}) watthours
UCE	Unit cost of energy
VSC	Voltage source converter
WF	Wound field
Wp	Watt peak

PREFIXES

k	kilo (10^3)
M	Mega (10^6)
G	Giga (10^9)
T	Trillion (10^{12})

Glossary

CONVERSION OF UNITS

The information contained in the book comes from many sources and many countries using different units in their reports. The data are kept in the form they were received by the authors and were not converted to a common system of units. The following is the conversion table for the most commonly used units in the book:

To Change From	Into	Multiply By
mph	m/sec	0.447
kn	m/sec	0.514
mi	km	1.609
ft	m	0.3048
in.	cm	2.540
lb	kg	0.4535
Btu	W·sec	1054.4
Btu	kW·h	2.93×10^{-4}
Btu/ft^2	W·sec/m^2	11357
Btu/h	W	0.293
Btu/ft^2/h	kW/m^2	3.15×10^{-7}
Btu/h/ft^2/°F	W/m^2/°C	5.678
hp	W	746
gallon of oil (U.S)	kWh	42
gal (U.S.)	liter	3.785
barrel of oil	Btu	6×10^6
barrel	gal (U.S.)	42

Part A

Wind Power Systems

1 Introduction

1.1 INDUSTRY OVERVIEW

Our world has been powered primarily by carbon fuels for more than two centuries, with some demand met by nuclear power plants over the last few decades. The increasing environmental concerns in recent years about global warming and the harmful effects of carbon emissions have created a new demand for clean and sustainable energy sources, such as wind, sun, sea, hydropower, and geothermal power. Among these, wind and solar power have experienced remarkably rapid growth in the past decade. Both are pollution-free sources of abundant power. Additionally, they generate power near load centers; hence, they eliminate the need of running high-voltage transmission lines through rural and urban landscapes. Deregulation, privatization, and consumer preferences for green power in many countries are expanding the wind and photovoltaic (PV) energy markets at an increasing pace.

The new capacity installation decisions for power plants of any kind today are becoming complicated in many parts of the world because of the difficulty in finding sites for new generation and transmission facilities. Given the potential for cost overruns, safety-related design changes during the construction and operation, and local opposition to new plants for safety and environmental concerns, most utility executives have been reluctant to plan on new nuclear power plants during the last three decades. Globally, the growth, decline, and stagnation of nuclear power plants over the last five decades is shown in Figure 1.1. At present, the nuclear power is going through stagnation all over the world and is declining in Western Europe. If no new nuclear plants are built and the existing plants are not relicensed at the expiration of their 40-year terms, the nuclear power output worldwide is expected to decline. This decline must be replaced by other means.

Alternatives to nuclear and fossil fuel power are renewable energy technologies (hydroelectric, in addition to those previously mentioned). Large-scale hydroelectric projects have become increasingly difficult to carry through in recent years because of the competing use of land and water. Relicensing requirements of existing hydroelectric plants may even lead to removal of some dams to protect or restore wildlife habitats. Among the other renewable power sources, wind and solar have recently experienced rapid growth around the world. Having wide geographical spread, they can be generated near the load centers, thus simultaneously eliminating the need for high-voltage transmission lines running through rural and urban landscapes.

The present status and benefits of renewable power sources are compared with conventional ones in Table 1.1 and Table 1.2, respectively. The renewables compare well with the conventional power in economy.

Global Electrical Energy Production TWh per Year

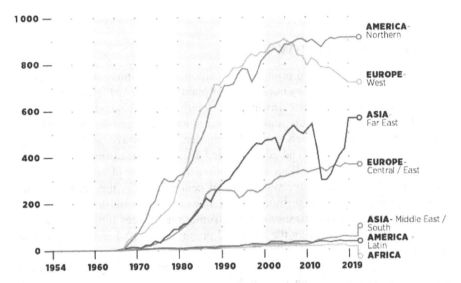

FIGURE 1.1 Growth, stagnation, and decline of nuclear power. (Source: International Atomic Energy Agency)

TABLE 1.1
Status of Conventional and Renewable Power Sources

Conventional	Renewable
Coal, nuclear, oil, and natural gas	Wind, solar, biomass, geothermal, and ocean
Fully matured technologies	Rapidly developing technologies
Numerous tax and investment subsidies embedded in national economies	Some tax credits and grants available from federal and some state governments
Accepted in society under a "grandfather clause" as a necessary evil	Being accepted on their own merit, even with the limited valuation of their environmental and other social benefits

1.2 HISTORY OF RENEWABLE ENERGY DEVELOPMENT

A great deal of renewable energy development in the U.S. occurred in the 1980s, and the prime stimulus was passage in 1978 of the Public Utility Regulatory Policies Act (PURPA). It created a class of nonutility power generators known as the *qualified facilities* (QFs). The QFs were defined to be small power generators utilizing renewable energy sources or cogeneration systems utilizing waste energy. For the first time, PURPA required electric utilities to interconnect with QFs and to purchase QFs' power generation at "avoided cost," which the utility would have incurred by generating that power by itself. PURPA also exempted QFs from certain federal and state utility regulations. Furthermore, significant federal investment tax credit, research and development tax credit, and energy tax credit—liberally available up to the mid-1980s—created a

TABLE 1.2
Benefits of Using Renewable Electricity

Traditional Benefits	Nontraditional Benefits per Million kWh Consumed
Monetary value of kWh consumed U.S. average 12 cents/kWh U.K. average 7.5 pence/kWh	Reduction in emission 750–100 t of CO_2 7.5–10 t of SO_2 3–5 t of NOx 50,000 kWh reduction in energy loss in power lines and equipment Life extension of utility power distribution equipment Lower capital cost as lower-capacity equipment can be used (such as transformer capacity reduction of 50 kW per MW installed)

wind energy rush in California, the state that also gave liberal state tax incentives. As of now, the financial incentives in the U.S. are reduced but are still available.

Many energy scientists and economists believe that the renewables would get many more federal and state incentives if their social benefits were given full credit. For example, the value of not generating 1 t of CO_2, SO_2, and NOx, and the value of not building long high-voltage transmission lines through rural and urban areas are not adequately reflected in the present evaluation of the renewables. If the renewables get due credit for pollution elimination of 600 t of CO_2 per million kWh of electricity consumed, they would get a further boost with greater incentives than those presently offered by the U.S. government.

For the U.S. wind and solar industries, there is additional competition in the international market. Other governments support green-power industries with well-funded research, low-cost loans, favorable tax-rate tariffs, and guaranteed prices not generally available to their U.S. counterparts. Under such incentives, the growth rate of wind power in Germany and India has been phenomenal over the last decade.

In Canada, the federal and provincial governments offer different programs to incentivize the renewable energy. To promote use of renewable power the Ontario provincial government introduced a Feed-In Tariff (FIT) program in 2009. The FIT program, managed by Independent Electricity System Operator (IESO), includes incentives for energy production from wind, solar, biomass, biogas, landfill gas, and hydroelectricity within the province of Ontario. The participants of the FIT program can be homeowners, communities, municipalities, Aboriginal communities, business owners, and private developers. Through FIT the Ontario government procures renewable energy from plants that have a capacity of up to and including 500 kilowatts (kW). For smaller producers, the Ontario offers a microFIT program, where it procures from plants with a capacity of 10 kW and smaller. Table 1.3 presents prices of different types of electricity generation in the FIT program.

From Table 1.3, it is seen that the price of the renewable energies through the FIT program is reduced from 2016 to 2017. This is due to the cost of technology, installation, and operation continuing to decline over time. Figure 1.2 shows the price of installation of commercial solar PV system in U.S. dollars published by IESO, Canada.

TABLE 1.3
Ontario FIT Program Prices for the Year 2017

Type of Energy	Size	Pricing 2016 (¢/kWh)	Pricing 2017 (¢/kWh)
Solar—Rooftop	≤ 6 kW	31.3	31.1
	> 6 kW ≤ 10 kW	29.4	28.8
	> 10 kW ≤ 100 kW	24.2	22.3
	> 100 kW ≤ 500 kW	22.5	20.7
Solar—Non-rooftop	≤ 10 kW	21.4	21.0
	> 10 kW ≤ 500 kW	20.9	19.2
On-farm Biomass	≤ 100 kW	26.3	25.8
	> 100 kW ≤ 250 kW	20.4	20.0
Biogas	> 250 kW ≤ 500 kW	16.8	16.5
Renewable Biomass	≤ 500 kW	17.5	17.2
Landfill gas	≤ 500 kW	17.1	16.8
On-shore wind	≤ 500 kW	12.8	12.5
Waterpower	≤ 500 kW	24.6	24.1

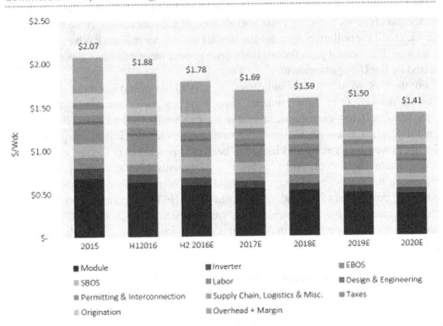

H1 2016: First half of 2016
H2 2016E: Second half of 2016 (estimated)
SBOS: Structural Balance of System
EBOS: Electrical Balance of System

FIGURE 1.2 Cost of installation of solar power modules. (From Independent Electricity System Operator (IESO), Canada).

1.3 UTILITY PERSPECTIVE

Until the late 1980s, interest in the renewables was confined primarily to private investors. However, as the considerations of fuel diversity, the environment, and market uncertainties are becoming important factors in today's electric utility resource planning, renewable energy technologies are beginning to find their place in the utility resource portfolio. Wind and solar power, in particular, have the following advantages over power supplied by electric utilities:

- Both are highly modular, in that their capacity can be increased incrementally to match gradual load growth.
- Their construction lead time is significantly shorter than that of conventional plants, thus reducing financial and regulatory risks.
- They use diverse fuel sources that are free of both cost and pollution.

Because of these benefits, many utilities and regulatory bodies have become increasingly interested in acquiring hands-on experience with renewable energy technologies in order to plan effectively for the future. These benefits are discussed in the following text in further detail.

1.3.1 MODULARITY FOR GROWTH

Both wind and solar PV power are highly modular. They allow installations in stages as needed without losing economy of size in the first installation. PV power is even more modular than wind power. It can be sized to any capacity, as solar arrays are priced directly by the peak generating capacity in watts and indirectly by square feet. Wind power is modular within the granularity of turbine size. Standard wind turbines come in different sizes ranging from several kW to several MW. For utility-scale installations, standard wind turbines in the recent past had been a few MW, and are moving into several MW ranges. Currently, the available wind turbine capacity exceeds 10 MW with a blade diameter of over 165 m. Prototypes of even larger wind turbines have been tested and made commercially available in Europe and Americas. For example, the largest wind turbine now available from GE, Haliade-X-12-MW, has each blade 107 m long (longer than a football field), 220 m rotor diameter, and electrical output rating of 12 MW.

For small grids, modularity of PV and wind systems is even more important. Slowly increasing demand, as is the case now, may be more economically met by adding small increments of green-power capacity. A wind farm may begin with the required number and size of wind turbines for the initial needs, adding more towers with no loss of economy when the plant needs to grow. Expanding or building a new conventional power plant in such cases may be neither economical nor free from market risks. Even when a small grid is linked by a transmission line to the main network, installing a wind or PV plant to serve growing demand may be preferable to laying another transmission line. Local renewable power plants can also benefit small power systems by moving generation near the load, thus reducing the voltage drop at the end of a long, overloaded line.

FURTHER READING

1. U.S. Department of Energy, International Energy Outlook 2004 with Projections to 2020, DOE Office of Integrated Analysis and Forecasting, April 2004.
2. Felix, F., State of the Nuclear Economy, *IEEE Spectrum*, November 1997, pp. 29–32.
3. Rahman, S., Green Power, *IEEE Power and Energy*, January–February 2003, pp. 30–37.
4. Independent Electricity System Operator (IESO) Report, 2017 FIT Price Review, August 2016, pp. 1–22.
5. Omid Beik, An HVDC Off-shore Wind Generation Scheme with High Voltage Hybrid Generator, Ph.D. thesis, McMaster University, 2016.
6. Omid Beik, Ahmad S. Al-Adsani, *DC Wind Generation Systems, Design, Analysis, and Multiphase Turbine Technology*, Springer, 2020.
7. Omid Beik, Ahmad S. Al-Adsani, *Wind Energy Systems*, pp. 1–9, Springer, 2020.
8. Omid Beik, Ahmad S. Al-Adsani, *DC Wind Generation System*, pp. 33–69, Springer, 2020.

2 Wind Power

In historical review, the first windmill to generate electricity in the rural U.S. was installed in 1890. An experimental grid-connected turbine with as large a capacity as 2 MW was installed in 1979 on Howard Knob Mountain near Boone, NC, and a 3-MW turbine was installed in 1988 on Berger Hill in Orkney, Scotland. Today, even larger wind turbines are routinely installed, commercially competing with electric utilities in supplying economical, clean power in many parts of the world.

The average turbine size of wind installations was 300 kW until the early 1990s. New machines being installed are in the 1 MW to 10 MW capacity range. Wind turbines over 10-MW capacity have been fully developed, installed, and in operation in the U.S. and most other countries. Figure 2.1 is a conceptual layout of a modern multi-megawatt wind tower suitable for utility-scale applications.[1]

Improved turbine designs and plant utilization have contributed to a decline in large-scale wind energy generation costs. The wind energy has become the least expensive new source of electric power in the world, less expensive than coal, oil, nuclear, and most natural-gas-fired plants, competing with these traditional sources on its own economic merit. Hence, it has become economically attractive to utilities and electric cooperatives.

Major factors that have accelerated the development of wind power technology are as follows:

- High-strength fiber composites for constructing large, low-cost blades
- Falling prices of the power electronics associated with wind power systems
- Variable-speed operation of electrical generators to capture maximum energy
- Improved plant operation, pushing the availability up to 95%
- Economies of scale as the turbines and plants are getting larger in size
- Accumulated field experience (the learning-curve effect) improving the capacity factor over 50 %

2.1 WIND POWER IN THE WORLD

Because wind energy has become the least expensive source of new electric power that is also compatible with environment preservation programs, many countries promote wind power technology by means of national programs and market incentives.

The international renewable energy agency (IRENA) reported that the global installed wind-generation capacity onshore and offshore has increased by a factor of almost 75 in the past two decades, jumping from 7.5 gigawatts (GW) in 1997 to some 564 GW by 2018. Production of wind electricity doubled between 2009 and 2013, and in 2016 wind energy accounted for 16% of the electricity generated by

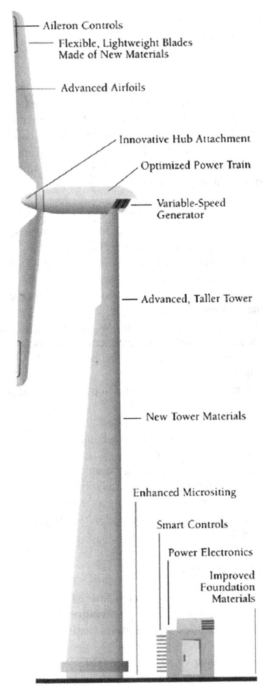

An artist's concept of a next-generation utility wind turbine is shown here.

FIGURE 2.1 Modern wind turbine for utility-scale power generation.

renewables. Many parts of the world have strong wind speeds, but the best locations for generating wind power are sometimes remote ones. Offshore and far-shore wind power offers tremendous potential.

Figure 2.2 shows the world total installed wind power capacity from 2010 to 2019 in MW per year: rising from 180,000 MW in 2010 to over 600,000 MW in 2019 amount to an average annual growth rate of 13 %.

Since 2010 more than half of all new wind power was added outside Europe and North America, mainly in China and India. China has an installed capacity of 221 GW. It has the world's largest onshore wind farm with a capacity of 7,965 megawatt (MW), which is five times larger than its nearest rival. The US comes second with 96.4 GW of installed capacity. The country has six of the 10 largest onshore wind farms. These include the Alta Wind Energy Centre in California, the world's second-largest onshore wind farm with a capacity of 1,548 MW. With 59.3 GW, Germany has the highest installed wind capacity in Europe. Its largest offshore wind farms are the Gode Windfarms, which have a combined capacity of 582 MW. India has the second-highest wind capacity in Asia, with a total capacity of 35 GW. The country has the third- and fourth-largest onshore wind farms in the world, namely the 1,500-MW Muppandal wind farm in Tamil Nadu and the 1,064-MW Jaisalmer Wind Park in Rajasthan.

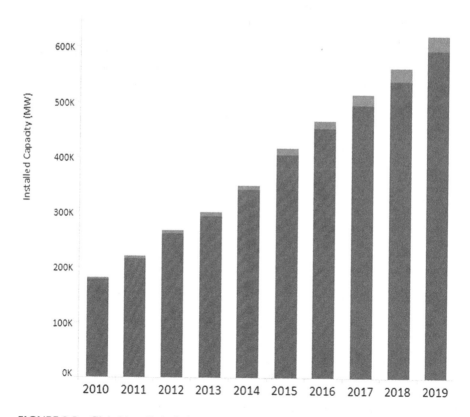

FIGURE 2.2 Global installed wind capacity. (From IRENA)

Much of the new wind power development around the world can be attributed to government policies to promote renewable energy sources. The renewable energy production and investment tax credits in the USA, the non-fossil fuel obligation of the U.K., Canada's wind power production incentive, and India's various tax rebates and exemptions are examples of such programs.

2.2 U.S. WIND POWER DEVELOPMENT

U.S. Energy Information Administration (EIA) reported that the total annual Electricity generation from wind electricity generation in the United States increased from about 6 billion kilowatthours (kWh) in 2000 to about 300 billion kWh in 2019. In 2019, wind turbines in the United States were the source of about 7.3% of total U.S. utility-scale electricity generation. Utility-scale includes facilities with at least one megawatt (1,000 kilowatts) of electricity generation capacity.

Figure 2.3 plots the U.S. wind electricity generation and its share compared to the total U.S. electricity generation from 1990 to 2019. In the 1990s the wind had less than 0.5% of total electricity generation. This has significantly increased to over 7% in 2019. It is expected that over the next decade this will significantly increase.

Figure 2.4 shows U.S. wind power versus hydroelectric. In 2019, U.S. annual wind generation exceeded hydroelectric generation for the first time, according to the U.S. Energy Information Administration's Electric Power Monthly. Wind is now the top renewable source of electricity generation in the country, a position previously held by hydroelectricity.

Technology development and the resulting price decline have caught the interest of a vast number of electric utilities that are now actively developing wind energy as one element of the balanced resource mix.[4] Projects are being built in most U.S. cities.

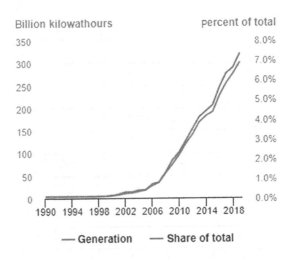

FIGURE 2.3 Wind generation and share of total U.S. electricity generation. (From EIA)

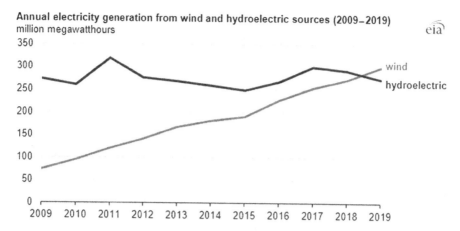

FIGURE 2.4 U.S. wind versus hydroelectric. (From EIA)

REFERENCES

1. U.S. Department of Energy, Wind Energy Programs Overview, NREL Report No. DE-95000288, March 1995.
2. International Energy Agency, Wind Energy Annual Report, International Energy Agency Report by NREL, March 1995.
3. Gipe, P., The BTM Wind Report: World Market Update, *Renewable Energy World*, July–August 2003, pp. 66–83.
4. Utility Wind Interest Group, Utilities Move Wind Technology Across America, November 1995.
5. Anson, S., Sinclair, K., and Swezey, B., Profiles in Renewables Energy, Case Studies of Successful Utility-Sector Projects, DOE/NREL Report No. DE-930000081, National Renewable Energy Laboratory, Golden, Colorado, August 1994.
6. Chabot, B. and Saulnier, B., Fair and Efficient Rates for Large-Scale Development of Wind Power: the New French Solution, Canadian Wind Energy Association Conference, Ottawa, October 2001.
7. Belhomme, R., Wind Power Developments in France, *IEEE Power Engineering Review*, October 2002, pp. 21–24.
8. Pane, E. D., Wind Power Developments in France, *IEEE Power Engineering Review*, October 2002, pp. 25–28.
9. Gupta, A. K., *Power Generation from Renewables in India*, Ministry of Non-Conventional Energy Sources, New Delhi, India, 1997.
10. Hammons, T. J., Ramakumar, R., Fraser, M., Conners, S. R., Davies, M., Holt, E. A., Ellis, M., Boyers, J., and Markard, J., Renewable Energy Technology Alternatives for Developing Countries, *IEEE Power Engineering Review*, December 1997, pp. 10–21.

3 Wind Speed and Energy

The wind turbine captures the wind's kinetic energy in a rotor consisting of two or more blades mechanically coupled to an electrical generator. The turbine is mounted on a tall tower to enhance the energy capture. Numerous wind turbines are installed at one site to build a wind farm of the desired power generation capacity. Obviously, sites with steady high wind produce more energy over the year.

Two distinctly different configurations are available for turbine design: the horizontal-axis configuration (Figure 3.1) and the vertical-axis configuration (Figure 3.2). The horizontal-axis machine has been the standard in Denmark from the beginning of the wind power industry. Therefore, it is often called the *Danish wind turbine*. The vertical-axis machine has the shape of an egg beater and is often called the *Darrieus rotor* after its inventor. It has been used in the past because of its specific structural advantage. However, most modern wind turbines use a horizontal-axis design. Except for the rotor, most other components are the same in both designs, with some differences in their placements.

3.1 SPEED AND POWER RELATIONS

The kinetic energy in air of mass m moving with speed V is given by the following in joules:

$$\text{kinetic energy} = \frac{1}{2}mV^2 \quad (3.1)$$

The power in moving air is the flow rate of kinetic energy per second in watts:

$$\text{power} = \frac{1}{2}(\text{mass flow per second})V^2 \quad (3.2)$$

If
  P = mechanical power in the moving air (watts),
  ρ = air density (kg/m³),
  A = area swept by the rotor blades (m²), and
  V = velocity of the air (m/sec),

then the volumetric flow rate is AV, the mass flow rate of the air in kilograms per second is ρAV, and the mechanical power coming in the upstream wind is given by the following in watts:

$$P = \frac{1}{2}(\rho AV)V^2 = \frac{1}{2}\rho AV^3 \quad (3.3)$$

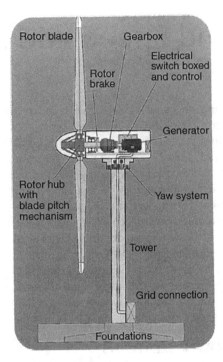

FIGURE 3.1 Horizontal-axis wind turbine showing major components. (Courtesy: Energy Technology Support Unit, DTI, U.K. with permission)

FIGURE 3.2 Vertical-axis 33-m-diameter wind turbine built and tested by DOE/Sandia National Laboratory during 1994 in Bushland, TX.

Wind Speed and Energy

Two potential wind sites are compared in terms of the specific wind power expressed in watts per square meter of area swept by the rotating blades. It is also referred to as the power density of the site, and is given by the following expression in watts per square meter of the rotor-swept area:

$$\text{specific power of the site} = \frac{1}{2}\rho V^3 \tag{3.4}$$

This is the power in the upstream wind. It varies linearly with the density of the air sweeping the blades and with the cube of the wind speed. The blades cannot extract all of the upstream wind power, as some power is left in the downstream air that continues to move with reduced speed.

3.2 POWER EXTRACTED FROM THE WIND

The actual power extracted by the rotor blades is the difference between the upstream and downstream wind powers. Using Equation 3.2, this is given by the following equation in units of watts:

$$P_o = \frac{1}{2}(\text{mass flow per second})\{V^2 - V_o^2\} \tag{3.5}$$

where

P_o = mechanical power extracted by the rotor, i.e., the turbine output power,
V = upstream wind velocity at the entrance of the rotor blades, and
V_o = downstream wind velocity at the exit of the rotor blades.

Let us leave the aerodynamics of the blades to the many excellent books available on the subject, and take a macroscopic view of the airflow around the blades. Macroscopically, the air velocity is discontinuous from V to V_o at the "plane" of the rotor blades, with an "average" of $\frac{1}{2}(V + V_o)$. Multiplying the air density by the average velocity, therefore, gives the mass flow rate of air through the rotating blades, which is as follows:

$$\text{mass flow rate} = \rho A \frac{V + V_o}{2} \tag{3.6}$$

The mechanical power extracted by the rotor, which drives the electrical generator, is therefore:

$$P_o = \frac{1}{2}\left[\rho A \frac{(V + V_o)}{2}\right](V^2 - V_o^2) \tag{3.7}$$

The preceding expression is algebraically rearranged in the following form:

$$P_o = \frac{1}{2}\rho A V^3 \frac{\left(1 + \frac{V_o}{V}\right)\left[1 - \left(\frac{V_o}{V}\right)^2\right]}{2} \tag{3.8}$$

The power extracted by the blades is customarily expressed as a fraction of the upstream wind power in watts as follows:

$$P_o = \frac{1}{2}\rho A V^3 C_p \tag{3.9}$$

where

$$C_p = \frac{\left(1+\frac{V_o}{V}\right)\left[1-\left(\frac{V_o}{V}\right)^2\right]}{2} \tag{3.10}$$

Comparing Equation 3.3 and Equation 3.9, we can say that C_p is the fraction of the upstream wind power that is extracted by the rotor blades and fed to the electrical generator. The remaining power is dissipated in the downstream wind. The factor C_p is called the *power coefficient* of the rotor or the *rotor efficiency*.

For a given upstream wind speed, Equation 3.10 clearly shows that the value of C_p depends on the ratio of the downstream to the upstream wind speeds (V_o/V). A plot of power vs. (V_o/V) shows that C_p is a single-maximum-value function (Figure 3.3). It has the maximum value of 0.59 when the V_o/V ratio is one third. The maximum power is extracted from the wind at that speed ratio, i.e., when the downstream wind speed equals one third of the upstream speed. Under this condition (in watts):

$$P_{max} = \frac{1}{2}\rho A V^3 \times 0.59 \tag{3.11}$$

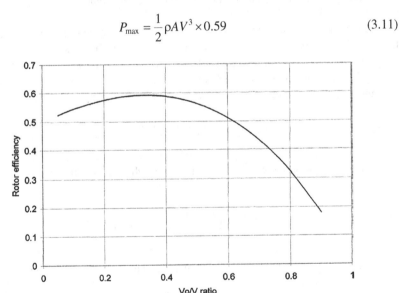

FIGURE 3.3 Rotor efficiency vs. V_o/V ratio has a single maximum. Rotor efficiency is the fraction of available wind power extracted by the rotor and fed to the electrical generator.

Wind Speed and Energy

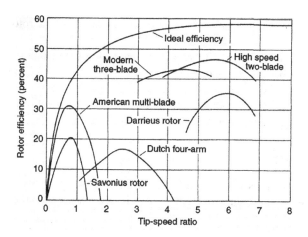

FIGURE 3.4 Rotor efficiency vs. V_o/V ratio for rotors with different numbers of blades. Two-blade rotors have the highest efficiency.

The theoretical maximum value of C_p is 0.59. C_p is often expressed as a function of the rotor tip-speed ratio (TSR) as shown in Figure 3.4. TSR is defined as the linear speed of the rotor's outermost tip to the upstream wind speed. The aerodynamic analysis of the wind flow around the moving blade with a given pitch angle establishes the relation between the rotor tip speed and the wind speed. In practical designs, the maximum achievable C_p ranges between 0.4 and 0.5 for modern high-speed two-blade turbines, and between 0.2 and 0.4 for slow-speed turbines with more blades. If we take 0.5 as the practical maximum rotor efficiency, the maximum power output of the wind turbine becomes a simple expression (in watts per square meter of swept area):

$$P_{max} = \frac{1}{4}\rho V^3 \qquad (3.12)$$

3.3 ROTOR-SWEPT AREA

As seen in the preceding power equation, the output power of the wind turbine varies linearly with the rotor-swept area. For the horizontal-axis turbine, the rotor-swept area is:

$$A = \frac{\pi}{4}D^2 \qquad (3.13)$$

where D is the rotor diameter.

For the Darrieus vertical-axis machine, the determination of the swept area is complex as it involves elliptical integrals. However, approximating the blade shape as a parabola leads to the following simple expression for the swept area:

$$A = \frac{2}{3}(\text{maximum rotor width at the center})(\text{height of the rotor}) \quad (3.14)$$

The wind turbine efficiently intercepts the wind energy flowing through the entire swept area even though it has only two or three thin blades with solidity between 5 and 10%. The solidity is defined as the ratio of the solid area to the swept area of the blades. The modern two-blade turbine has a low solidity ratio. It is more cost-effective as it requires less blade material to sweep large areas.

3.4 AIR DENSITY

Wind power also varies linearly with the air density sweeping the blades. The air density ρ varies with pressure and temperature in accordance with the gas law:

$$\rho = \frac{p}{RT} \quad (3.15)$$

where
p = air pressure,
T = temperature on the absolute scale, and
R = gas constant.

The air density at sea level at 1 atm (14.7 psi) and 60°F is 1.225 kg/m³. Using this as a reference, ρ is corrected for the site-specific temperature and pressure. The temperature and the pressure both vary with the altitude. Their combined effect on the air density is given by the following equation, which is valid up to 6,000 m (20,000 ft) of site elevation above sea level:

$$\rho = \rho_o e^{-\left(\frac{0.297 H_m}{3048}\right)} \quad (3.16)$$

where H_m is the site elevation in meters.
Equation 3.16 is often written in a simple form:

$$\rho = \rho_o - \left(1.194 \times 10^{-4} H_m\right) \quad (3.17)$$

The air density correction at high elevations can be significant. For example, the air density at 2000-m elevation would be 0.986 kg/m³, 20% lower than the 1.225 kg/m³ value at sea level.
For ready reference, the temperature varies with the elevation as follows in °C:

$$T = 15.5 - \frac{19.83 H_m}{3048} \quad (3.18)$$

3.5 WIND SPEED DISTRIBUTION

Having a cubic relation with power, wind speed is the most critical data needed to appraise the power potential of a candidate site. The wind is never steady at any site. It is influenced by the weather system, the local land terrain, and its height above the ground surface. Wind speed varies by the minute, hour, day, season, and even by the year. Therefore, the annual mean speed needs to be averaged over 10 years or more. Such a long-term average gives a greater confidence in assessing the energy-capture potential of a site. However, long-term measurements are expensive and most projects cannot wait that long. In such situations, the short-term data, for example, over 1 yr, is compared with long-term data from a nearby site to predict the long-term annual wind speed at the site under consideration. This is known as the *measure, correlate, and predict* (mcp) technique.

Because wind is driven by the sun and the seasons, the wind pattern generally repeats over a period of 1 yr. The wind site is usually described by the speed data averaged over calendar months. Sometimes, the monthly data is aggregated over the year for brevity in reporting the overall "windiness" of various sites. Wind speed variations over the period can be described by a probability distribution function.

3.5.1 WEIBULL PROBABILITY DISTRIBUTION

The variation in wind speed is best described by the Weibull probability distribution function h with two parameters, the shape parameter k, and the scale parameter c. The probability of wind speed being v during any time interval is given by the following:

$$h(v) = \left(\frac{k}{c}\right)\left(\frac{v}{c}\right)^{(k-1)} e^{-\left(\frac{v}{c}\right)^k} \quad \text{for } 0 < v < \infty \qquad (3.19)$$

In the probability distribution chart, h over a chosen time period is plotted against v, where h is defined as follows:

$$h = \frac{\text{fraction of time wind speed is between } v \text{ and } (v + \Delta v)}{\Delta v} \qquad (3.20)$$

By definition of the probability function, the probability that the wind speed will be between zero and infinity during the entire chosen time period is unity, i.e.:

$$\int_0^\infty h \, dv = 1 \qquad (3.21)$$

Because we often choose a time period of 1 yr, we express the probability function in terms of the number of hours in the year, such that:

$$h = \frac{\text{number of hours per year the wind is between } v \text{ and } (v + \Delta v)}{\Delta v} \qquad (3.22)$$

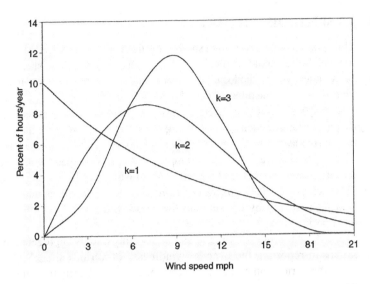

FIGURE 3.5 Weibull probability distribution function with scale parameter $c = 10$ and shape parameters $k = 1, 2,$ and 3.

The unit of h is hours per year per meter per second, and the integral (Equation 3.21) now becomes 8760 (the total number of hours in the year) instead of unity.

Figure 3.5 is the plot of h vs. v for three different values of k in Equation 3.19. The curve on the left with $k = 1$ has a heavy bias to the left, where most days are windless ($v = 0$). The curve on the right with $k = 3$ looks more like a normal bell-shaped distribution, where some days have high wind and an equal number of days have low wind. The curve in the middle with $k = 2$ is a typical wind speed distribution found at most sites. In this distribution, more days have speeds lower than the mean speed, whereas a few days have high wind. The value of k determines the shape of the curve and hence is called the *shape parameter*.

The Weibull distribution with $k = 1$ is called the *exponential distribution*, which is generally used in reliability studies. For $k > 3$, it approaches the normal distribution, often called the *Gaussian* or the *bell-shaped distribution*.

Figure 3.6 shows the distribution curves corresponding to $k = 2$ with different values of c ranging from 8 to 16 mph (1 mph = 0.446 m/sec). For greater values of c, the curves shift right to the higher wind speeds. That is, the higher the c is, the greater the number of days that have high winds. Because this shifts the distribution of hours at a higher speed scale, c is called the *scale parameter*.

At most sites, wind speed has the Weibull distribution with $k = 2$, which is specifically known as the *Rayleigh distribution*. The actual measurement data taken at most sites compare well with the Rayleigh distribution, as seen in Figure 3.7. The Rayleigh distribution is then a simple and accurate enough representation of the wind speed with just one parameter, the scale parameter c.

Wind Speed and Energy

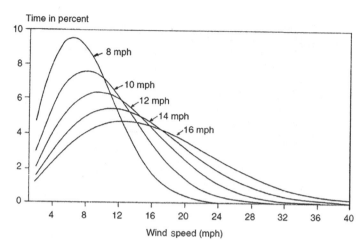

FIGURE 3.6 Weibull probability distribution with shape parameter $k = 2$ and scale parameters ranging from 8 to 16 mph.

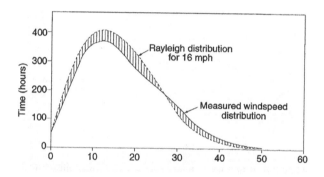

FIGURE 3.7 Rayleigh distribution of hours per year compared with measured wind speed distribution at St. Annes Head, Pembrokeshire, U.K.

Summarizing the characteristics of the Weibull probability distribution function:

$k = 1$ makes it the exponential distribution, $h = \lambda e^{-\lambda V}$ where $\lambda = 1/c$,

$k = 2$ makes it the Rayleigh distribution, $h = 2\lambda^2 v e^{-(\lambda v)^2}$, and (3.23)

$k > 3$ makes it approach a normal bell-shaped distribution.

Because most wind sites would have the scale parameter ranging from 10 to 20 mph (about 5–10 m/sec) and the shape parameter ranging from 1.5 to 2.5 (rarely 3.0), our discussion in the following subsections will center around those ranges of c and k.

Figure 3.8 displays the number of hours on the vertical axis vs. the wind speed on the horizontal axis. The three graphs (a), (b), and (c) are the distributions with

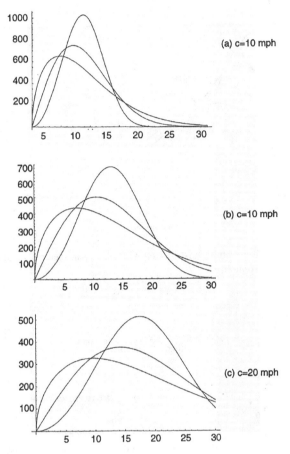

FIGURE 3.8 Weibull distributions of hours per year with three different shape parameters $k = 1.5, 2,$ and 3, each with scale parameters $c = 10, 15,$ and 20 mph.

different scale parameters $c = 10, 15,$ and 20 mph, each with three values of the shape parameters $k = 1.5, 2,$ and 3. The values of h in all three sets of curves are the number of hours in a year the wind is in the speed interval v and $(v + \Delta v)$. Figure 3.9 depicts the same plots in three-dimensional h–v–k space. It shows the effect of k in shifting the shape from the bell shape in the front right-hand side ($k = 3$) to the Rayleigh and to flatten shapes as the value of k decreases from 3.0 to 1.5. It is also observed from these plots that as c increases, the distribution shifts to higher speed values.

3.5.2 MODE AND MEAN SPEEDS

We now define the following terms applicable to wind speed:

Mode speed is defined as the speed corresponding to the hump in the distribution function. This is the speed of the wind most of the time.

Mean speed over the period is defined as the total area under the h–v curve integrated from $v = 0$ to ∞ and divided by the total number of hours in the period (8760 if the period is 1 yr). The annual mean speed is therefore the weighted average speed and is given by:

Wind Speed and Energy

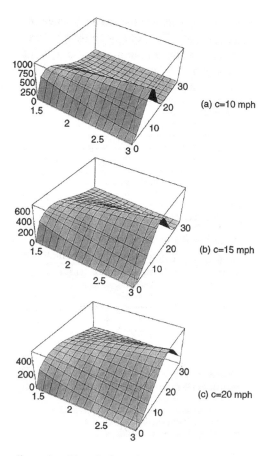

FIGURE 3.9 Three-dimensional h–v–k plots with c ranging from 10 to 20 mph and k ranging from 1.5 to 3.0.

$$V_{mean} = \frac{1}{8760} \int_0^\infty hv\,dv \qquad (3.24)$$

For c and k values in the range found at most sites, the integral expression can be approximated to the Gamma function:

$$V_{mean} = c\sqrt{\left(1+\frac{1}{k}\right)} \qquad (3.25)$$

For the Rayleigh distribution with $k = 2$, the Gamma function can be further approximated to the following:

$$V_{mean} = 0.90c \qquad (3.26)$$

This is a very simple relation between the scale parameter c and V_{mean}, which can be used with reasonable accuracy. For example, most sites are reported in terms of their mean wind speeds. The c parameter in the corresponding Rayleigh distribution is then $c = V_{mean}/0.9$. The k parameter is, of course, 2.0 for the Rayleigh distribution. Thus, we have the Rayleigh distribution of the site using the generally reported mean speed as follows:

$$h(v) = \frac{2v}{c^2} e^{-\left(\frac{v}{c}\right)^2} = \frac{2v}{(V_{mean})^2} e^{-\left(\frac{v}{V_{mean}}\right)^2} \quad (3.27)$$

3.5.3 Root Mean Cube Speed

Wind power is proportional to the cube of the speed, and the energy collected over the year is the integral of $hv^3 \cdot dv$. We, therefore, define the root mean cube or the rmc speed in a manner similar to the root mean square (rms) value in alternating current (AC) electrical circuits:

$$V_{rmc} = \sqrt[3]{\frac{1}{8760} \int_0^\infty hv^3 dv} \quad (3.28)$$

The rmc speed is useful in quickly estimating the annual energy potential of the site. Using V_{rmc} in Equation 3.12 gives the annual average power generation in W/m²:

$$P_{rmc} = \frac{1}{4} \rho V_{rmc}^3 \quad (3.29)$$

Then, we obtain the annual energy production potential of the site by simply multiplying the P_{rmc} value by the total number of hours in the year.

The importance of the rmc speed is highlighted in Table 3.1. It compares the wind power density at three sites with the same annual average wind speed of 6.3 m/sec. The San Gorgonio site in California has 66% greater power density than the Culebra site in Puerto Rico. The difference comes from having different shape factors k and, hence, different rmc speeds, although all have the same annual mean speed.

TABLE 3.1
Comparison of Three Wind Farm Sites with the Same Mean Wind Speed but Significantly Different Specific Power Densities

Site	Annual Mean Wind Speed (m/sec)	Annual Average Specific Power (W/m²)
Culebra, PR	6.3	220
Tiana Beach, NY	6.3	285
San Gorgonio, CA	6.3	365

Wind Speed and Energy

3.5.4 Mode, Mean, and RMC Speeds

The important difference between the mode, the mean, and the rmc speeds is illustrated in Table 3.2. The values of the three speeds are compiled for four shape parameters ($k = 1.5, 2.0, 2.5,$ and 3.0) and three scale parameters ($c = 10, 15,$ and 20 mph). The upstream wind power densities are calculated using the respective speeds in the wind power equation $P = \frac{1}{2}\rho V^3$ W/m² using the air mass density of 1.225 kg/m³.

We observe the following from the $c = 15$ rows:

- For $k = 1.5$, the power density using the mode speed is 230 as against the correct value of 4134 W/m² using the rmc speed. The ratio of the incorrect to correct value of the power density is 1–18, a huge difference.
- For $k = 2$, the power densities using the mode and rmc speeds are 731 and 2748 W/m², respectively, in the ratio of 1–3.76. The corresponding power densities with the mean and the rmc speeds are 1439 and 2748 W/m² in the ratio of 1–1.91.
- For $k = 3$, the power densities using the mode and rmc speeds are 1377 and 2067 W/m², respectively, in the ratio of 1–1.50. The corresponding power densities with the mean and the rmc speeds are 1472 and 2067 W/m², in the ratio of 1–1.40.

The last column in Table 3.2 gives the yearly energy potentials of the corresponding sites in kilowatthours per year per square meter of the blade area for the given k and c values. These values are calculated for a rotor efficiency C_p of 50%, which is the maximum that can be practically achieved.

TABLE 3.2
Influence of Shape and Scale Parameters on the Mode, Mean, and RMC Speeds and the Energy Density

c	k	Mode Speed	Mean Speed	RMC Speed	Pmode (W/m²)	Pmean (W/m²)	Prmc (W/m²)	Ermc (KWh/yr)
10	1.5	3.81	9.03	12.60	68	451	1225	5366
	2.0	7.07	8.86	11.00	216	426	814	3565
	2.5	8.15	8.87	10.33	331	428	675	2957
	3.0	8.74	8.93	10.00	409	436	613	2685
15	1.5	7.21	13.54	18.90	230	1521	4134	18107
	2.0	10.61	13.29	16.49	731	1439	2748	12036
	2.5	12.23	13.31	15.49	1120	1444	2278	9978
	3.0	13.10	13.39	15.00	1377	1472	2067	9053
20	1.5	9.61	18.05	25.19	544	3604	9790	42880
	2.0	13.14	17.72	22.00	1731	3410	6514	28531
	2.5	16.30	17.75	20.66	2652	3423	5399	23648
	3.0	17.47	17.86	20.00	3266	3489	4900	21462

Note: P = upstream wind power density in watts per square meter of the blade-swept area = $0.5 \rho V^3$, where $\rho = 1.225$ kg/m³; the last column is the energy potential of the site in kWh per year per m² of the blade area, assuming a rotor efficiency C_p of 50% (i.e., the maximum power that can be converted into electric power is $0.25 \rho V^3$).

Thus, regardless of the shape and the scale parameters, use of the mode or the mean speed in the power density equation would introduce a significant error in the annual energy estimate, sometimes off by several folds, making the estimates completely useless. Only the rmc speed in the power equation always gives the correct average power over a period.

The following example illustrates this point. A site has ¼ the number of hours in the year at 0, 10, 20, and 30 mph, with an annual average speed of 15 mph. The energy distribution at this site is shown in the following table:

Speed mph	0	10	20	30	(Avg. 15)
% h/yr	25	25	25	25	
% Energy/yr	0	3	22	75	

Annual wind energy using average speed = 3375 units
Annual wind energy using rmc speed = 9000 units
Energy potential estimated correctly using rmc speed = 2.67 × energy potential estimated incorrectly using average speed.

3.5.5 Energy Distribution

If we define the energy distribution function:

$$e = \frac{\text{kWh contribution in the year by the wind between } v \text{ and } (v + \Delta v)}{\Delta v} \quad (3.30)$$

then, for the Rayleigh speed distribution ($k = 2$), the energy distribution would look like the shaded curve in Figure 3.10. The wind speed curve has the mode at 5.5 m/sec and the

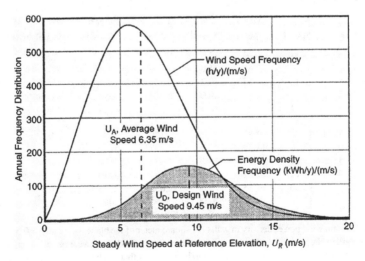

FIGURE 3.10 Annual frequency distributions of hours vs. wind speed and energy density per year with $c = 10$ and $k = 2$ (Rayleigh distribution).

Wind Speed and Energy

mean at 6.35 m/sec. However, because of the cubic relation with speed, the maximum energy contribution comes from the wind speed at 9.45 m/sec. Above this speed, although V^3 continues to increase in a cubic manner, the number of hours at those speeds decreases faster than V^3. The result is an overall decrease in the yearly energy contribution. For this reason, it is advantageous to design the wind power system to operate at variable speeds to capture the maximum energy available during high-wind periods.

Figure 3.11 is a similar chart showing the speed and energy distribution functions for a shape parameter of 1.5 and a scale parameter of 15 mph. The mode speed is 10.6 mph, the mean speed is 13.3 mph, and the rmc speed is 16.5 mph. The energy distribution function has the mode at 28.5 mph. That is, the most energy is captured at 28.5-mph wind speed, although the probability of wind blowing at that speed is low.

Comparing Figure 3.10 and Figure 3.11, we see that as the shape parameter value decreases from 2.0 to 1.5, the speed and the energy modes move farther apart. On the other hand, as the speed distribution approaches the bell shape for $k > 3$, the speed and the energy modes get closer to each other.

Figure 3.12 compares the speed and the energy distributions with $k = 2$ (Rayleigh) and $c = 10$, 15, and 20 mph. As seen here, the relative spread between the speed mode and the energy mode remains about the same, although both shift to the right as c increases.

3.5.6 Digital Data Processing

The mean wind speed over a period of time is obtained by adding numerous readings taken over that period and dividing the sum by the number of readings. Many digital data loggers installed over the last few decades collected average wind speed data primarily for meteorological purposes, as opposed to assessing wind power. They logged the speed every hour, and then averaged over the day, which, in turn, was averaged over the month and over the year. The averaging was done as follows:

$$V_{avg} = \frac{1}{n}\sum_{i=1}^{n} V_i \tag{3.31}$$

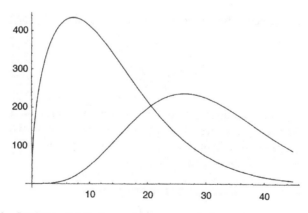

FIGURE 3.11 Rayleigh distributions of hours vs. wind speed and energy per year with $c = 15$ and $k = 1.5$. (Left curve is wind speed and right curve is wind energy)

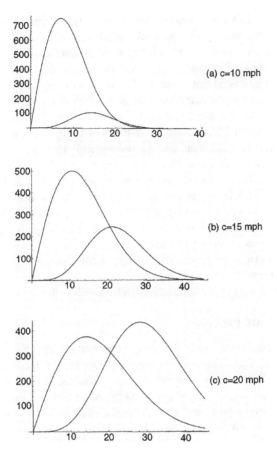

FIGURE 3.12 Rayleigh distributions of hours vs. wind speed and energy per year with $k = 2$ and $c = 10$, 15, and 20 mph.

The month-to-month wind speed variation at a typical site over the year can be ±30–35% over the annual average. As seen earlier, for assessing the wind power, the rmc speed is what matters. The rmc equivalent of the digital data logging is as follows:

$$V_{rmc} = \sqrt[3]{\frac{1}{n}\sum_{i=1}^{n} V_i^3} \qquad (3.32)$$

The preceding equation does not take into account the variation in the air mass density, which is also a parameter (although of second order) in the wind power density. Therefore, a better method of processing wind speed data for power calculations is to digitize the yearly average power density as

$$P_{rmc} = \frac{1}{2n}\sum_{i=1}^{n} \rho_i V_i^3 \qquad (3.33)$$

where

n = number of observations in the averaging period,
ρ_i = air density (kg/m³), and
V_i = wind speed (m/sec) at the ith observation time.

3.5.7 EFFECT OF HUB HEIGHT

The wind shear at a ground-level surface causes the wind speed to increase with height in accordance with the following expression:

$$V_2 = V_1 \left(\frac{h_2}{h_1} \right)^\alpha \tag{3.34}$$

where

V_1 = wind speed measured at the reference height h_1,
V_2 = wind speed estimated at height h_2, and
α = ground surface friction coefficient.

The friction coefficient α is low for smooth terrain and high for rough ones. The values of α for typical terrain classes are given in Table 3.3, and their effects on the wind speed at various heights are plotted in Figure 3.13. It is noteworthy that the offshore wind tower, being in low-α terrain, always sees a higher wind speed at a given height and is less sensitive to tower height.

Wind speed does not increase with height indefinitely, even at a slower rate. The data collected at Merida airport in Mexico show that typically wind speed increases with height up to about 450 m and then decreases (Figure 3.14).[1] The wind speed at 450-m height can be 4–5 times greater than that near the ground surface.

Modern wind turbines operate on increasingly taller towers to take advantage of the increased wind speeds at higher altitudes. Very little is known about the turbulent wind patterns at these heights, which can damage the rotor. Of particular interest on the Great Plains, where many wind farms are located, are the high-level wind flows called *nocturnal jets* that dip close to the ground at night, creating violent turbulence. Engineers at the National Wind Technology Center (NWTC) have been measuring

TABLE 3.3
Friction Coefficient α of Various Terrains

Terrain Type	Friction Coefficient α
Lake, ocean, and smooth, hard ground	0.10
Foot-high grass on level ground	0.15
Tall crops, hedges, and shrubs	0.20
Wooded country with many trees	0.25
Small town with some trees and shrubs	0.30
City area with tall buildings	0.40

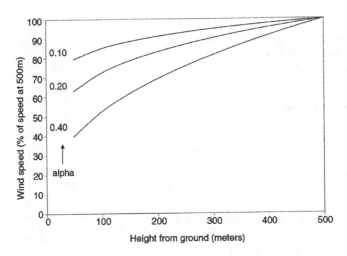

FIGURE 3.13 Wind speed variations with height over different terrain. Smooth, low-friction terrain with low α develops a thinner layer of slow wind near the surface and high wind at heights.

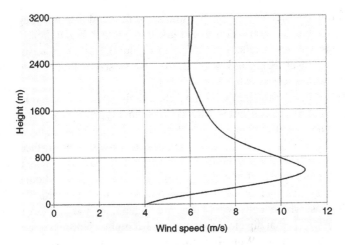

FIGURE 3.14 Wind speed variations with the height measured at Merida airport, Yucatan, in Mexico. (From Schwartz, M.N. and Elliott, D.L., "Mexico Wind Resource Assessment Project," DOE/NREL Report No. DE95009202, National Renewable Energy Laboratory, Golden, Colorado, March 1995.)

higher-altitude wind patterns and developing simulation models of the turbine's interaction with turbulent wind patterns to develop designs that can prevent potential damage to the rotor.

3.5.8 Importance of Reliable Data

Some of the old wind speed data around the world may have been collected primarily for meteorological use with rough instruments and relatively poor exposure to the

TABLE 3.4
Comparison of Calculated Average Wind Power Density between 1983 OLADE Atlas and 1995 NREL Analysis for Several Locations in Mexico

		Wind Power Density (in W/m^2)	
Region in Mexico	Data Site	OLADE Atlas (1983)	NREL Data (1995)
Yucatan Peninsula	Merida	22	165
	Campeche	23	120
	Chetumal	28	205
Northern Gulf Plain	Tampico	8	205
	Cuidad Victoria	32	170
	Matamoros	32	165
Central Highlands	Durango	8	140
	San Luis Potosi	35	155
	Zacatecas	94	270
Northwest	Chihuahua	27	120
	Hermosillo	3	80
	La Paz	10	85

Note: Data at 10-m height.

wind. This is highlighted by a recent wind resource study in Mexico.[1] Significant differences between the old and the new data have been found, as listed in Table 3.4. The 1983 Organización Latinoamericana de Energía (OLADE) Atlas of Mexico indicates very low energy potential, whereas the 1995 NREL data reports energy potentials several times greater. The values from the OLADE Atlas are from a few urban locations where anemometers could be poorly exposed to the prevailing winds. In contrast, the new NREL wind data come from a large number of stations, including open airport locations, and have incorporated the terrain effect in the final numbers. The message here is clear: it is important to have reliable data on the annual wind speed distribution over at least a few years before sitting a wind farm with a high degree of confidence.

The most widely available wind speed data in the past came from weather stations at airports. Unfortunately, that is not very useful to wind farm planners, as airports are not necessarily located at favorable wind sites. Other widely used locations in towns and cities for wind measurements have been existing structures such as water towers. Such data can also be unreliable due to shear effect. Near the tower, the wind speed can be 10–15% higher or lower than the actual speed. This results in an energy estimate up to 52% on the high side or 60% on the low side, a difference of (1.52/0.60) 2.5–1. Such an enormous error could make a site look lucrative, whereas in fact it may not be even marginally economical.

Because airports were intentionally not built on high-wind sites, the good wind sites were left unmonitored for a long time. The new data from such sites are, therefore, limited to a few recent years. For this reason, one must verify the data in more than one way to ascertain high reliability and confidence in its use.

3.6 WIND SPEED PREDICTION

Because the available wind energy at any time depends on the wind speed at that time, which is a random variable, knowing the average annual energy potential of a site is one thing and the ability to accurately predict when the wind will blow is quite another thing. For the wind farm operator, this poses difficulties in system scheduling and energy dispatching as the schedule of wind power availability is not known in advance. However, a reliable forecast of wind speed several hours in advance can give the following benefits:

- Generating schedule can efficiently accommodate wind generation in a timely manner
- Allows the grid-connected wind farm to commit to power purchase contracts in advance for a better price
- Allows investors to proceed with new wind farms and avoid the penalties they must pay if they do not meet their hourly generation targets

Therefore, development of short-term wind-speed-forecasting tools helps wind energy producers. NWTC researchers work in cooperation with the National Oceanic and Atmospheric Administration (NOAA) to validate the nation's wind resource maps and develop methods of short-term (1–4 h) wind forecasting. Alexiadis et al.[2] have also proposed a new technique for forecasting wind speed and power output up to several hours in advance. Their technique is based on cross-correlation at neighboring sites and artificial neural networks and is claimed to significantly improve forecasting accuracy compared to the persistence-forecasting model.

REFERENCES

1. Schwartz, M.N. and Elliott, D.L., Mexico Wind Resource Assessment Project, DOE/NREL Report No. DE95009202, National Renewable Energy Laboratory, Golden, CO, March 1995.
2. Alexiadis, M.C., Dokopoulos, P.S., and Sahsamanogdou, H.S., Wind Speed and Power Forecasting Based on Spatial Correlation Models, IEEE Paper No. PE-437-EC-0-03-1998.
3. Elliott, D.L., Holladay, C.G., Barchet, W.R., Foote, H.P., and Sandusky, W.F., Wind Energy Resources Atlas of the United States, DOE/Pacific Northwest Laboratory Report No. DE-86004442, April 1991.
4. Elliot, D.L., Synthesis of National Wind Assessment, DOE/Pacific Northwest Laboratory, NTIS Report No. BNWL-2220 Wind-S, 1997.
5. Rory, A., *Minnesota Wind Resources Assessment Program*, Minnesota Department of Public Service Report, 1994.
6. Freris, L.L., *Wind Energy Conversion Systems*, Prentice Hall, London, 1990.
7. Elliott, D.L. and Schwartz, M.N., Wind Energy Potential in the United States, Pacific Northwest Laboratory PNL-SA-23109, NTIS No. DE94001667, September 1993.
8. Elliott, D.L., Holladay, C.G., Barchet, W.R., Foote, H.P., and Sandusky, W.F. *Wind Energy Resource Atlas of the United States, Report DOE/CH 10093-4*, Solar Energy Research Institute, Golden, CO, 1987.
9. Elliott, D.L., Wendell, L.L., and Gower, G.L., *An Assessment of the Available Windy Land Area and Wind Energy Potential in the Contiguous United States, Report PNL-7789*, Pacific Northwest Laboratory, Richland, WA, 1991.

10. Schwartz, M.N., Elliott, D.L., and Gower, G.L., Gridded State Maps of Wind Electric Potential. Proceedings of the American Wind Energy Association's WindPower '92 Conference, October 19–23, Seattle, WA, 1992.
11. Gupta, A.K., *Power Generation from Renewables in India*, Ministry of Non-Conventional Energy Sources, New Delhi, India, 1997.
12. Omid Beik, An HVDC Off-shore Wind Generation Scheme with High Voltage Hybrid Generator, Ph.D. thesis, McMaster University, 2016.
13. Omid Beik, Ahmad S. Al-Adsani, *DC Wind Generation Systems, Design, Analysis, and Multiphase Turbine Technology*, Springer, 2020.
14. Omid Beik, Ahmad S. Al-Adsani, *Wind Energy Systems*, pp. 1–9, Springer, 2020.
15. Omid Beik, Ahmad S. Al-Adsani, *DC Wind Generation System*, pp. 33–69, Springer, 2020.
16. Beik, O., Dekka A., Narimani, M., *A new modular neutral point clamped converter with space vector modulation control,' Proc. IEEE Int. Conf. Ind. Tech*, Lyon, France, 2018, pp. 591–595.
17. O. Beik and A. S. Al-Adsani, "Parallel Nine-Phase Generator Control in a Medium-Voltage DC Wind System," in *IEEE Transactions on Industrial Electronics*, vol. 67, no. 10, pp. 8112–8122, Oct. 2020.
18. O. Beik and N. Schofield, "High-Voltage Hybrid Generator and Conversion System for Wind Turbine Applications," in *IEEE Transactions on Industrial Electronics*, vol. 65, no. 4, pp. 3220–3229, April 2018.
19. Beik, O., Schofield, N., 'High voltage generator for wind turbines', 8th IET International Conference on Power Electronics, Machines and Drives (PEMD 2016), 2016.
20. O. Beik and N. Schofield, "An Offshore Wind Generation Scheme With a High-Voltage Hybrid Generator, HVDC Interconnections, and Transmission," in *IEEE Transactions on Power Delivery*, vol. 31, no. 2, pp. 867–877, April 2016.
21. O. Beik and N. Schofield, "Hybrid generator for wind generation systems," 2014 IEEE Energy Conversion Congress and Exposition (ECCE), Pittsburgh, PA, 2014, pp. 3886–3893.

4 Wind Power Systems

The wind power system is fully covered in this and the following two chapters. This chapter covers the overall system-level performance, design considerations, and trades.

4.1 SYSTEM COMPONENTS

The wind power system comprises one or more wind turbine units operating electrically in parallel. Each turbine is made of the following basic components:

- Tower structure
- Rotor with two or three blades attached to the hub
- Shaft with mechanical gear
- Electrical generator
- Yaw mechanism, such as the tail vane
- Sensors and control

Because of the large moment of inertia of the rotor, design challenges include starting, speed control during the power-producing operation, and stopping the turbine when required. The eddy current or another type of brake is used to halt the turbine when needed for emergency or for routine maintenance.

In a modern wind farm, each turbine must have its own control system to provide operational and safety functions from a remote location (Figure 4.1). It also must have one or more of the following additional components:

- Anemometers, which measure the wind speed and transmit the data to the controller.
- Numerous sensors to monitor and regulate various mechanical and electrical parameters. A 1-MW turbine may have several hundred sensors.
- Stall controller, which starts the machine at set wind speeds of 8–15 mph and shuts off at 50–70 mph to protect the blades from overstressing and the generator from overheating.
- Power electronics to convert and condition power to the required standards.
- Control electronics, usually incorporating a computer.
- Battery for improving load availability in a stand-alone plant.
- Transmission link for connecting the plant to the area grid.

The following are commonly used terms and terminology in the wind power industry:

Low-speed shaft: The rotor turns the low-speed shaft at 30–60 rotations per minute (rpm).

FIGURE 4.1 Baix Ebre wind farm and control center, Catalonia, Spain. (From *Wind Directions*, Magazine of the European Wind Energy Association, London, October 1997. With permission.)

High-speed shaft: It drives the generator via a speed step-up gear.

Brake: A disc brake, which stops the rotor in emergencies. It can be applied mechanically, electrically, or hydraulically.

Gearbox: Gears connect the low-speed shaft to the high-speed shaft and increase the turbine speed from 30–60 rpm to the 1200–1800 rpm required by most generators to produce electricity in an efficient manner. Because the gearbox is a costly and heavy part, design engineers are exploring slow-speed, direct-drive generators that need no gearbox.

Generator: It is usually an off-the-shelf induction generator that produces 50- or 60-Hz AC power.

Nacelle: The rotor attaches to the nacelle, which sits atop the tower and includes a gearbox, low- and high-speed shafts, generator, controller, and a brake. A cover protects the components inside the nacelle. Some nacelles are large enough for technicians to stand inside while working.

Pitch: Blades are turned, or pitched, out of the wind to keep the rotor from turning in winds that have speeds too high or too low to produce electricity.

Upwind and downwind: The upwind turbine operates facing into the wind in front of the tower, whereas the downwind runs facing away from the wind after the tower.

Vane: It measures the wind direction and communicates with the yaw drive to orient the turbine properly with respect to the wind.

Yaw drive: It keeps the upwind turbine facing into the wind as the wind direction changes. A yaw motor powers the yaw drive. Downwind turbines do not require a yaw drive, as the wind blows the rotor downwind.

The design and operating features of various system components are described in the following subsections.

Wind Power Systems

4.1.1 TOWER

The wind tower supports the rotor and the nacelle containing the mechanical gear, the electrical generator, the yaw mechanism, and the stall control. Figure 4.2 depicts the component details and layout in a large nacelle, and Figure 4.3 shows the installation on the tower. The height of the tower in the past has been in the 20–50 m range, but in newer installations it can be 100 m or higher. For medium and large-sized turbines, the tower height is approximately equal to the rotor diameter, as seen in the dimension drawing of a 600-kW wind turbine (Figure 4.4). Small turbines are generally mounted on the tower a few rotor diameters high. Otherwise, they would suffer fatigue due to the poor wind speed found near the ground surface. Figure 4.5 shows tower heights of various-sized wind turbines relative to some known structures.

Both steel and concrete towers are available and are being used. The construction can be tubular or lattice. Towers must be at least 25–30 m high to avoid the turbulence caused by trees and buildings. Utility-scale towers are typically twice or thrice as high to take advantage of the swifter winds at those heights.

1. Nacelle
2. Heat Exchanger
3. Offshore Container
4. Small Gantry Crane
5. Oil Cooler
6. Control Pane
7. Generator
8. Impact Noise Reduction
9. Hydraulic Parking Brake
10. Main Frame
11. Swiveling Crane
12. Gearbox
13. Rotor Lock
14. Rotor Shaft
15. Yaw Drive
16. Rotor Hub
17. Pitch Drive
18. Nose Cone

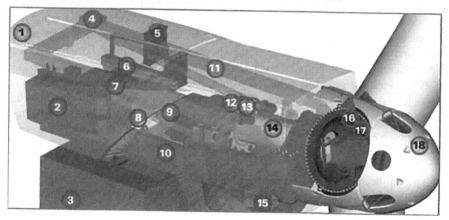

FIGURE 4.2 Nacelle details of a 3.6-MW/104-m-diameter wind turbine. (From GE Wind Energy. With permission.)

FIGURE 4.3 A large nacelle under installation. (From Nordtank Energy group, Denmark. With permission.)

The main issue in the tower design is the structural dynamics. The tower vibration and the resulting fatigue cycles under wind speed fluctuation are avoided by the design. This requires careful avoidance of all resonance frequencies of the tower, the rotor, and the nacelle from the wind fluctuation frequencies. Sufficient margin must be maintained between the two sets of frequencies in all vibrating modes.

The resonance frequencies of the structure are determined by complete modal analyses, leading to the eigenvectors and eigenvalues of complex matrix equations representing the motion of the structural elements. The wind fluctuation frequencies are found from the measurements at the site under consideration. Experience on a similar nearby site can bridge the gap in the required information.

Big cranes are generally required to install wind towers. Gradually increasing tower height, however, is bringing a new dimension in the installation (Figure 4.6). Large rotors add to the transportation problem as well. Tillable towers to nacelle and rotors moving upwards along with the tower are among some of the newer developments in wind tower installation. The offshore installation comes with its own challenge that must be met.

The top head mass (THM) of the nacelle and rotor combined has a significant bearing on the dynamics of the entire tower and the foundation. Low THM is generally a measure of design competency, as it results in reduced manufacturing and installation costs. The THMs of Vestas' 3-MW/90-m turbine is 103 t, NEG Micon's new 4.2-MW/110-m machine is 214 t, and Germany's REpower's 5-MW/125-m machine is about 350 t, which includes extra 15–20% design margins.

Wind Power Systems

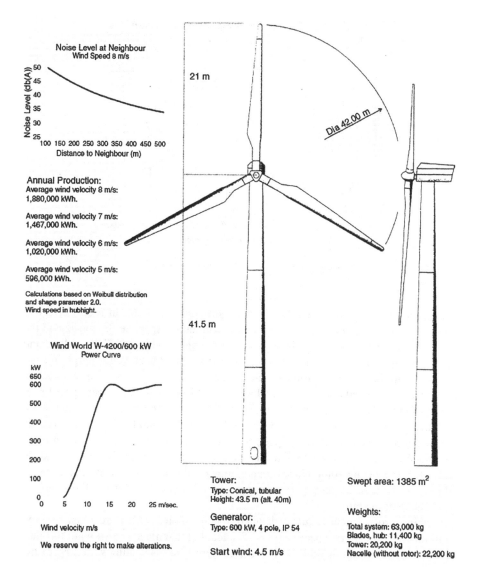

FIGURE 4.4 A 600-kW wind turbine and tower dimensions with specifications. (From Wind World Corporation, Denmark. With permission.)

4.1.2 TURBINE

The turbine size has been steadily increasing, ranging from a few kW for small stand-alone remote applications to several MW each for utility-scale power generation. By the end of 2003, about 1200 1.5-MW turbines made by GE Wind Energy alone were installed and in operation. Today, larger machines are being routinely installed on a large commercial scale for offshore wind farms both in Europe and in the U.S. For example, GE's 3.6-MW turbines offers lighter variable-speed,

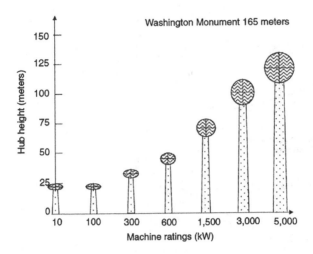

FIGURE 4.5 Tower heights of various capacity wind turbines.

pitch-controlled blades on a softer support structure, resulting in a cost-effective foundation. Its rated wind speed is 14 m/sec with cut-in speed at 3.5 m/sec and the cutout at 25 m/sec. The blade diameter is 104 m with hub height 100 m on land and 75 m offshore. Enercon's 4.5-MW wind turbine installed near Magdeburgh in eastern Germany has a 113-m rotor diameter, 124-m hub height, and an egg-shaped nacelle. Its reinforced concrete tower diameter is 12 m at the base, tapering to 4 m at the top. On the larger size turbines, *Vestas V164*-8.0 has 164 m diameter blades and the rated power of 8 MW. At a wind speed of 4 m/s, the wind turbine starts working (cut-in speed). The cut-out wind speed is 25 m/s to protect the generator from overheating. As for the mass, a 5-MW turbine mass can vary from 150 to 250 t in nacelle and 70–100 t in the rotor blades, depending on the manufacturing technologies adopted at the time of design. The most modern designs would naturally be on the lighter side of the range.

Turbine procurement requires detailed specifications, which are often tailored from the manufacturers' specifications. The leading manufacturers of wind turbines in the world are listed in Table 4.1, with Denmark's Vestas leading with 20% of the world's market share.

4.1.3 Blades

Modern wind turbines have two or three blades, which are carefully constructed airfoils that utilize aerodynamic principles to capture as much power as possible. The airfoil design uses a longer upper-side surface whereas the bottom surface remains somewhat uniform. By the Bernoulli principle, a "lift" is created on the airfoil by the pressure difference in the wind flowing over the top and bottom surfaces of the foil. This aerodynamic lift force flies the plane high, but rotates the wind turbine blades about the hub. In addition to the lift force on the blades, a drag force is created, which acts perpendicular to the blades, impeding the lift effect and slowing the rotor down.

Wind Power Systems

FIGURE 4.6 WEG MS-2 wind turbine installation at Myers Hill. (From Wind Energy Group, a Taylor Woodrow subsidiary and ETSU/DTI, U.K.)

The design objective is to get the highest lift-to-drag ratio that can be varied along the length of the blade to optimize the turbine's power output at various speeds.

The rotor blades are the foremost visible part of the wind turbine and represent the forefront of aerodynamic engineering. The steady mechanical stress due to centrifugal forces and fatigue under continuous vibrations make the blade design the weakest mechanical link in the system. Extensive design effort is needed to avoid premature fatigue failure of the blades. A swift increase in turbine size has been recently made possible by the rapid progress in rotor blade technology, including emergence of the

TABLE 4.1
World's Major Wind Turbine Suppliers in 2020

Supplier	% Share of World Market
Vestas (Denmark)	20
Siemens-Gamesa (Spain)	14
Goldwind (China)	14
GE Wind (USA).	10
Envision (China)	8
All others	34

carbon- and glass-fiber-based epoxy composites. The turbine blades are made of high-density wood or glass fiber and epoxy composites.

The high pitch angle used for stall control also produces a high force. The resulting load on the blade can cause a high level of vibration and fatigue, possibly leading to a mechanical failure. Regardless of the fixed- or variable-speed design, the engineer must deal with the stall forces. Researchers are moving from the 2-D to 3-D stress analyses to better understand and design for such forces. As a result, the blade design is continually changing, particularly at the blade root where the loading is maximum due to the cantilever effect.

The aerodynamic design of the blade is important, as it determines the energy-capture potential. The large and small machine blades have significantly different design philosophies. The small machine sitting on a tower relatively taller than the blade diameter, and generally unattended, requires a low-maintenance design. On the other hand, a large machine tends to optimize aerodynamic performance for the maximum possible energy capture. In either case, the blade cost is generally kept below 10% of the total installed cost.

4.1.4 Speed Control

The wind turbine technology has changed significantly in the last 25 yr.[1] Large wind turbines being installed today tend to be of variable-speed design, incorporating pitch control and power electronics. Small machines, on the other hand, must have simple, low-cost power and speed control. The speed control methods fall into the following categories:

No speed control whatsoever: In this method, the turbine, the electrical generator, and the entire system are designed to withstand the extreme speed under gusty winds.

Yaw and tilt control: The yaw control continuously orients the rotor in the direction of the wind. It can be as simple as the tail vane or more complex on modern towers. Theoretical considerations dictate free yaw as much as possible. However, rotating blades with large moments of inertia produce high gyroscopic torque during yaw, often resulting in loud noise. A rapid

Wind Power Systems 45

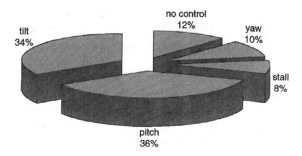

FIGURE 4.7 Speed control methods used in small- to medium-sized turbines.

yaw may generate noise exceeding the local ordinance limit. Hence, a controlled yaw is often required and used, in which the rotor axis is shifted out of the wind direction when the wind speed exceeds the design limit.

Pitch control: This changes the pitch of the blade with changing wind speed to regulate the rotor speed. Large-scale power generation is moving towards variable-speed rotors with power electronics incorporating a pitch control.

Stall control: Yaw and tilt control gradually shifts the rotor axis in and out of the wind direction. But, in gusty winds above a certain speed, blades are shifted (profiled) into a position such that they stall and do not produce a lift force. At stall, the wind flow ceases to be smooth around the blade contour, but separates before reaching the trailing edge. This always happens at a high pitch angle. The blades experience a high drag, thus lowering the rotor power output. This way, the blades are kept under the allowable speed limit in gusty winds. This not only protects the blades from mechanical overstress, but also protects the electrical generator from overloading and overheating. Once stalled, the turbine has to be restarted after the gust has subsided.

Figure 4.7 depicts the distribution of the control methods used in small wind turbine designs. Large machines generally use the power electronic speed control, which is covered in later chapters.

4.2 TURBINE RATING

The method of assessing the nominal rating of a wind turbine has no globally accepted standard. The difficulty arises because the power output of the turbine depends on the square of the rotor diameter and the cube of the wind speed. The rotor of a given diameter, therefore, would generate different power at different wind speeds. A turbine that can generate 300 kW at 7 m/sec would produce 450 kW at 8 m/sec wind speed. What rating should then be assigned to this turbine? Should we also specify the rated speed? Early wind turbine designers created a rating system that specified the power output at some arbitrary wind speed. This method did not work well because everyone could not agree on one speed for specifying the power rating. The "rated" wind speeds varied from 10 to 15 m/sec under this practice. Manufacturers quoted on the higher side to claim a greater output from the same design.

Such confusion in quoting the rating was avoided by some European manufacturers who quoted only the rotor diameter. But the confusion continued as to the maximum power the machine can generate under the highest wind speed in which the turbine can continuously and safely operate. Many manufacturers have, therefore, adopted the combined rating designations x/y, the wind turbine diameter followed by the generator's maximum electrical capacity. For example, a 30–300 kW wind turbine means 30-m diameter rotor with 300-kW rated electrical generator, and *V164-8.0* turbine means Vesta's 164-m diameter rotor with 8-MW generator. The specific rated capacity (SRC) is often used as a comparative index of the wind turbine designs. It measures the power generation capacity per square meter of the blade-swept area, and is defined as follows in units of kW/m^2:

$$SRC = \frac{\text{Generator electrical capacity}}{\text{Rotor-swept area}} \qquad (4.1)$$

The SRC for a 300/30 wind turbine is $300/\pi \times 15^2 = 0.42$ kW/m^2. It increases with diameter, giving favorable economies of scale for large machines. For example, it is approximately 0.2 kW/m^2 for a 10-m diameter rotor and 0.5 kW/m^2 for a 150-m diameter rotor. Some aggressively rated turbines have an SRC of 0.7 kW/m^2, and some reach as high as 1 kW/m^2. The higher-SRC rotor blades have higher operating stresses, which result in a shorter fatigue life. All stress concentration regions are carefully identified and eliminated in high-SRC designs. Modern design tools, such as the finite element stress analysis and the modal vibration analysis, can be of great value in rotor design.

Turbine rating is important as it indicates to the system designer how to size the electrical generator, the plant transformer, and the connecting cables to the substation and the transmission link interfacing the grid. The power system must be sized on the peak capacity of the generator. Because turbine power depends on the cube of the wind speed, the system-design engineer matches the turbine and the generator performance characteristics. This means selecting the rated speed of the turbine to match with the generator. As the gearbox and generator are manufactured only in discrete sizes, selecting the turbine's rated speed can be complex. The selection process goes through several iterations, trading the cost with benefit of the available speeds. Selecting a low rated speed would result in wasting much energy at high winds. On the other hand, if the rated speed is high, the rotor efficiency will suffer most of the time.

Figure 4.8 is an example of the technical summary datasheet of the 550/41-kW/m wind turbine manufactured by Nordtank Energy Group of Denmark. Such data is used in the preliminary design of the overall system. The SRC of this machine is 0.414. It has a cut-in wind speed of 5 m/sec, a cutout speed of 25 m/sec, and it reaches the peak power at 15 m/sec.

Wind Power Systems

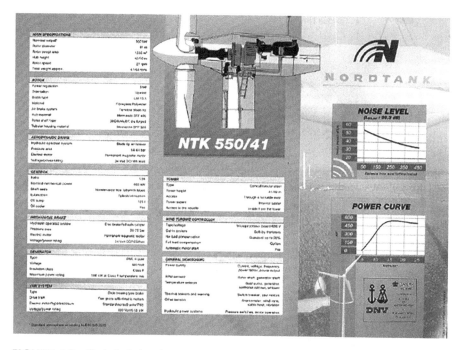

FIGURE 4.8 Technical datasheet of a 550-kW/41-m diameter wind turbine, with power level and noise level. (From Nordtank Energy Group, Denmark. With permission.)

4.3 POWER VS. SPEED AND TSR

The typical turbine torque vs. rotor speed is plotted in Figure 4.9. It shows a small torque at zero speed, rising to a maximum value before falling to nearly zero when the rotor just floats with the wind. Two such curves are plotted for different wind speeds V_1 and V_2, with V_2 being higher than V_1. The corresponding power vs. rotor speed at the two wind speeds are plotted in Figure 4.10. As the mechanical power converted into the electric power is given by the product of the torque T and the angular speed, the power is zero at zero speed and again at high speed with zero torque. The maximum power is generated at a rotor speed somewhere in between, as marked by P_{1max} and P_{2max} for speeds V_1 and V_2, respectively. The speed at the maximum power is not the same speed at which the torque is maximum. The operating strategy of a well-designed wind power system is to match the rotor speed to generate power continuously close to the P_{max} points. Because the P_{max} point changes with the wind speed, the rotor speed must, therefore, be adjusted in accordance with the wind speed to force the rotor to work continuously at P_{max}. This can be done with a variable-speed system design and operation.

At a given site, the wind speed varies over a wide range from zero to high gust. We define tip speed ratio (TSR) as follows:

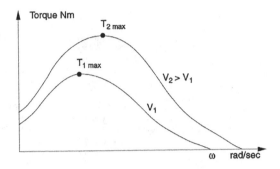

FIGURE 4.9 Wind turbine torque vs. rotor speed characteristic at two wind speeds, V_1 and V_2.

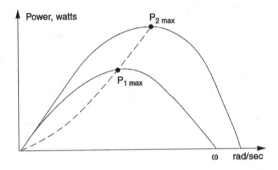

FIGURE 4.10 Wind turbine power vs. rotor speed characteristic at two wind speeds, V_1 and V_2.

$$TSR = \frac{\text{Linear speed of the blade's outermost tip}}{\text{Free upstream wind velocity}} = \frac{\omega R}{V} \qquad (4.2)$$

where R and ω are the rotor radius and the angular speed, respectively.

For a given wind speed, the rotor efficiency C_p varies with TSR as shown in Figure 4.11. The maximum value of C_p occurs approximately at the same wind speed that gives peak power in the power distribution curve of Figure 4.10. To capture high power at high wind, the rotor must also turn at high speed, keeping TSR constant at the optimum level. However, the following three system performance attributes are related to TSR:

1. The maximum rotor efficiency C_p is achieved at a particular TSR, which is specific to the aerodynamic design of a given turbine. As was seen in Figure 3.4, the TSR needed for maximum power extraction ranges from nearly one for multiple-blade, slow-speed machines to nearly six for modern high-speed, two-blade machines.

Wind Power Systems

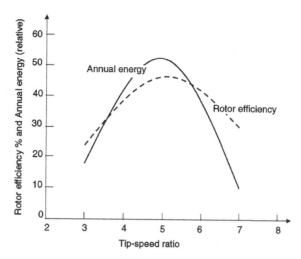

FIGURE 4.11 Rotor efficiency and annual energy production vs. rotor TSR.

2. The centrifugal mechanical stress in the blade material is proportional to the TSR. The machine working at a higher TSR is necessarily stressed more. Therefore, if designed for the same power in the same wind speed, the machine operating at a higher TSR would have slimmer rotor blades.
3. The ability of a wind turbine to start under load is inversely proportional to the design TSR. As this ratio increases, the starting torque produced by the blade decreases.

A variable-speed control is needed to maintain a constant TSR to keep the rotor efficiency at its maximum. At the optimum TSR, the blades are oriented to maximize the lift and minimize the drag on the rotor. The turbine selected for a constant TSR operation allows the rotational speed of both the rotor and generator to vary up to 60% by varying the pitch of the blades.

4.4 MAXIMUM ENERGY CAPTURE

The wind power system design must optimize the annual energy capture at a given site. The only operating mode for extracting the maximum energy is to vary the turbine speed with varying wind speed such that at all times the TSR is continuously equal to that required for the maximum power coefficient C_p. The theory and field experience indicate that the variable-speed operation yields 20–30% more power than with the fixed-speed operation. Nevertheless, the cost of variable-speed control is added. In the system design, this trade-off between energy increase and cost increase has to be optimized. In the past, the added costs of designing the variable-pitch rotor, or the speed control with power electronics, outweighed the benefit of the increased energy capture. However, the falling prices of power electronics for speed control and the availability of high-strength fiber composites for constructing high-speed rotors have made it economical to capture more energy when the speed is high.

TABLE 4.2
Advantages of Fixed- and Variable-Speed Systems

Fixed-Speed System	Variable-Speed System
Simple and inexpensive electrical system	Higher rotor efficiency, hence, higher energy capture per year
Fewer parts, hence, higher reliability	Low transient torque
Lower probability of excitation of mechanical resonance of the structure	Fewer gear steps, hence, inexpensive gear box
No frequency conversion, hence, no current harmonics present in the electrical system	Mechanical damping system not needed; the electrical system could provide damping if required
Lower capital cost	No synchronization problems
	Stiff electrical controls can reduce system voltage sags

The variable-speed operation has an indirect advantage. It allows controlling the active and reactive powers separately in the process of automatic generation control. In fixed-speed operation, on the other hand, the rotor is shut off during high wind speeds, losing significant energy. The pros and cons of fixed- and variable-speed operations are listed in Table 4.2.

Almost all major suppliers now offer variable-speed systems in combination with pitch regulation. Potential advantages of the variable-speed system include active grid support, peak-power-tracking operation, and cheaper offshore foundation structure. The doubly fed induction generator is being used in some large wind turbines such as NEG Micon's 4.2-MW, 110-m diameter machines, and multimegawatt GE machines. It is an emerging trendsetting technology in the variable-speed gear-driven systems, primarily because only the slip frequency power (20–30% of the total) has to be fed through the frequency converter. This significantly saves power electronics cost.

4.5 MAXIMUM POWER OPERATION

As seen earlier, operating the wind turbine at a constant TSR corresponding to the maximum power point at all times can generate 20–30% more electricity per year. However, this requires a control scheme to operate with a variable speed to continuously generate the maximum power. Two possible schemes for such an operation are as follows:

4.5.1 Constant-TSR Scheme

In this scheme, the machine is continuously operated at its optimum TSR, which is a characteristic of the given wind turbine. This optimum value is stored as the reference TSR in the control computer. The wind speed is continuously measured and compared with the blade tip speed. The error signal is then fed to the control system, which changes the turbine speed to minimize the error (Figure 4.12). At this time

Wind Power Systems

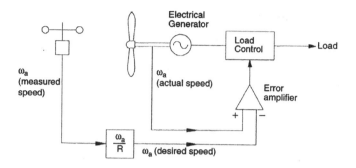

FIGURE 4.12 Maximum power operation using rotor tip speed control scheme.

the rotor must be operating at the reference TSR, generating the maximum power. This scheme has the disadvantage of requiring the local wind speed measurements, which could have a significant error, particularly in a large wind farm with shadow effects. Being sensitive to the changes in the blade surface, the optimum TSR gradually changes with age and environment. The computer reference TSR must be changed accordingly many times, which is expensive. Besides, it is difficult to determine the new optimum TSR with changes that are not fully understood or easily measured.

4.5.2 Peak-Power-Tracking Scheme

The power vs. speed curve has a single well-defined peak. If we operate at the peak point, a small increase or decrease in the turbine speed would result in no change in the power output, as the peak point locally lies in a flat neighborhood. In other words, a necessary condition for the speed to be at the maximum power point is as follows:

$$\frac{dP}{d\omega} = 0 \qquad (4.3)$$

This principle is used in the control scheme (Figure 4.13). The speed is increased or decreased in small increments, the power is continuously measured, and $\Delta P/\Delta\omega$ is continuously evaluated. If this ratio is positive—meaning we get more power by increasing the speed—the speed is further increased. On the other hand, if the ratio is negative, the power generation will reduce if we change the speed any further. The speed is maintained at the level where $\Delta P/\Delta\omega$ is close to zero. This method is insensitive to errors in local wind speed measurement, and also to wind turbine design. It is, therefore, the preferred method. In a multiple-machine wind farm, each turbine must be controlled by its own control loop with operational and safety functions incorporated.

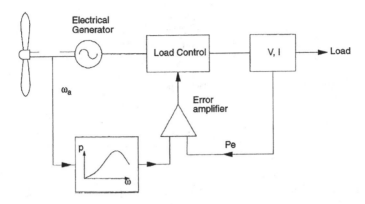

FIGURE 4.13 Maximum power operation using power control scheme.

4.6 SYSTEM-DESIGN TRADE-OFFS

When the land area is limited or is at a premium price, one optimization study that must be conducted in an early stage of the wind farm design is to determine the number of turbines, their size, and the spacing for extracting the maximum energy from the farm annually. The system trade-offs in such a study are as follows:

4.6.1 Turbine Towers and Spacing

Large turbines cost less per megawatt of capacity and occupy less land area. On the other hand, fewer large machines can reduce the megawatthour energy crop per year, as downtime of one machine would have larger impact on the energy output. A certain turbine size may stand out to be the optimum for a given wind farm from the investment and energy production cost points of view.

Tall towers are beneficial, but the height must be optimized with the local regulations and constrains of the terrain and neighborhood. Nacelle weight and structural dynamics are also important considerations.

When installing a cluster of machines in a wind farm, certain spacing between the wind towers must be maintained to optimize the energy crop over the year. The spacing depends on the terrain, wind direction, wind speed, and turbine size. The optimum spacing is found in rows 8–12 rotor diameters apart in the wind direction, and 2–4 rotor diameters apart in the crosswind direction (Figure 4.14). A wind farm consisting of 20 towers, rated at 500 kW each, needs 1–2 km^2 of land area. Of this, less than 5% of the land is actually required for turbine towers and access roads. The remaining land could continue its original use. Thus, wind turbines can co-exist with grazing, farming, fishing, and recreational use (Figure 4.15). The average number of machines in wind farms varies greatly, ranging from several to hundreds depending on the required power capacity of the farm. The preceding spacing rules would

Wind Power Systems

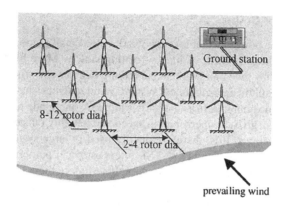

FIGURE 4.14 Optimum tower spacing in wind farms in flat terrain.

FIGURE 4.15 Original land use continues in a wind farm in Germany. (From Vestas Wind Systems, Denmark. With permission.)

ensure that the turbines do not shield those further downwind. Some wind farms have used narrow spacing of five to six rotor diameters in the wind direction. One such farm in Mackinaw City, MI, has reported the rotors in downwind direction running slower due to the wake effect of the upwind rotors.

The wind power fluctuations and electrical transients on fewer large machines would cost more in the filtering of power and voltage fluctuations, or would degrade the quality of power, inviting penalty from the grid.

The optimization method presented by Roy[2] takes into account the preceding trades. Additionally, it includes the effect of tower height that goes with the turbine diameter, available standard ratings, cost at the time of procurement, and wind speed. The wake interaction and tower shadow are ignored for simplicity. Such optimization leads to a site-specific number and size of the wind turbines that will minimize the energy cost.

4.6.2 Number of Blades

One can extract the power available in the wind with a small number of blades rotating quickly, or a large number of blades rotating slowly. More blades do not give more power, but they give more torque and require heavier construction. A few fast-spinning blades result in an economical system. Wind machines have been built with the number of blades ranging from 1 to 40 or more. A one-blade machine, although technically feasible, gives a supersonic tip speed and a highly pulsating torque, causing excessive vibrations. It is, therefore, hardly used in large systems. A very high number of blades were used in old low-TSR rotors for water pumping and grain milling, the applications requiring high starting torque. Modern high-TSR rotors for generating electric power have two or three blades, many of them with just two, although the Danish standard is three blades. The major factors involved in deciding the number of blades are as follows:

- The effect on power coefficient
- The design TSR
- The means of limiting yaw rate to reduce the gyroscopic fatigue

Compared to the two-blade design, the three-blade machine are more common, as they give smoother power output and a balanced gyroscopic force. There is no need to teeter the rotor, allowing the use of a simple rigid hub. Large machines up to several MW are being built using the three-blade configuration. Some machines in the past, especially in the U.S., have been two-blade design. Adding the third blade increases the power coefficient only by about 5%, thus giving a diminished rate of return for the 50% more blade weight and cost. The two-blade rotor is also simpler to erect, because it can be assembled on the ground and lifted to the shaft without complicated maneuvers during the lift. The number of blades is often viewed as the blade solidity. Higher solidity ratio gives higher starting torque and leads to low-speed operation. For electric power generation, the turbine must run at high speeds as the electrical generator weighs less and operates more efficiently at high speeds. That is why all large-scale wind turbines have low solidity ratio, with just two or three blades.

4.6.3 Rotor Upwind or Downwind

Operating the rotor upwind of the tower produces higher power as it eliminates the tower shadow on the blades. This results in lower noise, lower blade fatigue, and smoother power output. A drawback is that the rotor must constantly be turned into the wind via the yaw mechanism. The heavier yaw mechanism of an upwind turbine requires a heavy-duty and stiffer rotor compared to a downwind rotor.

The downwind rotor has the wake (wind shade) of the tower in the front and loses some power from the slight wind drop. On the other hand, it allows the use of a free yaw system. It also allows the blades to deflect away from the tower when loaded. Its drawback is that the machine may yaw in the same direction for a long period of time, which can twist the cables that carry current from the turbines.

Both types have been used in the past with no clear trend. However, the upwind rotor configuration has recently become more common.

4.6.4 Horizontal vs. Vertical Axis

In the horizontal-axis Danish machine, considered to be classical, the axis of blade rotation is horizontal with respect to the ground and parallel to the wind stream. Most wind turbines are built today with the horizontal-axis design, which offers a cost-effective turbine construction, installation, and control by varying the blade pitch.

The vertical-axis Darrieus machine has different advantages. First of all, it is omnidirectional and requires no yaw mechanism to continuously orient itself toward the wind direction. Secondly, its vertical drive shaft simplifies the installation of the gearbox and the electrical generator on the ground, making the structure much simpler. On the negative side, it normally requires guy wires attached to the top for support. This could limit its applications, particularly at offshore sites. Overall, the vertical-axis machine has not been widely used, primarily because its output power cannot be easily controlled in high winds simply by changing the blade pitch. With modern low-cost variable-speed power electronics emerging in the wind power industry, the Darrieus configuration may revive, particularly for large-capacity applications.

The Darrieus has structural advantages compared to a horizontal-axis turbine because it is balanced. The blades only "see" the maximum lift torque twice per revolution. Seeing maximum torque on one blade once per revolution excites many natural frequencies, causing excessive vibrations. Also a vertical-axis wind turbine configuration is set on the ground. Therefore, it is unable to effectively use higher wind speeds using a higher tower, as there is no tower here.

4.7 SYSTEM CONTROL REQUIREMENTS

Both the speed and the rate of change must be controlled in a good system design.

4.7.1 Speed Control

The rotor speed must be controlled for three reasons:

- To capture more energy, as seen before.
- To protect the rotor, generator, and power electronic equipment from overloading during high-gust winds.
- When the generator is disconnected from the electrical load, accidentally or for a scheduled event. Under this condition, the rotor speed may run away, destroying it mechanically, if it is not controlled.

The speed control requirement of the rotor has five separate regions as shown in Figure 4.16:

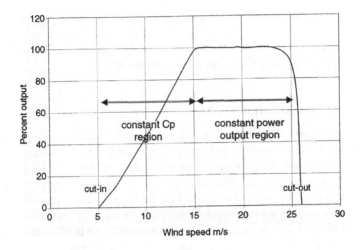

FIGURE 4.16 Five regions of turbine speed control.

1. The cut-in speed at which the turbine starts producing power. Below this speed, it is not worthwhile, nor efficient, to turn the turbine on.
2. The constant maximum C_p region where the rotor speed varies with the wind speed variation to operate at the constant TSR corresponding to the maximum C_p value.
3. During high winds, the rotor speed is limited to an upper constant limit based on the design limit of the system components. In the constant-speed region, the C_p is lower than the maximum C_p, and the power increases at a lower rate than that in the first region.
4. At still higher wind speeds, such as during a gust, the machine is operated at a controlled constant power to protect the generator and power electronics from overloading. This can be achieved by lowering the rotor speed. If the speed is decreased by increasing the electrical load, then the generator will be overloaded, defeating the purpose. To avoid generator overloading, some sort of a brake (eddy current or another type) must be installed on the rotor.
5. The cutout speed, at which the rotor is shut off to protect the blades, the electrical generator, and other components of the system beyond a certain wind speed.

4.7.2 Rate Control

The inertia of large rotors must be taken into account in controlling the speed. The acceleration and deceleration must be controlled to limit the dynamic mechanical stresses on the blades and hub, and the electrical load on the generator and power electronics. The instantaneous difference between the mechanical power produced by the blades and the electric power delivered by the generator will change the rotor speed as follows:

$$J \frac{d\omega}{dt} = \frac{P_m - P_e}{\omega} \tag{4.4}$$

Wind Power Systems

where

J = polar moment of inertia of the rotor
ω = angular speed of the rotor
P_m = mechanical power produced by the blades
P_e = electric power delivered by the generator

Integrating Equation 4.4, we obtain:

$$\frac{1}{2}J\left(\omega_2^2 - \omega_1^2\right) = \int_{t_1}^{t_2}\left(P_m - P_e\right)dt \tag{4.5}$$

Let us examine this aspect for a sample rotor with the moment of inertia $J = 7500$ kg.m². Changing this rotor speed from 100 to 95 rpm in 5 sec requires ΔP of 800 kW. The resulting torque of 80 Nm would produce torsional stress on the rotor structure and hub components. If the same speed change were made in 1 sec, the required power would be 4000 kW, and the torque, 400 Nm. Such high torque can overstress and damage the rotor parts or shorten its life significantly. For this reason, the acceleration and deceleration must be kept within design limits, with adequate margins.

The strategy for controlling the speed of the wind turbine varies with the type of electrical machine used, i.e., the induction, synchronous, or DC machine.

4.8 ENVIRONMENTAL ASPECTS

The following aspects of environmental considerations enter the wind farm setting:

4.8.1 AUDIBLE NOISE

The wind turbine is generally quiet. It poses no objectionable noise disturbance in the surrounding area. The wind turbine manufacturers supply the machine noise level data in dB(A) vs. the distance from the tower. A typical 600-kW machine noise level is shown in Figure 4.8. This machine produces 55-dB(A) noise at 50-m distance from

TABLE 4.3
Noise Levels of Some Commonly Known Sources Compared with Wind Turbine

Source	Noise Level (dB)
Elevated train	100
Noisy factory	90
Average street	70
Average factory	60
Average office	50
Quiet conversation	30

the turbine and 40-dB(A) at 250-m distance. Table 4.3 compares the turbine noise level with other generally known noise levels. The table indicates that the turbine at a 50-m distance produces no noise higher than the average factory. This noise, however, is a steady noise. Additionally, the turbine makes a louder noise while yawing under the changing wind direction. In either case, the local noise ordinance must be complied with. In some instances, there have been cases of noise complaints reported by the nearby communities. Although noise pollution is not a major problem with offshore wind farms, it depends on the size whether or not one can hear the turbines while operating. It has also been suggested that the noise from the turbines travels underwater and disturbs sea life as well.

In general, there are two main sources of noise emitted from the wind turbine. One is mechanical, which is inherent in the gearing system. The other is created by the aerodynamics of the rotating blade, which emits a noise when passing the tower, known as the tower thump or simply the aerodynamic noise. The first may be at a somewhat low level, generally uniform over the year. The other—the tower thump—can be loud. It varies with the speed of blade rotation and may cause most of the problems and complaints. Some residents describe the tower thump noise as being like a boot in a tumble dryer. A large wind turbine can produce an aggregate noise level of up to 100 dB(A), which weakens to a normal level within a 1.5-km distance. The worst conditions are when the wind is blowing lightly and the background noise is minimal. Residents up to 1-km radius have complained to the Environmental Health Department about noise from such turbines.

4.8.2 Electromagnetic Interference (EMI)

Any stationary or moving structure in the proximity of a radio or TV tower interferes with the signals. The wind turbine tower, being a large structure, can cause objectionable EMI in the performance of a nearby transmitter or a receiver. Additionally, the rotating blades of an operating wind turbine may reflect impinging signals so that the electromagnetic signals in the neighborhood may experience interference at the blade passage frequency. The exact nature and magnitude of such EMIs depend on a number of parameters. The primary parameters are the location of the wind turbine tower relative to the radio or TV tower, physical and electrical properties of the rotor blades, the signal frequency modulation scheme, and the high-frequency electromagnetic wave propagation characteristics in the local atmosphere.[3]

EMI may be a serious issue with wind farm planning. For example, 5 of the 18 offshore wind farms planned around the U.K. coasts were blocked by the U.K. Ministry of Defense due to concerns that they may interfere with radar and flight paths to airfields close to the proposed sites. Detailed studies on the precise effects of wind turbines on radar and possible modifications in radar software may mitigate the concerns. The potential cost of such studies and legal appeals should be factored into the initial planning of large wind farms.

The University of Liverpool in the U.K. has investigated the effects of electromagnetic fields generated by offshore wind farm cables on the marine environment.

4.8.3 EFFECTS ON BIRDS

The effect of wind farms on wild life and avian population—including endangered species protected by federal laws—has created controversy and confusion within the mainstream environmental community. The breeding and feeding patterns of some birds may be disturbed. The wind turbine blade can weigh few tons and the blade tip speed can be up to 200 mph, a lethal weapon against any airborne creature. The birds may be killed or at least injured if they collide with a blade. Often the suction draft created by the wind flowing to a turbine draws the birds into the airstream headed for the blades. Although less usual, birds are attracted by the tower hum (music!) and simply fly into the towers. On the other hand, studies at an inshore site near Denmark have determined that birds alter their flight paths 200 m around the turbine. Thus, a wind farm can significantly alter the flight paths of large avian populations. In another study, the population of water fowl declined 75–90% within 3 yr after installing an offshore wind farm in Denmark. Such a large decline could have a massive impact on the ecosystem of the surrounding area.

The initially observed high bird-kill rate of the 1980s has significantly declined with larger turbines used at present having longer slower-moving blades, which are easier for the birds to see and avoid. Tubular towers have a lower bird-hit rate compared to lattice towers, which attract birds to nest. The turbines are now mounted on either solid tubular towers or towers with diagonal bracing, eliminating the horizontal supports that attracted the birds for nesting. New wind farms are also sited away from avian flight paths.

Historical data from various sources on the birds killed follow:

- In Altamont Pass, CA, 200–300 red-tailed hawks and 40–60 golden eagles were killed annually by early wind farms in the 1980s. Bill Evans of Cornell University estimated that the bird kills due to wind farms may have exceeded 5,000,000/yr during the 1980s. Up to 10,000 have been killed by one turbine in one night. A 1994 estimate in Europe showed a significant number of 13 bird species, protected by European law, killed by turbines at Tarifa, Spain.
- Forest City School District in Iowa had installed wind turbines for self-sufficiency in energy. Their turbine sites had the most wind in the city, and were in the migratory path of many geese. Even then, neither bird kills nor noise complaints were reported over the first several years of operation.
- By way of comparison, U.S. hunters annually kill over 120 million birds, and about one billion birds are killed by flying into glass windows on their own each year in the U.S.

It is generally agreed that migration paths and nesting grounds of rare species of birds should be protected against the threat of wind farms. Under these concerns, obtaining permission from the local planning authorities can take considerable time and effort.

4.8.4 OTHER IMPACTS

The visual impact of the wind farm may be unpleasant to the property owners around the wind farm. This is more so for offshore wind farms, where the property owners are usually affluent people with significant influence on the area's policymaking organizations. In the early 2000s, the most publicized proposed project in Nantucket Sound had become highly controversial by high-profiled owners of the waterfront estates.

Wind farm designers can minimize aesthetic complaints by installing identical turbines and spacing them uniformly.

Because wind is a major transporter of energy across the globe, the impact of the energy removed by many large wind farms on a grand scale may possibly impact the climate. This may be a subject of future studies.

4.9 POTENTIAL CATASTROPHES

A fire or an earthquake can be a major catastrophe for a wind power plant.

4.9.1 FIRE

Fire damage amounts to 10–20% of the wind plant insurance claims. A fire on a wind turbine is rare but difficult to fight. Reaching the hub height is slow, and water pressure is always insufficient to extinguish the fire. This generally leads to a total loss of the turbine, leading to 9–12 months of downtime and lost revenue. The cost of replacing a single 30-m blade can exceed $100,000, and that of replacing the whole 3-MW turbine can exceed $2 million.

The following are some causes of fire in wind turbines:

Lightning strike: Lightning arresters are used to protect the turbine blades, nacelle, and tower assembly. However, if lightning is not properly snubbed, it can lead to local damage or total damage if it leads to sparks and subsequent turbine fire. Lightning occurrences depend on the location. Offshore turbines are more prone to lightning than land turbines. On land, lightning is rare in Denmark, whereas it is frequent in northern Germany and the Alps regions, and even more frequent in parts of Japan and the U.S., particularly in Florida and Texas. The growing trends of using electrically conducting carbon fiber-epoxy composites for their high strength and low weight in the blade construction make the blades more vulnerable to lightning. For this reason, some manufacturers avoid carbon fibers in their blades, more so in large, tall turbines for offshore installations.

Internal fault: Any electrical or mechanical fault leading to a spark with the transmission fluids or other lubricants is a major risk. The flammable plastic used in the construction, such as the nacelle covers, is also a risk.

Wind Power Systems

Typical internal faults that can cause excessive heat leading to a fire are as follows:

- Bearings running dry and failing
- Failing cooling system
- Brakes becoming hot under sustained braking
- Oil and grease spills
- Short circuit in the battery pack of the pitch-control system
- Cables running against rotating or vibrating components

Frequent physical checks of the entire installation, servicing and maintenance, and a condition-monitoring system, accessed remotely by computers used in modern installations, can detect potential fire hazards and avoid fires.

4.9.2 Earthquake

Lateral loads resulting from an earthquake are important data to consider in designing a tall structure in many parts of the world. The wind tower, being always tall, is especially vulnerable to seismic events. The seismic energy is concentrated in the 1- to 10-Hz frequency band. A dynamic analysis is required, as a dynamic response amplification is expected. However, because of the complexities in modeling and performing such analyses, it has been standard practice to represent seismic loads with equivalent static loads. The severity of the seismic loads, the potential failure modes, and the resulting effects require that design engineers make reasonable tradeoffs between potential safety concerns and economics during the design phase. To alleviate this difficulty, the recent trend in the U.S. has been to require dynamic analyses to estimate seismic stresses. For example, all primary components of nuclear power plants in the U.S. must be dynamically analyzed for the specified seismic loadings. This is required even for plants located in seismically inactive areas.

4.10 SYSTEM-DESIGN TRENDS

Significant research and development work is underway at the NREL and the National Wind Technology Center in Golden, CO. The main areas of applied research conducted at the NWTC are as follows:

- Aerodynamics to increase energy capture and reduce acoustic impacts.
- Inflow and turbulence to understand the nature of wind.
- Structural dynamics models to minimize the need of prototypes.
- Controls to enhance energy capture, reduce loads, and maintain stable closed-loop behavior of these flexible systems are an important design goal.

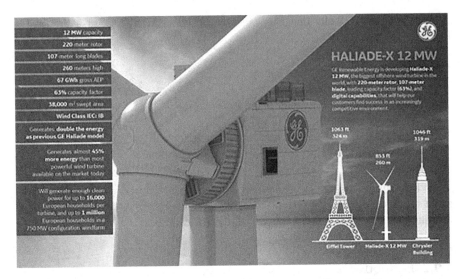

FIGURE 4.17 A large 12 MW wind turbine design data sheet. (Source: GE's product catalog on public internet site)

- The wind turbine design progress includes larger turbines on taller towers to capture higher wind speed; the design difficulty increases as these machines become larger and the towers become taller. The GE's 12 MW turbine (Figure 4.17) is an example of the progress made in the wind turbine design technology.

REFERENCES

1. Van Kuik, G., Wind turbine technology—25 years' progress, *Wind Directions*, April 1998.
2. Roy, S., Optimal planning of wind energy conversion systems over an energy scenario, *IEEE Transactions on Energy Conversion*, September 1997.
3. Spera, D. A., *Wind Turbine Technology*, American Society of Mechanical Engineers, New York, 1994.

5 Electrical Generators

The mechanical power of a wind turbine is converted into electric power by usually an alternating current (AC) generator. The AC generator can be either a synchronous machine or an induction machine.

The electrical machine works on the principle of action and reaction of electromagnetic induction. The resulting electromechanical energy conversion is reversible. The same machine can be used as a motor for converting electric power into mechanical power or as a generator for converting mechanical power into electric power.

Figure 5.1 depicts common construction features of electrical machines. Typically, there is an outer stationary member (stator) and an inner rotating member (rotor). The rotor is mounted on bearings fixed to the stator. Both the stator and the rotor carry cylindrical iron cores, which are separated by an air gap. The cores are made of magnetic iron of high permeability and have conductors embedded in slots distributed on the core surface. Alternatively, the conductors are wrapped in the coil form around salient magnetic poles. Figure 5.2 is a cross-sectional view of a rotating electrical machine with the stator made of salient poles and the rotor with distributed conductors. The magnetic flux, created by the excitation current in one of the two coils, passes from one core to the other in a combined magnetic circuit always forming a closed loop. Electromechanical energy conversion is accomplished by interaction of the magnetic flux produced by one coil with the electrical current in the other coil. The current may be externally supplied or electromagnetically induced. The induced current in a coil is proportional to the rate of change in the flux linkage of that coil.

The various types of machines differ fundamentally in the distribution of the conductors forming the windings, and by their elements: whether they have continuous slotted cores or salient poles. The electrical operation of any given machine depends on the nature of the voltage applied to its windings. The narrow annular air gap between the stator and the rotor is the critical region of the machine operation, and the theory of performance is mainly concerned with the conditions in or near the air gap.

5.1 TURBINE CONVERSION SYSTEMS

The early wind turbines were what is today called windmills that were used by the Persians (Iranians) in 640s for mechanical power. The first wind turbines used for electrical power generation used DC generators in 1890s. Over the decades the wind turbine conversion systems developed in terms of aerodynamics and mechanics, generator and power electronics, and the control schemes. In its modern form, one of the early turbines used an induction machine connected directly to the grid, and referred to direct online (DOL) as schematically shown in Figure 5.3. An issue with the DOL conversion is the lack of a control by the generator and dependency on a variable

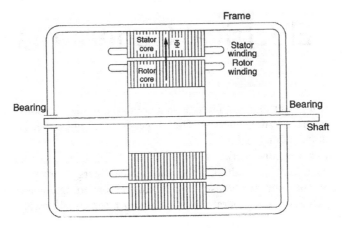

FIGURE 5.1 Common constructional features of rotating electrical machines.

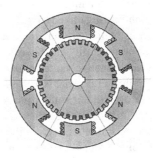

FIGURE 5.2 Cross-section of the electrical machine stator and rotor.

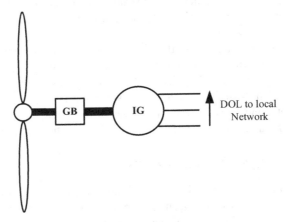

FIGURE 5.3 Direct online (DOL) turbine.

speed gearbox that is mechanically controlled. This made these turbines unpopular. In addition, the reactive current for the induction generator in the DOL scheme is supplied by the grid. To improve this a switched capacitance bank may be connected to the terminals of DOL scheme, as shown in Figure 5.4, where the capacitor bank supplies the required induction generator reactive power.

Electrical Generators

Introducing power electronics to the wind turbine power system, the wind-to-electrical power conversion efficiency gets significantly improved. Figure 5.5 shows a wind power system with doubly fed induction generator (DFIG), and two power electronics converters, an AC/DC and a DC/AC converter. The AC/DC and DC/AC converters in Figure 5.5 are connected in back-to-back arrangement, where they decouple the variable-speed variable-frequency rotor of the DFIG from the fixed-frequency fixed-voltage 3-phase grid. The rotor and back-to-back converters process a portion of the power while the DFIG stator is directly connected to the grid. In the past decade, the wind turbines with DFIGs have been widely used in the wind farms. However, the DFIG-based systems are limited in the power and speed operation as the converters are not fully rated.

To obtain full operational speed, the generator is decoupled from the grid using fully rated power electronics converters as shown in Figure 5.6. In this scheme, the

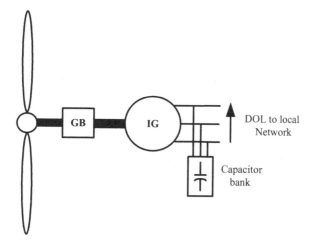

FIGURE 5.4 Direct online (DOL) with a capacitor bank to supply the excitation power.

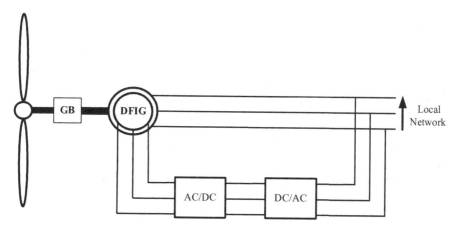

FIGURE 5.5 Turbine conversion system with doubly fed induction generator (DFIG).

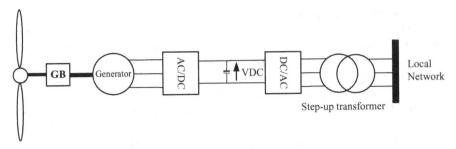

FIGURE 5.6 Fully rated wind turbine power conversion system with power electronics for maximum power extraction.

generator outputs are rectified to DC using an AC/DC converter. A DC/AC converter is then used to convert the DC to a 3-phase AC output voltage with fixed magnitude and frequency. The two power electronics converters are arranged in back-to-back and usually use a voltage source converter (VSC) form. The generator may be permanent magnet generator, induction generator, or conventional synchronous generator.

As the wind velocity changes, the output voltage, frequency, and power of the generator vary. The AC/DC converter controls the generator to extract maximum power from the wind at each wind velocity, a concept referred to as maximum power point tracking (MPPT). The generator's variable voltage and frequency output is then converted to DC. The DC/AC converter controls the DC-link around a nominal fixed value, which is essential for the correct operation of the AC/DC converter. The frequency and voltage at the output of DC/AC converters are fixed. As the conversion scheme in this scheme is rated at low voltage, a step-up transformer is used to increase the voltage suitable for transmission to a substation that collects the power from all turbines.

5.2 SYNCHRONOUS GENERATOR

The synchronous generator produces most of the electrical power consumed in the world. For this reason, the synchronous machine is technically matured and hence widely used machine in utility power plants. The machine works at a constant speed related to the fixed supply frequency. Therefore, it is not well suited for variable-speed operation in wind power plants without power electronic frequency converters. Moreover, the conventional synchronous machine requires DC current to excite the rotor field, which has traditionally used sliding carbon brushes on slip rings on the rotor shaft. This introduces a routine maintenance and some unreliability in its use. The modern synchronous machines are made brushless by generating the required DC field current on the rotor itself. Reliability is greatly improved while reducing the cost. In small synchronous machine for wind power, the DC field current need can be eliminated altogether by using a reluctance rotor,[1] in which the synchronous operation is achieved by the reluctance torque. The reluctance machine rating, however, is

Electrical Generators

limited to below 100 kW. It is being investigated at present for small wind generators.

The synchronous machine is ideally suited to constant-speed systems such as solar thermal power plants. It is, therefore, covered further in detail in Chapter 18.

The synchronous machine, when used in a grid-connected system, has some advantages over the induction machine. It does not require reactive power from the grid. This results in a better quality of power at the grid interface. This advantage is more pronounced when the wind farm is connected to a small-capacity grid using a long low-voltage transmission link. For this reason, California plants in the early 1980s used synchronous generators. Present-day wind plants connected to large grids using short lines almost universally use the induction generator.

The synchronous generator is rarely used in gear-driven wind systems. However, the low-speed design of the synchronous generator is often found advantageous in the direct-drive variable-speed wind turbine. In such a design, the generator is completely decoupled from the grid by a voltage source power electronic converter connected to the stator, and the rotor is excited by an excitation winding or a permanent magnet. GE Wind's 2.X series of 2- to 3-MW generators have switched from the doubly fed induction generator to the synchronous generator, whereas the variable-speed pitch-control operating principle has remained unchanged.

5.2.1 Equivalent Circuit

The synchronous generator single-phase equivalent circuit is formed by induced back-EMF, phase resistance, and synchronous reactance as shown in Figure 5.7. The phase back-EMF is a function of flux-linkage (λ), speed (ω), and the number of stator turns:

$$EMF = N\omega \frac{d\lambda}{dt} \tag{5.1}$$

The synchronous reactance (X_s) includes the effect of mutual inductance between the three phases, and the resistance (R_s) represents the per-phase stator winding resistance. On the equivalent circuit, the phase current (I_s) and terminal voltage (V_t) are also shown.

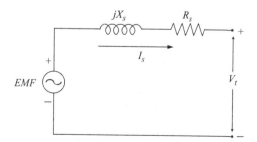

FIGURE 5.7 Single-phase equivalent circuit for synchronous generator.

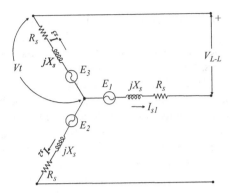

FIGURE 5.8 3-phase equivalent circuit for synchronous generator.

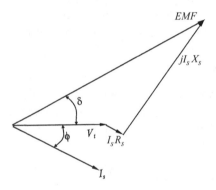

FIGURE 5.9 Phasor diagram of phase voltage and phase current in synchronous generator.

Figure 5.8 shows the 3-phase equivalent of a synchronous generator where the phases are connected in star (also called Y). The line voltage (V_L) is defined as the terminal voltage between the two phase lines.

Figure 5.9 shows a per-phase phasor diagram for the synchronous generator using the equivalent circuit diagram. In this case, the phase current lags the terminal voltage indicating an inductive-reactance dominant load. The angle between the phase current and terminal voltage is called the power factor angle, while the angle between the back-EMF and terminal voltage is called the load angle. If the load angle is zero, the synchronous generator provides net zero active power. The reactive power, however, is defined by the amplitude of the voltage vectors. If the amplitude of the back-EMF is greater than that of terminal voltage, the synchronous generator injects reactive power to the load. The synchronous generator received reactive power if the amplitude of terminal voltage is greater than back-EMF.

5.2.2 Synchronous Generators in Wind Turbines

The synchronous generators have been used in wind turbines. ENERCON (a turbine manufacturer) uses low-speed synchronous generators in a direct drive (no gearbox) wind turbine conversion schemes. The stator windings are closed sign-layer basket with class F (155 °C) insulation. The winding uses round wires and the stator is designed for low voltages (<1000 Vac). Figure 5.10 shows ENERCON direct-drive generator. The stator has a large diameter compared to its axial length. This is due to the generator operating at low speeds (up to 20 RPM), a common condition for direct-drive generators in wind turbines.

The power-angle characteristic for synchronous generator in a wind turbine is shown in Figure 5.11. The generator power depends on the load angle controlled by AC/DC converter shown in Figure 5.6. The maximum power is generated when the load angle (δ in Figure 5.9) is 90°. The plots in Figure 5.11 also show the variation of synchronous generator output power with respect to synchronous inductance. As noted, the lower the inductance the higher the output power. A feature that is taken into account in the generator design.

The output power of ENERCON wind turbine using synchronous generator is shown in Figure 5.12. The turbine power increases with wind velocity from a cut-in velocity of 2 m/s to a rated velocity of 13 m/s. The wind turbine reaches its nominal power 2 MW at the rated velocity. Above the rated velocity and up to cut-out velocity the wind turbine delivers its rated output power while the blade pitch control adjusts the blade angle. The synchronous generator in the turbine is controlled to deliver the power at each wind velocity as its rotational speed is related to the wind velocity. The turbine power coefficient (C_p) is a feature that defines the amount of power that turbine is capable of extracting from the kinetic wind energy. This coefficient is also plotted in Figure 5.12 with its values shown on the right y-axis.

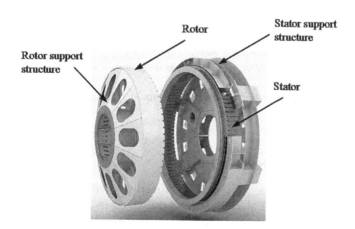

FIGURE 5.10 ENERCON synchronous generator for wind turbine applications. (From ENERCON, Germany).

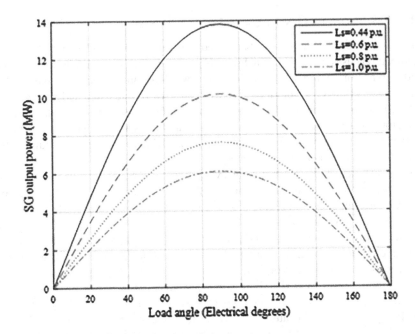

FIGURE 5.11 Synchronous generator power-angle characteristics.

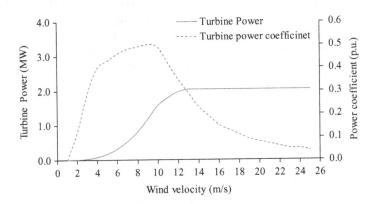

FIGURE 5.12 ENERCON wind turbine power-speed characteristics using synchronous generator.

5.3 INDUCTION GENERATOR

The electric power in industry is consumed primarily by induction machines working as motors driving mechanical loads. For this reason, the induction machine, invented by Nikola Tesla and financed by George Westinghouse in the late 1880s, represents a well-established technology. The primary advantage of the induction machine is the rugged brushless construction that does not need a separate DC field power. The

Electrical Generators

disadvantages of both the DC machine and the synchronous machine are eliminated in the induction machine, resulting in low capital cost, low maintenance, and better transient performance. For these reasons, the induction generator is extensively used in small and large wind farms and small hydroelectric power plants. The machine is available in numerous power ratings up to several megawatts capacity, and even larger.

For economy and reliability, many wind power systems use induction machines as electrical generators. The remaining part of this chapter is devoted to the construction and the theory of operation of the induction generator.

5.3.1 Construction

In the electromagnetic structure of the induction generator, the stator is made of numerous coils wound in three groups (phases), and is supplied with three-phase current. The three coils are physically spread around the stator periphery and carry currents, which are out of time phase. This combination produces a rotating magnetic field, which is a key feature in the working of the induction machine. The angular speed of the rotating magnetic field is called the *synchronous speed*. It is denoted by N_s and is given by the following in rpm:

$$N_s = 60 \frac{f}{p} \quad (5.2)$$

where
f = frequency of the stator excitation
p = number of magnetic pole pairs

The stator coils are embedded in slots in a high-permeability magnetic core to produce the required magnetic field intensity with a small exciting current.

The rotor, however, has a completely different structure. It is made of solid conducting bars, also embedded in slots in a magnetic core. The bars are connected together at both ends by two conducting end rings (Figure 5.13). Because of its

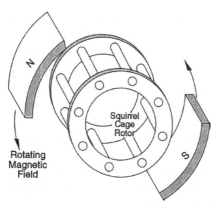

FIGURE 5.13 Squirrel cage rotor of the induction machine under rotating magnetic field.

resemblance, the rotor is called a *squirrel cage rotor*, or the *cage rotor*, for short, and the machine is called the *squirrel cage induction machine*.

5.3.2 Working Principle

The stator magnetic field is rotating at the synchronous speed determined by Equation 5.2. This field is conceptually represented by the rotating magnets in Figure 5.13. The relative speed between the rotating field and the rotor induces the voltage in each closed loop of the rotor conductors linking the stator flux ϕ. The magnitude of the induced voltage is given by Faraday's law of electromagnetic induction, namely:

$$e = -\frac{d\phi}{dt} \qquad (5.3)$$

where ϕ = the magnetic flux of the stator linking the rotor loop.

This voltage in turn sets up the circulating current in the rotor. The electromagnetic interaction of the rotor current and stator flux produces the torque. The magnitude of this torque is given by the following:

$$T = k\Phi I_2 \cos\phi_2 \qquad (5.4)$$

where
k = constant of proportionality
Φ = magnitude of the stator flux wave
I_2 = magnitude of induced current in the rotor loops
ϕ_2 = phase angle by which the rotor current lags the rotor voltage

The rotor accelerates under this torque. If the rotor were on frictionless bearings in a vacuum with no mechanical load attached, it would be completely free to rotate with zero resistance. Under this condition, the rotor would attain the same speed as the stator field, namely, the synchronous speed. At this speed, the current induced in the rotor is zero, no torque is produced, and none is required. Under these conditions, the rotor finds equilibrium and will continue to run at the synchronous speed.

If the rotor is now attached to a mechanical load such as a fan, it will slow down. The stator flux, which always rotates at a constant synchronous speed, will have a relative speed with respect to the rotor. As a result, electromagnetically induced voltage, current, and torque are produced in the rotor. The torque produced must equal that needed to drive the load at this speed. The machine works as a motor in this condition.

If we attach the rotor to a wind turbine and drive it faster than its synchronous speed via a step-up gear, the induced current and the torque in the rotor reverse the direction. The machine now works as the generator, converting the mechanical power of the turbine into electric power, which is delivered to the load connected to the stator terminals. If the machine were connected to a grid, it would feed power into the grid. Thus, the induction machine can work as an electrical generator only at speeds higher than the synchronous speed. The generator operation, for this reason, is often called the *supersynchronous operation* of the induction machine.

Electrical Generators

As described in the preceding text, an induction machine needs no electrical connection between the stator and the rotor. Its operation is entirely based on electromagnetic induction; hence, the name. The absence of rubbing electrical contacts and simplicity of its construction make the induction generator a very robust, reliable, and low-cost machine. For this reason, it is widely used in numerous industrial applications.

Engineers familiar with the theory and operation of the electrical transformer would see the working principle of the induction machine can be seen as the transformer, in which the fixed high-voltage primary coil on the stator is excited, and the low-voltage secondary coil on the rotor is shorted on itself and is free to rotate. The electrical or mechanical power from one to the other can flow in either direction. The theory and operation of the transformer, therefore, holds true when modified to account for the relative motion between the stator and the rotor. This motion is expressed in terms of the slip of the rotor relative to the synchronously rotating magnetic field.

5.3.3 Rotor Speed and Slip

The slip of the rotor is defined as the ratio of the speed of rotating magnetic field sweeping past the rotor and the synchronous speed of the stator magnetic field as follows:

$$s = \frac{N_s - N_r}{N_s} \qquad (5.5)$$

where

s = slip of the rotor in a fraction of the synchronous speed
N_s = synchronous speed = $60 f/$
pN_r = rotor speed

The slip is positive in the motoring mode and negative in the generating mode. In both modes, a higher rotor slip induces a proportionally higher current in the rotor, which results in greater electromechanical power conversion. In both modes, the value of slip is generally a few to several percent. Higher slips, however, result in greater electrical loss, which must be effectively dissipated from the rotor to keep the operating temperature within the allowable limit.

The heat is removed from the machine by the fan blades attached to one end-ring of the rotor. The fan is enclosed in a shroud at the end. The forced air travels axially along the machine exterior, which has fins to increase the dissipation area. Figure 5.14 is an exterior view of a 150-kW induction machine showing the end shroud and the cooling fins running axially. Figure 5.15 is a cutaway view of the machine interior of a 2-MW induction machine.

The induction generator feeding a 60-Hz grid must run at a speed higher than 3600 rpm in a 2-pole design, 1800 rpm in a 4-pole design, and 1200 rpm in a 6-pole design. The wind turbine speed, on the other hand, varies from a few hundred rpm in kW-range machines to a few tens of rpm in MW-range machines. The wind turbine, therefore, must interface the generator via a mechanical gear. As this somewhat degrades efficiency and reliability, many small stand-alone plants operate with custom-designed generators operating at lower speeds without any mechanical gear.

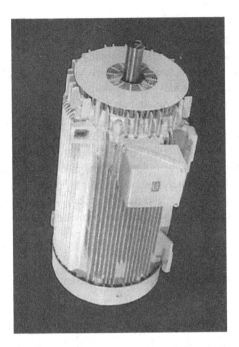

FIGURE 5.14 A 150-kW induction machine. (From General Electric Company, Fort Wayne, IN.)

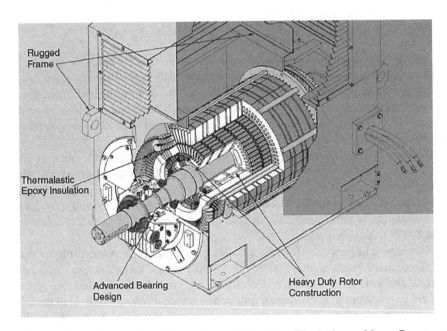

FIGURE 5.15 A 2-MW induction machine. (From Teco Westinghouse Motor Company, Round Rock, TX. With permission.)

Electrical Generators

Under the steady-state operation at slip "s," the induction generator has the following operating speeds in rpm:

- Stator flux wave speed $\quad N_s$
- Rotor mechanical speed $\quad N_r = (1-s)N_s$
- Stator flux speed with respect to rotor $\quad sN_s$ (5.6)
- Rotor flux speed with respect to stator $\quad N_r + sN_s = N_s$

Thus, the squirrel cage induction machine is essentially a constant-speed machine, which runs slightly slipping behind the rotating magnetic field of the three-phase stator current. The rotor slip varies with the power converted, and the rotor speed variations are within a few percent. It always consumes reactive power — undesirable when connected to a weak grid — which is often compensated by capacitors to achieve the systems power factor closed to one. Changing the machine speed is difficult. It can be designed to run at two different but fixed speeds by changing the number of poles of the stator winding.

The voltage usually generated in the induction generator is 690-V AC. It is not economical to transfer power at such a low voltage over a long distance. Therefore, the machine voltage is stepped up to a higher value between 10,000 V and 30,000 V via a step-up transformer to reduce the power losses in the lines.

5.3.4 Equivalent Circuit

The theory of operation of the induction machine is represented by the equivalent circuit shown in Figure 5.16(a). It is similar to that of the transformer. The left-hand side of the circuit represents the stator and the right-hand side, the rotor. The stator and rotor currents are represented by I_1 and I_2, respectively. The vertical circuit branch at the junction carries the magnetizing (also known as excitation or no-load) current I_0, which sets the magnetic flux required for the electromagnetic operation of the machine. The total stator current is then the sum of the rotor current and the

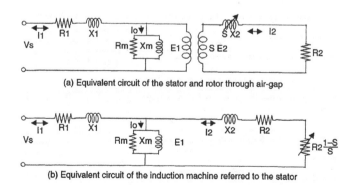

FIGURE 5.16 Equivalent electrical circuit of induction machine for performance calculations.

excitation current. The air-gap separation is not shown, nor is the difference in the number of turns in the stator and rotor windings. This essentially means that the rotor is assumed to have the same number of turns as the stator and has an ideal 100% magnetic coupling. We calculate the performance parameters taking the stator winding as the reference. The actual rotor voltage and current would be related to the calculated values through the turn ratio between the two windings. Thus, the calculations are customarily performed in terms of the stator, as we shall do in this chapter. This matches the practice, as the performance measurements are always done on the stator side. The rotor is inaccessible for any routine measurements.

Most of the flux links both the stator and the rotor coils. The flux that does not link both the coils is called the leakage flux, and is represented by the leakage reactance in ohms per phase. One half of the total leakage reactance is attributed to each side, namely the stator leakage reactance X_1 and the rotor leakage reactance X_2 in Figure 5.16(b). The stator and rotor conductor resistance is represented by R_1 and R_2, respectively. The magnetizing parameters X_m and R_m represent the permeability and losses (hysteresis and eddy currents) in the magnetic circuit of the machine.

The slip-dependent electrical resistance $R_2(1-s)/s$ represents the equivalent mechanical resistance on the shaft. Therefore, the electromechanical power conversion per phase is given by $I_2^2 R_2(1-s)/s$, and the three-phase power conversion is then given by the following in watts:

$$P_{em} = 3 I_2^2 R_2 (1-s)/s \tag{5.7}$$

If the machine is not loaded and has zero friction, it runs at the synchronous speed, the slip is zero, and the value of $R_2(1-s)/s$ becomes infinite. The rotor current is then zero and P_{em} is also zero, as it should be. On the other speed extreme, when the rotor is standing still, the slip is unity, and the value of $R_2(1-s)/s$ is zero. The rotor current is not zero, but the P_{em} is zero again, as the mechanical power delivered by the standstill rotor is zero. These two extreme operating points briefly validate the electromechanical power conversion given by Equation 5.7.

At any slip other than zero or unity, neither the rotor current nor the speed is zero, resulting in a nonzero value of P_{em}.

The machine's capacity rating is the power developed under rated conditions, that is:

$$\text{Machine rating} = \frac{P_{em\,rated}}{1000}(\text{kW}) \quad \text{or} \quad \frac{P_{em\,rated}}{746}(\text{hp}) \tag{5.8}$$

The mechanical torque is given by the power divided by the angular speed as follows:

$$T_{em} = P_{em}/\omega \tag{5.9}$$

where

T_{em} = electromechanical torque developed in the rotor in Nm
ω = angular speed of the rotor = $2\pi \cdot N_s(1-s)/60$ in mechanical rad/sec

Electrical Generators

Combining Equations 5.7 and 5.9, we obtain the torque at any slip s, as follows in units of Nm:

$$T_{em} = (180 / 2\pi N_s) I_2^2 R_2 / s \qquad (5.10)$$

The value of I_2 in Equation 5.10 is determined by the equivalent circuit parameters, and is slip-dependent. The torque developed by the induction machine rotor is, therefore, highly slip-dependent, as is discussed later in this chapter.

We take a note here that the performance of the induction machine is completely determined by the equivalent circuit parameters. The circuit parameters are supplied by the machine manufacturer, but can be determined by two basic tests on the machine. The full-speed test under no-load and the zero-speed test with blocked-rotor determine the complete equivalent circuit of the machine.[2-3]

Equivalent circuit parameters are generally expressed in fractions (per unit) of their respective rated values per phase. The rated impedance per phase is defined as follows in ohms:

$$Z_{rated} = \frac{\text{Rated voltage per phase}}{\text{Rated current per phase}} \qquad (5.11)$$

For example, the per unit (pu) stator resistance is expressed in the following way:

$$R_{1\,per\,unit} = \frac{R_1 (\text{ohms})}{Z_{rated} (\text{ohms})} \qquad (5.12)$$

and similar expressions for all other circuit parameters. When expressed as such, X_1 and X_2 are equal, each a few to several percent, and R_1 and R_2 are also approximately equal, each a few percent of the rated impedance. The magnetizing parameters X_m and R_m are usually large, in several hundred percent of Z_{rated}, hence, drawing negligible current compared to the rated current. For this reason, the magnetizing branch of the circuit is often ignored in making approximations of the machine performance calculations.

All of the preceding performance equations hold true for both the induction motor and the induction generator by taking the proper sign of the slip. In the generator mode, the value of the slip is negative in the performance equations wherever it appears. We must also remember that the real mechanical power output is negative, meaning that the shaft receives the mechanical power instead of delivering it. The reactive electric power drawn from the stator terminals remains leading with respect to the line voltage; hence, we say that the induction generator delivers a leading reactive power. Both of these mean that the magnetizing volt-amperes are supplied by an external source.

5.3.5 Efficiency and Cooling

The values of R_1 and R_2 in the equivalent circuit represent electrical losses in the stator and rotor, respectively. As seen later, for a well-designed machine, the magnetic core loss must equal the conductor loss. Therefore, with R_1 and R_2 expressed in pu of the base impedance, the induction machine efficiency is approximately equal to the following:

$$\eta = 1 - 2(R_1 + R_2) \qquad (5.13)$$

For example, in a machine with R_1 and R_2 each 2%, we write $R_1 = R_2 = 0.02$ pu. The efficiency is then simply $1 - 2(0.02 + 0.02) = 0.92$ pu or 92%. Of the input power in this machine, 8% is lost in the electromechanical conversion process.

Losses taking place in the machine are minimized by providing adequate cooling. Small machines are generally air cooled. Large generators located inside the nacelle can be difficult to cool by air. Water cooling, being much more effective than air cooling in such situations, can be advantageous in three ways:

- Water cooling reduces the generator weight on the nacelle, thus benefiting the structural design of the tower.
- It absorbs and thus reduces noise and vibrations.
- It eliminates the nacelle opening by mounting the heat exchanger outside, making the nacelle more weatherproof.
- Overall, it reduces the maintenance requirement, a significant benefit in large machines usually sitting on tall towers in inclement weather.

5.3.6 Self-Excitation Capacitors

The induction machine needs AC excitation current. The machine is either self-excited or externally excited. Because the excitation current is mainly reactive, a stand-alone system is self-excited by shunt capacitors. An induction generator connected to the grid draws the excitation power from the network. The synchronous generators connected to the base-load network must be capable of supplying this reactive power.

As the generator, the induction machine has a drawback in requiring a leading reactive power for the excitation. For a stand-alone, self-excited induction generator, the exciting power can be provided by an external capacitor connected to the generator terminals (Figure 5.17). No separate AC supply is needed in this case. In the grid-connected generator, the reactive power is supplied from the base synchronous generators working at the other end of the grid. Where the grid capacity of supplying the reactive power is limited, local capacitors can be used to partly supply the needed reactive power.

The induction generator can self-excite using the external capacitor only if the rotor has an adequate remnant magnetic field. In the self-excited mode, the speed, load, and capacitance value affect the generator output frequency and voltage. The operating voltage and frequency are determined in the following text in terms of the approximate equivalent circuit of Figure 5.17.

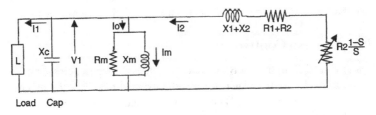

FIGURE 5.17 Self-excited induction generator with external capacitor.

Electrical Generators

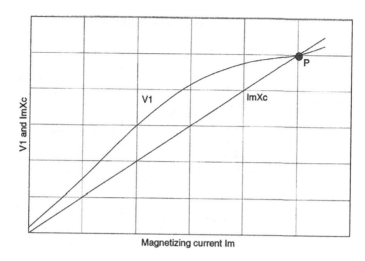

FIGURE 5.18 Determination of stable operation of self-excited induction generator.

With no load on the machine terminals, the capacitor current $I_c = V_1/X_c$ must be equal to the magnetizing current $I_m = V_1/X_m$. The voltage V_1 is a function of I_m, linearly rising until the saturation point of the magnetic core is reached (Figure 5.18). The stable operation requires the line $I_m X_c$ to intersect the V_1 vs. I_m curve. The operating point is fixed where V_1/X_c equals V_1/X_m, that is, when $1/X_c = 1/X_m$, where $X_c = 1/\omega C$. This settles the operating frequency in hertz. With the capacitor value C, the output frequency of the self-excited generator is therefore given by the following:

$$f = \frac{1}{2\pi C X_m} \text{ or } \omega = \frac{1}{2\pi \sqrt{CL_m}} \tag{5.14}$$

Under load conditions, the generated power $V_1 I_2 \cos \phi_2$ provides for the power in the load resistance R and the loss in R_m, and the reactive currents must sum to zero, i.e.,

$$\frac{V_1}{X} + \frac{V_1}{X_m} + I_2 \sin \phi_2 = \frac{V_1}{X_c} \tag{5.15}$$

Equation 5.15 determines the output voltage under load.

Equations 5.14 and 5.15 determine the induction generator output frequency and the voltage with a given value of the capacitance. Inversely, they can be used to determine the required value of the capacitance for the desired frequency and voltage.

5.3.7 TORQUE-SLIP CHARACTERISTIC

If we vary the slip over a wide range in the equivalent circuit, we get the torque-slip characteristic as shown in Figure 5.19. In the region of negative slips, the machine works as the generator powering the electrical load connected to its terminals. In the

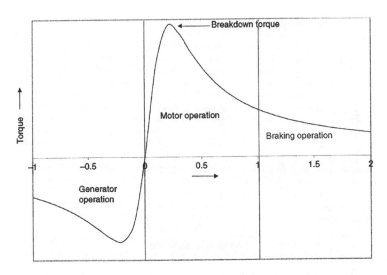

FIGURE 5.19 Torque vs. slip characteristic of the induction machine in three operating modes (slip in multiples of flux speed on horizontal axis).

region of positive slips, the machine works as the motor turning the mechanical load connected to its shaft. In addition to the motoring and generating regions, the induction machine has yet another operating mode, and that is the braking mode. If the machine is operated at slips greater than 1.0 by turning it backward, it absorbs power with a motoring torque. That is, it works as a brake. The power, in this case, is converted into I^2R loss in the rotor conductors, which must be dissipated as heat. The eddy current brake used with the wind turbine rotor works on this principle. As such, in case of emergencies, the grid-connected induction generator itself as a whole can be used as an eddy current brake by reversing the three-phase grid voltage sequence at the stator terminals. This reverses the direction of rotation of the magnetic flux wave with respect to the rotor. The torsional stress on the turbine blades and the hub, however, must be considered, which may limit such a high-torque braking operation only for emergencies.

The torque-slip characteristic in the generating mode is separately shown in Figure 5.20. If the generator is loaded at a constant load torque T_L, it has two possible points of operation, P_1 and P_2. Only one of these two points, P_1, is stable. Any perturbation in speed around P_1 will produce stabilizing torque to bring it back to P_1. The figure also shows the limit to which the generator can be loaded. The maximum torque it can support is called the *breakdown torque*, which is shown as T_{max}. If the generator is loaded under a constant torque above T_{max}, it will become unstable, stall, draw excessive current, and destroy itself thermally if not properly protected.

5.3.8 Transients

The induction generator may experience the following three types of transient currents:

Electrical Generators

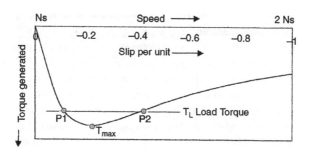

FIGURE 5.20 Torque vs. slip characteristic of induction generator under load.

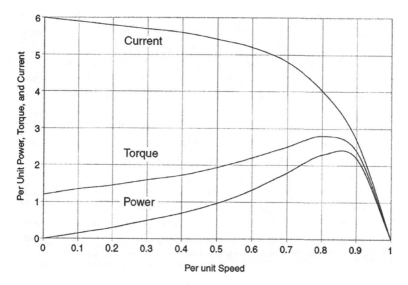

FIGURE 5.21 Induction machine starting and accelerating characteristics.

Starting transient: In a grid-connected system, the induction generator is used as the motor for starting the turbine from rest to supersynchronous speeds. Only then is the machine switched to the generating mode, feeding power to the grid. If the full-rated voltage is applied for starting, the motor draws high starting current at zero speed when the slip is 1.0, and the rotor resistance is the least. Therefore, the starting inrush current can be 5–7 times the rated current, causing overheating problems, particularly in large machines. Moreover, as seen in Figure 5.21, the torque available to accelerate the rotor may be low, resulting in a slow start. This also adds to the heating problem. For this reason, the large induction motor is often started with a soft-start circuit, such as the voltage-reducing autotransformer or the star-delta starter. The modern method of starting is to apply a reduced voltage at a reduced frequency, maintaining a constant volts-to-hertz ratio. This method starts the motor with the least mechanical and thermal stress.

Reswitching transient: A severe transient current can flow in the system if the induction generator operating in a steady-state suddenly gets disconnected due to a system fault or some other reason, and then gets reconnected by automatic reswitching. The magnitude of the current depends on the phase angle of the voltage wave when the generator gets reconnected to the grid. The physical appreciation of this transient current comes from the constant flux linkage theorem. A coil having no resistance keeps its flux linkage constant. Because the winding resistance is small compared to the inductance in most electrical machines, the theorem of constant flux linkage holds after the fault. If the reswitching was done when the stator and rotor voltages were in phase opposition, large transient currents are established to maintain the flux linkage, which then decreases slowly to a small value after tens of milliseconds. Meanwhile, the transient electromechanical torque may be large enough to give the machine and the tower a severe jolt. The actual amplitude and sign of the first peak of the transient torque are closely dependent on the rotor speed and duration of the interruption. In the worst case, the first peak may reach 15 times the rated full-load torque. Frequent faults of this nature can cause shaft breakage due to fatigue stresses, particularly at the coupling with the wind turbine.

Short circuit: When a short-circuit fault occurs at or near the generator terminals, the machine significantly contributes to the system fault current, particularly if it is running on a light load. The short-circuit current is always more severe for a single-phase fault than a three-phase fault. The most important quantity is the first peak current as it determines the rating of the protective circuit breaker needed to protect the generator against such faults. The short-circuit current has a slowly decaying DC component superimposed on an AC component. The latter is larger than the direct online starting inrush current, and may reach 10–15 times the full-loadrated current.

The transient current and torque, in any case, are calculated using the generalized equivalent circuit of the machine in terms of the d-axis and q-axis transient and subtransient reactance and time constants.[4–7] The q-axis terms in the generalized theory do not enter in the induction generator transient analysis, as the d-axis and q-axis terms are identical due to the perfect circular symmetry in electromagnetic structure.

5.4 DOUBLY FED INDUCTION GENERATOR

The doubly fed induction generator is an emerging trendsetting technology, presently used in some large wind turbines with variable-speed operation. NEG Micon's 4.2-MW/100-m diameter machine and 2- to 3-MW GE machines are two examples. In the doubly fed induction motor, a voltage source frequency converter feeds slip frequency power to the three-phase rotor. Thus, the motor is fed from both sets of terminals, the stator as well as the rotor, hence the name. In the grid-connected wind turbine applications, the rotor speed and grid frequency are decoupled by using power electronic frequency converters.

Figure 5.5 shows a doubly fed induction generator connected to a grid using the two back-to-back converters. The per-phase equivalent circuit for the doubly fed

Electrical Generators

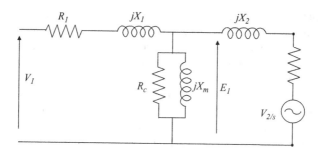

FIGURE 5.22 Doubly fed induction generator equivalent circuit.

induction generator is shown in Figure 5.22. Compared to conventional induction generator the equivalent circuit of doubly fed induction generator has a second controlled source (rotor). The second source is controlled by the DC/AC converter shown in Figure 5.5, which injects a controlled current to the rotor winding.

The most attractive feature of the doubly fed induction generator is that only 20–30% of the power needs to pass through frequency conversion as compared with 100% in the variable-speed synchronous generator. This gives a substantial cost advantage in the power electronics cost. However, the fractional power conversion of the doubly fed induction generator introduces a grid-related problem. A new regulation requires that wind turbines remain connected to the grid in case of a voltage dip and that the built-in capacity of the wind turbine actively supports the grid. Instantly switching off a large wind-power-generating capacity during an emergency could lead to catastrophic grid blackout.[7]

5.5 DIRECT-DRIVEN GENERATOR

The wind turbine typically runs at tens of rpm, whereas the generator runs near 1800 or 3600 rpm in a 60-Hz system or near 1500 rpm or 3000 rpm in a 50-Hz system. In the conventional gear-driven system, a turbine running at 36 rpm, driving a four-pole generator must use a 1:50 gearbox. The cost, vibration, and noise associated with the gearbox can be eliminated by using a gearless direct-driven generator, which also improves the conversion efficiency.

Designing an electrical generator running at tens of rpm and yet having a comparable efficiency is a challenging task. It must have a very high-rated torque to convert the required power at such low speeds. Because the machine size and power losses depend on the rated torque, the low-speed design is inherently heavy and less efficient. This disadvantage is partly overcome by designing the machine with a large diameter and small pole pitch. Many types of direct-driven generators have been considered. For such applications, they are as follows:

- The conventional synchronous, switched-reluctance, and sector-induction generators among the electrically excited designs
- The radial-flux synchronous, axial-flux synchronous, and transverse-flux synchronous generators among the permanent-magnet designs

A variety of direct-driven generator designs in the 30-kW to 3-MW range have been compared with the conventional gear-driven generator designs by Grauers.[8] His design methodology is based on well-known analytical methods and lumped-parameter thermal models. The Grauers study concludes that a radial-flux permanent-magnet generator fits well with the direct-driven wind turbine. The design can be small and efficient for a grid-connected generator with a frequency converter if the rated torque is allowed to be close to the pull-out torque. Such design could be as efficient as the conventional gear-driven four-pole induction generator. That is, the generator must have a large number of poles. Such a machine must have a short pole pitch, resulting in a poor magnetic design. To circumvent such limitations, the permanent-magnet and wound-rotor synchronous machines are being considered for multi-megawatts direct-driven generators.

Another possible solution is the axial-gap induction machine. It can be designed with a large number of poles with less difficulty compared to the conventional radial-gap induction machine. The axial-gap machine is being considered for direct-drive marine propulsion, which is also inherently a low-speed system. For small gearless wind drives with the wind turbine, the axial-flux permanent-magnet generator may find some interest for its simplicity. A 5-kW, 200-rpm laboratory prototype of the axial-gap permanent-magnet design has been recently tested.[9] However, a significant research and development effort is needed before the variable-speed direct-driven systems can be commercially made available for large wind power systems.

The direct-drive is increasingly used along with the permanent-magnet generators. The simplified generator design and higher partial load efficiency are the principal benefits of the permanent-magnet generator.

The use of direct-driven permanent magnet generators has been increasing. GE's 12 MW Haliade-X wind turbine uses a direct drive permanent magnet generator. Other major players in the wind turbine domain have also adopted permanent magnet generator technology for their higher power turbines such as Siemens Gamesa 11 MW SG 11.0-200 DD.

Figure 5.23 compares a 3 MW geared and direct drive wind turbine generator. The generator in a geared turbine operates at higher speeds (up to 2000 RPM), requires lower torque and it has a longer axial length than its diameter. The direct-drive

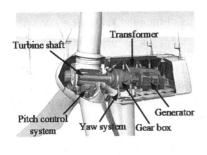
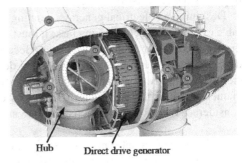

FIGURE 5.23 Comparison of geared and direct drive turbine generators. (From Vestas, Denmark and ENERCON, Germany).

Electrical Generators 85

generator operates at low speeds (up to 20 RPM), has high torque, and its diameter is larger than axial length (sometimes referred to as annular generator due to its shape).

5.6 UNCONVENTIONAL GENERATORS

Normally wind farms include one to several wind turbines installed in one site operating individually and independently of each other. Since the wind velocity in a wind farm varies from one location of wind turbine to the next, the output voltage and power of wind turbines in a wind farm differs from one turbine to another even when turbines use the same physical configurations. This implies need for sophisticated control schemes over wind farms to keep the same level of voltage and power for all turbines. Considerable research studies have investigated these issues with different point of view including amendments in the structure of wind turbines and electrical machines and their controls, power electronic converters control, and in some cases power system corrective actions to avoid power fluctuations coming from wind farms.

As previously discussed, the permanent magnet generators have attracted interest as alternative electric machines in wind energy systems particularly in direct-drive applications in which the gear-boxes are eliminated. The weight and size of electrical machines increases when the torque rating increases for the same active power. Therefore, it is essential for the machine designer to consider an electrical machine with high torque density to minimize the weight and the size. The permanent magnet machines have higher torque density compared with DC, induction, and switched reluctance machines. Therefore, permanent magnet generators are a candidate technology for the wind generation systems. However, permanent magnet machines do have some drawbacks, namely the constant permanent magnet flux linkage which implies the problem of uncontrollable flux which results in output voltage variation when the generator is subject to speed variations. Therefore, it requires some form of output voltage management and, as was mentioned before in the wind generation schemes, the power electronics converters need to be fully dimensioned.

Newly introduced hybrid permanent magnet generators can replace the conventional permanent magnet generator with simplified power electronic requirements. Figure 5.24 shows a hybrid permanent magnet generator incorporating advantages of switch reluctance and permanent magnet machines. In this design, a stator DC field winding is in series with permanent magnets while the rotor has neither permanent magnets nor field windings. The field winding on the stator is used to adjust the flux generated by permanent magnets to either weaken or strengthen it. Moreover, an airbridge is in shunt with each permanent magnet in which amplifies the effect of flux weakening or strengthening and with a proper design of the air-bridge width, a wide flux regulating range is achieved by using a small DC field excitation.

Another type of hybrid permanent magnet machine for wind turbine applications is shown in Figure 5.25. The generator in this topology has two stators, and inner and an outer stator, and a salient pole rotor. The outer stator has 36 salient poles and a three-phase winding while the inner stator had both permanent magnets and DC field windings, which act as excitation. The rotor has 24 salient poles with neither permanent magnets nor field windings. An air bridge is also considered locating in shunt

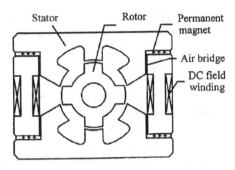

FIGURE 5.24 Example of a hybrid permanent magnet generator. (Chau, K.T.; Li, Y.B.; Jiang, J.Z.; Shuangxia Niu, "Design and Control of a PM Brushless Hybrid Generator for Wind Power Application," Magnetics, IEEE Transactions on, vol.42, no.10, pp.3497-3499, Oct. 2006).

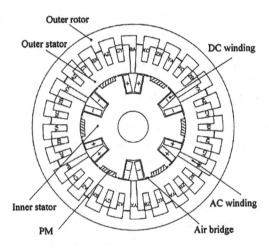

FIGURE 5.25 Example of a hybrid permanent magnet generator. (Chunhua Liu; Chau, K.T.; Jiang, J.Z.; Jian, L., "Design of a New Outer-Rotor Permanent Magnet Hybrid Machine for Wind Power Generation," Magnetics, IEEE Transactions on, vol.44, no.6, pp.1494-1497, June 2008).

with each permanent magnet so that the effect of flux weakening can be significantly amplified. By adjusting the DC field current, the machine can maintain a constant voltage output over 1/3 to 3 times the base wind speed suitable for direct drive applications.

Another type of hybrid permanent magnet generator for wind turbine applications is shown in Figure 5.26. There are two rotors, a permanent magnet and a wound rotor that has DC windings. Both rotors are on the same shaft, rotate with the same speed, and exist inside a single stator and housing. DC current for the WF rotor excitation is provided via a brushless exciter system common to industrial synchronous machine

Electrical Generators

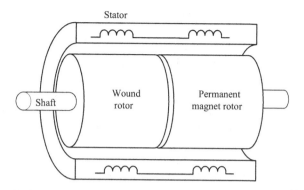

FIGURE 5.26 Example of a hybrid permanent magnet generator. (O. Beik and N. Schofield, High-Voltage Hybrid Generator and Conversion System for Wind Turbine Applications, IEEE Transactions on Industrial Electronics, vol. 65, no. 4, April 2018, pp. 3220–3229).

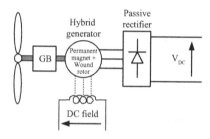

FIGURE 5.27 Hybrid generator in a wind turbine.

systems. The hybrid generator in Figure 5.26 combines the output voltage due to a fixed field from the permanent magnet rotor and a controlled variable voltage due to the variable field of the wound rotor. The total machine output voltage is the sum of the voltages due to these two excitation fields. The wound rotor therefore contributes a controllable variable voltage at the stator output over a limited range. The range of the total voltage variation depends on the design of the wound rotor. Figure 5.27 shows the hybrid generator in a wind turbine system.

5.7 MULTIPHASE GENERATORS

The research and application of multiphase machines (greater than 3-phase) have recently gained renewed attention. The multiphase machines and their associated multiphase power electronics converters bring a number of system advantages including increased reliability, enhanced power, and torque density, improved performance, and speed range capabilities. Multiphase generators, compared to their 3-phase counterparts, result in lower ratings for the power electronic converters, lower DC-link ripple, and relatively lower winding copper losses. The multiphase generators also result in improved fault tolerance, an important factor for wind generators when they connect to the local grid. On the other hand, the higher phase numbers increase the component count and complexity of the phase terminal connections.

The multiphase winding (> 3-ph) arrangements may include any number of phases; however, the phase numbers 5, 6, 7, and 9 are the most considered possibilities. The phase numbers such as 4 and 8 result in the same arrangement as the 2-phase system hence, not usually studied. A 3-phase and 9-phase representation of phases is shown in Figure 5.28. The electrical angle between two consecutive phases in the 9-phase system is 40° (2π/9) and the there are four line-to-line voltages, i.e. between phase 1 and phase 2 (40°); between phase 1 and phase 3 (80°); between phase 1 and phase 4 (120°); between phase 1 and phase 5 (160°). In the 3-phase system, the electrical angle between two consecutive phases is the same (120°); hence, there exists only one line-to-line voltage in the 3-phase system.

As mentioned earlier, the higher phase numbers reduce the ripple on the rectified DC voltage, and hence reduce the capacity of capacitor/inductor used on the DC-link. The rectified DC-link voltage is expressed as follows:

$$V_{dc} = \frac{4}{2\pi/n} \int_0^{\pi/n} E_p \cos(\omega t) d(\omega t) = E_p \frac{2n}{\pi} \sin\left(\frac{\pi}{n}\right) \quad (5.16)$$

where
n is the number of phases (p.u.),
E_p is the peak back-EMF phase voltage (V), and
ω is the rotational speed (rad/s).

For a 3-phase system, Equation 5.16 is simplified to:

$$V_{dc-3ph} = E_p \frac{5.20}{\pi} \quad (5.17)$$

while for a 9-phase system it is:

$$V_{dc-9ph} = E_p \frac{6.16}{\pi} \quad (5.18)$$

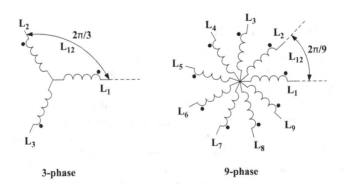

FIGURE 5.28 Electrical representation of 3-phase and 9-phase systems.

Electrical Generators

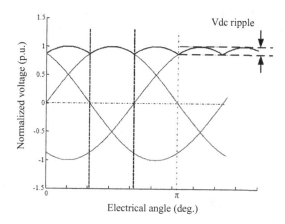

FIGURE 5.29 3-phase rectified voltage.

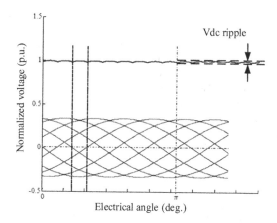

FIGURE 5.30 9-phase rectified voltage.

Note that Equations 5.16–5.18 denote average values for the DC-link voltage and the AC waveforms are considered to be sinusoidal. Figures 5.29 and 5.30 show the 3-phase and 9-phase rectified DC voltage respectively. It is seen that DC-voltage ripple is reduced in a 9-phase system. This reduction, in addition to the higher ripple frequency in 9-phase allows for a smoother rectified DC voltage, hence reduced capacitance. Note that the 9-phase power electronics rectifier has three times the 3-phase converter legs; however, the phase current is reduced in the 9-phase system. This distribution of current to more legs results in better use of rectifier which results in improved losses for the rectifier, and improved packaging.

REFERENCES

1. Rahim, Y.H.A., Controlled Power Transfer from Wind Driven Reluctance Generators, IEEE Winter Power Meeting, Paper No. PE-230-EC-1-09, New York, 1997.

2. Say, M.G., *Alternating Current Machines*, John Wiley & Sons, New York, 1983.
3. Alger, P.L., *The Nature of Induction Machines*, Gordon and Breach, New York, 1965.
4. Adkins, B., *The General Theory of Electrical Machines*, Chapman and Hall, London, 1964.
5. Kron, G., *Equivalent Circuits of Electrical Machines*, Dover Publications, New York, 1967.
6. Yamayee, Z.A. and Bala, J.L., *Electromechanical Devices and Power Systems*, John Wiley & Sons, New York, 1994.
7. Slootweg, J.G., Wind Power Modeling and Impact on Power System Dynamics, Ph.D. thesis in Electric Power Systems, TU Delft, The Netherlands, 2003.
8. Grauers, A., Design of Direct-Driven Permanent-Magnet Generators for Wind Turbines, Ph.D. thesis, Chalmers University of Technology, Goteborg, Sweden, October 1996.
9. Chalmers, B.J. and Spooner, E., Axial Flux Permanent Magnet Generator for Gearless Wind Energy Systems, IEEE Paper No. PE-P13-EC-02, July 1998.
10. Erich Hau, *Wind Turbines, Fundamentals, Technologies, Application, Economics*. Springer, 2006.
11. Omid Beik, An HVDC Off-shore Wind Generation Scheme with High Voltage Hybrid Generator, Ph.D. Thesis, McMaster University, 2016.
12. Omid Beik, Ahmad S. Al-Adsani, *DC Wind Generation Systems, Design, Analysis, and Multiphase Turbine Technology*, Springer, 2020.
13. A. Dekka, O. Beik and M. Narimani, Modulation and Voltage Balancing of a Five-Level Series-Connected Multilevel Inverter with Reduced Isolated DC Sources, *IEEE Transactions on Industrial Electronics*, 2020.
14. O. Beik, A. S. Al-Adsani, Parallel Nine-Phase Generator Control in a Medium Voltage DC Wind System, *IEEE Transactions on Industrial Electronics*, 2020.
15. O. Beik, A. Dekka and M. Narimani, A New Modular Neutral Point Clamped Converter with Space Vector Modulation Control, *IEEE International Conference on Industrial Technology (ICIT)*, Lyon, 2018, pp. 591–595.
16. O. Beik and N. Schofield, High-Voltage Hybrid Generator and Conversion System for Wind Turbine Applications, *IEEE Transactions on Industrial Electronics*, vol. 65, no. 4, April 2018, pp. 3220–3229.
17. O. Beik and N. Schofield, An Offshore Wind Generation Scheme With a High-Voltage Hybrid Generator, HVDC Interconnections, and Transmission, *IEEE Transactions on Power Delivery*, vol. 31, no. 2, April 2016, pp. 867–877.
18. O. Beik and N. Schofield, *Hybrid Generator for Wind Generation Systems, IEEE Energy Conversion Congress and Exposition (ECCE)*, Pittsburgh, PA, 2014, pp. 3886–3893.
19. Nigel Schofield, Omid Beik, *Wind Turbine High Voltage Generator and System, The 15th International Workshop on Large-Scale Integration of Wind Power into Power Systems*, 2016.

6 Generator Drives

The wind turbine speed is much lower than the optimum speed for the electrical generator. For this reason, the turbine speed is stepped up using a gear drive. The system can be fixed speed or variable speed, as described in this chapter.

The wind power equation as derived in Chapter 3 is as follows:

$$P = \frac{1}{2}\rho A V^3 C_p \qquad (6.1)$$

where C_p = rotor power coefficient.

As seen earlier, the value of C_p varies with the ratio of the rotor tip speed to the wind speed, termed as the tip speed ratio (TSR). Figure 6.1 depicts a typical relationship between the power coefficient and the TSR. The TSR and the power coefficient vary with the wind speed. The C_p characteristic has a single maximum at a specific value of TSR, around 5 in this case. Therefore, when operating the rotor at a constant speed, the power coefficient can be maximum at only one wind speed. However, for achieving the highest annual energy yield, the value of the rotor power coefficient must be maintained at the maximum level at all times, regardless of the wind speed.

The value of C_p varies not only with the TSR but also with the construction features of the rotor. The theoretical maximum value of C_p is 0.59, but the practical limit around 0.5. Attaining C_p above 0.4 is considered good. Whatever maximum value is attainable with a given wind turbine, it must be maintained constant at that value. Therefore, the rotor speed must change in response to the changing wind speed. This is achieved by incorporating a speed control in the system design to run the rotor at high speed in high wind and at low speed in low wind. Figure 6.2 illustrates this principle. For given wind speeds V_1, V_2, or V_3, the rotor power curves vs. the turbine speed are plotted in solid curves. For extracting the maximum possible energy over the year, the turbine must be operated at the peak power point at all wind speeds. In the figure, this happens at points P_1, P_2, and P_3 for wind speeds V_1, V_2, and V_3, respectively. The common factor among the peak power production points P_1, P_2, and P_3 is the constant high value of TSR around 5.

Operating the machine at the constant TSR corresponding to the peak power point means turning the rotor at high speed in gusty winds. The centrifugal forces produced in the rotor blades above a certain top speed can mechanically destroy the rotor. Moreover, the electrical machine producing power above its rated capacity may overheat and thermally destroy itself. For these reasons, the turbine speed and the generator power output must be controlled.

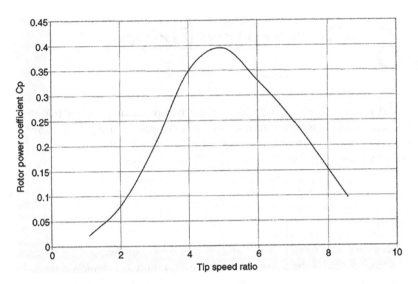

FIGURE 6.1 Rotor power coefficient vs. tip speed ratio has a single maximum.

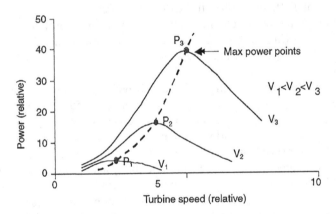

FIGURE 6.2 Turbine power vs. rotor speed characteristics at different wind speeds. The peak power point moves to the right at higher wind speed.

6.1 SPEED CONTROL REGIONS

The speed and power controls in wind power systems have three distinct regions as shown in Figure 6.3, where the solid curve is the power and the dotted curves, the rotor efficiency. They are as follows:

- The optimum constant C_p region, generating linearly increasing power with increasing wind speed
- The power-limited region, generating a constant power even at higher winds, by decreasing the rotor efficiency C_p
- The power-shut-off region, where the power generation is ramped down to zero as the wind speed approaches the top cutout limit

Generator Drives

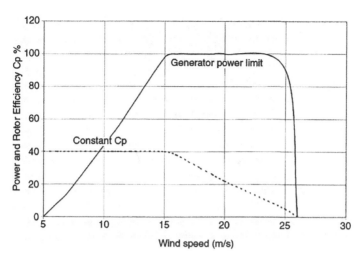

FIGURE 6.3 Three distinct rotor speed control regions of the system.

Typically, the turbine starts operating (cut-in) when the wind speed exceeds 4–5 m/sec, and is shut off at speeds exceeding 25–30 m/sec. In between, it operates in one of the above three regions. At a typical site, the wind turbine may operate about 70–80% of the time. Other times, it is off because wind speed is too low or too high.

The constant C_p region is the normal mode of operation, where the speed controller operates the system at the optimum constant C_p value stored in the system computer. Two alternative schemes of controlling the speed in this region were described in Section 4.4.

To maintain a constant C_p, the control system increases the rotor speed in response to the increasing wind speed only up to a certain limit. When this limit is reached, the control shifts into the speed-limiting region. The power coefficient C_p is no longer at the optimum value, and the rotor power efficiency suffers.

If the wind speed continues to rise, the system approaches the power limitation of the electrical generator. When this occurs, the turbine speed is reduced, and the power coefficient C_p moves farther away from the optimum value. The generator output power remains constant at the design limit. When the speed limit and power limit cannot be maintained under an extreme gust of wind, the machine is cut out of the power-producing operation.

Two traditional methods of controlling the turbine speed and generator power output are as follows:

1. *Pitch control:* The turbine speed is controlled by controlling the blade pitch by mechanical and hydraulic means. The power fluctuates above and below the rated value as the blade pitch mechanism adjusts with the changing wind speed. This takes some time because of the large inertia of the rotor. Figure 6.4 depicts the variation in wind speed, the pitch angle of the blades, the generator speed, and the power output with respect to time in a fluctuating wind based on the actual measurements on a Vestas 1.65-MW wind turbine with

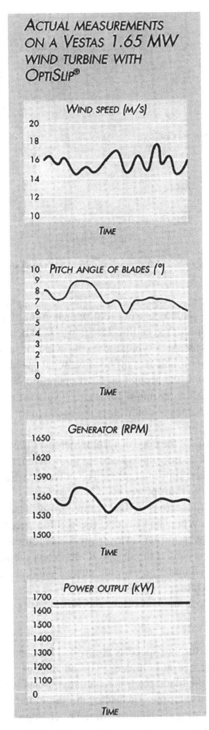

FIGURE 6.4 Wind speed, pitch angle, generator speed, and power output under fluctuating wind speed in 1650-kW turbine. (From Vestas Wind Systems, Denmark. With permission.)

Generator Drives

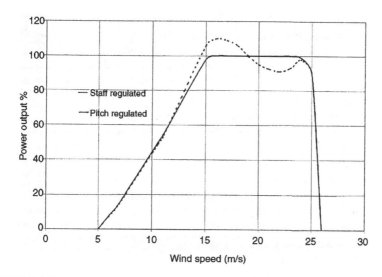

FIGURE 6.5 Generator output power variation with wind speed in blade pitch-regulated and stall-regulated turbines.

OptiSlip® (registered trade name of Vestas Wind Systems, Denmark). The generator power output is held constant even with a 10% fluctuation in the generator speed. This minimizes the undesired fluctuations on the grid. The "elasticity" of such a system also reduces the stress on the turbine and the foundation.

2. *Stall control:* The turbine uses the aerodynamic stall to regulate the rotor speed in high winds. The power generation peaks somewhat higher than the rated limit and then declines until the cutout wind speed is reached. Beyond that point, the turbine stalls and the power production drops to zero (Figure 6.5).

In both methods of speed regulation, the power output of most machines in practice is not as smooth. Theoretical considerations give only approximations of the powers produced at any given instant. For example, the turbine can produce different powers at the same speed depending on whether the speed is increasing or decreasing.

6.2 GENERATOR DRIVES

Selecting the operating speed of the generator and controlling it according to changing wind speed must be determined early in the system design. This is important, as it determines all major components and their ratings. The alternative generator drive strategies and the corresponding speed control methods fall in the following categories.

6.2.1 ONE FIXED-SPEED DRIVE

The fixed-speed operation of the generator offers a simple system design. It naturally fits well with the induction generator, which is inherently a fixed-speed machine.

However, the turbine speed is generally low, whereas the electrical generator works more efficiently at high speed. The speed match between the two is accomplished by the mechanical gear. The gearbox reduces the speed and increases the torque, thus improving the rotor power coefficient C_p. Under varying wind speed, the increase and decrease in the electromagnetically converted torque and power are accompanied by the corresponding increase or decrease in the rotor slip with respect to the stator. The wind generator generally works at a few percent slip. The higher value benefits the drive gear, but increases the electrical loss in the rotor, which leads to a cooling difficulty.

The annual energy yield for a fixed-speed wind turbine must be analyzed with the given wind speed distribution at the site of interest. Because the speed is held constant under this scheme, the turbine running above the rated speed is not a design concern. But, the torque at the generator shaft must be higher. Therefore, it is possible to generate electric power above the rated capacity of the generator. When this happens, the generator is shut off by opening the circuit breaker, thus shedding the load and dropping the system power generation to zero.

The major disadvantage of one fixed-speed operation is that it almost never captures the wind energy at the peak efficiency in terms of the rotor coefficient C_p. The wind energy is wasted when the wind speed is higher or lower than a certain value selected as the optimum.

With the generator operating at a constant speed, the annual energy production depends on the wind speed and the gear ratio. Figure 6.6 depicts the annual energy vs. gear ratio relation typical of such a system. It is seen that the annual energy yield is highly dependent on the selected gear ratio. For the given wind speed distribution in the figure, the energy production for this turbine would be maximum at the gear ratio of 25. When choosing the gear ratio, it is therefore important to consider the average wind speed at the specific site. The optimum gear ratio for the operation of the wind turbine varies from site to site.

Because of the low energy yield over the year, the fixed-speed drives are generally limited to small machines.

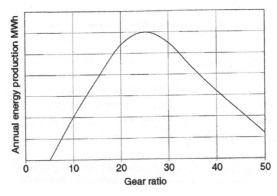

FIGURE 6.6 Annual energy production strongly varies with gear ratio for a given wind speed in one fixed-speed drive.

Generator Drives

6.2.2 TWO FIXED-SPEED DRIVE

The two-speed machine improves energy capture and reduces electrical loss in the rotor and gear noise. The speed is changed by changing the gear ratio. The two operating speeds are selected to optimize the annual energy production with the expected wind speed distribution at the site. Obviously, the wind speeds V_1 and V_2 that would generate peak powers with two gear ratios must be on the opposite side of the expected annual average wind speed. In the specific example of Figure 6.7, the system is operated on the low gear ratio for wind speeds below 10 m/sec, and on the high gear ratio for wind speeds above 10 m/sec. The gear ratio would be changed during operation at 10 m/sec in this example.

In some early American designs, two speeds were achieved by using two separate generators and switching between the generators with a belt drive. An economic and efficient method is to design the induction generator to operate at two speeds. The cage motor with two separate stator windings of different pole numbers can run at two or more integrally related speeds. The pole-changing motor, on the other hand, has a single stator winding, the connection of which is changed to give a different number of poles. Separate windings that match with the system requirement may be preferred where the speed change must be made without losing control of the machine. Separate windings are, however, difficult to accommodate.

In the pole-changing method with one winding, the stator is wound with coils that can be connected either in P or 2P number of poles. No changes are needed, nor are they possible, in the squirrel cage rotor. The stator connection, which produces a higher pole number for low-speed operation, is changed to one half as many poles for high-speed operation. This maintains the TSR near the optimum to produce a high rotor power coefficient C_p. The machine, however, operates with only one speed ratio of 2:1.

Figure 6.8 shows one phase of a pole-changing stator winding. For the higher pole number, the coils are in series. For the lower number, they are in series–parallel. The resulting magnetic flux pattern corresponds to eight and four poles, respectively. It is common to use a double-layer winding with 120° electrical span for the higher pole number. An important design consideration in such a winding is to limit the space harmonics,

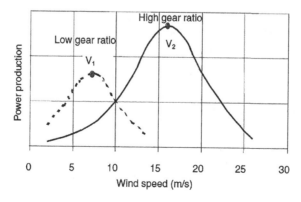

FIGURE 6.7 Power production probability distribution for various wind speeds with low and high gear ratios.

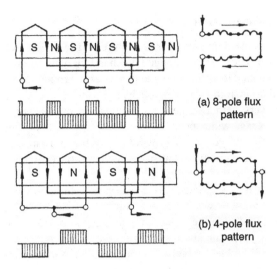

FIGURE 6.8 Pole-changing stator winding for a speed ratio of 2:1.

which may decrease the efficiency in the generating mode and may also produce a tendency to crawl when using the machine as a motor during the start-up operation.

The coil pitch of the stator winding is fixed once wound, but its electrical span depends on the number of poles. A coil pitch one-eighth of the circumference provides full-pitch coils for an 8-pole connection, two thirds for a 6-pole, and one half for a 4-pole connection. Too narrow a coil span must be avoided. For a 2:1 speed-ratio generator, a possible coil span is 1.33-pole pitch for the larger and 0.67 for the smaller pole number. In each case, the coil span factor would be 0.86. Using the spans near 1 and 0.5, with span factors of 1.0 and 0.71, one can avoid an excessive leakage reactance in the lower-speed operation.

Two-speed technology using a fixed-blade rotor with stalled type design is still available in small (<1 MW) machines.

6.2.3 Variable-Speed Gear Drive

The variable-speed operation using a variable gear ratio has been considered in the past but has been found to add more problems than benefits. Therefore, such drives are not generally used at present.

6.3 DRIVE SELECTION

The constant-speed system allows a simple, robust, and low-cost drive train. The variable-speed system, on the other hand, brings the following advantages:

- 20–30% higher energy yield
- Lower mechanical stress—a wind gust accelerating the blades instead of a torque spike
- Less fluctuation in electric power because the rotor inertia works as the energy buffer
- Reduced noise at lower wind speed

Generator Drives

The current and/or voltage harmonics introduced by the power electronics in the variable-speed design may be of concern.

The variable-speed operation can capture theoretically about a third more energy per year than the fixed-speed system.[2] The actual improvement reported by the variable-speed system operators in the field is lower, around 20 to 30%. However, an improvement of even 15 to 20% in the annual energy yield can make the variable-speed system commercially viable in a low-wind region. This can open an entirely new market for the wind power installations, and this is happening at present in many countries. Therefore, the newer installations are more likely to use the variable-speed systems.

6.4 CUTOUT SPEED SELECTION

In any case, it is important that the machine is operated below its top speed and power limits. Exceeding either one above the design limit can damage and even destroy the machine.

In designing the variable-speed system, an important decision must be made for the top limit of the operating speed. For the energy distribution shown in Figure 6.9, if the wind plant is designed to operate up to 18 m/sec, it can capture energy E_1 integrated over the year (area under the curve). On the other hand, if the system is designed to operate at a variable speed up to 25 m/sec, it can capture energy E_2 over the same period. The latter, however, comes with an added cost of designing the wind turbine and

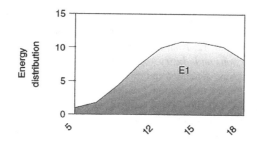

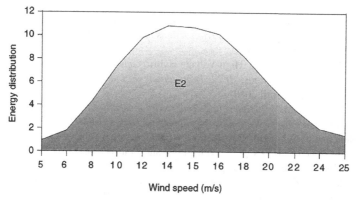

FIGURE 6.9 Probability distribution of annual energy production at two cutout speeds.

the generator to handle a higher speed and a higher power. The benefit and cost must be traded off for the given site to arrive at the optimum upper limit on the rotor speed.

On one side of the trade-off is additional energy ($E_2 - E_1$) that can be captured over the year. If the revenue of the generated electricity is valued at p \$/kWh, the added benefit per year is $p(E_2 - E_1)$ dollars. The present worth PW of this yearly benefit over the life of n years at the annual cost of capital i is as follows:

$$PW = p(E_2 - E_1)\left[\frac{(1+i)^n - 1}{i(1+i)^n}\right] \quad (6.2)$$

In the present example, if the initial capital cost is C_1 for the variable-speed system with cutout speed of 25 m/sec, and C_2 for that with 18 m/sec, then the variable-speed system with 25 m/sec cutout speed will be financially beneficial if the following is true:

$$PW > (C_2 - C_1) \quad (6.3)$$

Such trade-offs should account for other incidental but important issues, such as potential noise concerns at higher cutout speeds.

REFERENCES

1. Datta, R. and Ranganathan, V.T., Variable-Speed Wind Power Generation Using a Doubly Fed Wound Rotor Induction Machine: A Comparison with Alternative Schemes, IEEE Transactions on Energy Conversion, Paper No. PE-558EC, September 2002.
2. Zinger, D.S. and Muljadi, E., Annualized wind energy improvement using variable-speed, *IEEE Transactions on Industry Applications*, Vol. 33–6, pp. 1444–1447, 1996.
3. Gardner, P., Wind turbine generator and drive systems, *Wind Directions*, Magazine of the European Wind Energy Association, London, October 1996.
4. Omid Beik, An HVDC Off-shore Wind Generation Scheme with High Voltage Hybrid Generator, Ph.D. thesis, McMaster University, 2016.
5. Omid Beik, Ahmad S. Al-Adsani, *DC Wind Generation Systems, Design, Analysis, and Multiphase Turbine Technology*, Springer, 2020.
6. Omid Beik, Ahmad S. Al-Adsani, *Wind Energy Systems* pp. 1–9, Springer, 2020.
7. Omid Beik, Ahmad S. Al-Adsani, *DC Wind Generation System*, pp. 33–69, Springer, 2020.
8. O. Beik and A. S. Al-Adsani, "Parallel Nine-Phase Generator Control in a Medium-Voltage DC Wind System," *IEEE Transactions on Industrial Electronics*, vol. 67, no. 10, pp. 8112–8122, October 2020.
9. O. Beik and N. Schofield, "High-Voltage Hybrid Generator and Conversion System for Wind Turbine Applications," *IEEE Transactions on Industrial Electronics*, vol. 65, no. 4, pp. 3220–3229, April 2018.
10. Beik, O., Schofield, N., *"High voltage generator for wind turbines"*, *8th IET International Conference on Power Electronics, Machines and Drives (PEMD 2016)*, 2016.
11. O. Beik and N. Schofield, "An Offshore Wind Generation Scheme With a High-Voltage Hybrid Generator, HVDC Interconnections, and Transmission," *IEEE Transactions on Power Delivery*, vol. 31, no. 2, pp. 867–877, April 2016.
12. O. Beik and N. Schofield, *"Hybrid generator for wind generation systems,"* *2014 IEEE Energy Conversion Congress and Exposition (ECCE)*, Pittsburgh, PA, 2014, pp. 3886–3893.

7 Offshore Wind Farms

The oceans cover two-thirds of the earth's surface, contain energy resources far greater than the entire human race could possibly use, and offer open space for deploying new energy technologies on a grand scale without significant interference with the environment or normal human activities. As for wind farms, the wind speed in the ocean is higher than on land—30 to 40% higher in the open ocean and 15 to 20% higher near the shore. An offshore wind farm, therefore, can generate up to 50 to 70% more power and reduce its electricity costs, even with a higher cost of installation in water. For this reason, many large offshore wind farms have been installed in Europe (Figure 7.1), and more are under construction. Governments in many countries in Europe, Asia, and the Americas are evaluating new proposals every year.

The U.S. east coast, being thousands of miles long, has more promise as compared to many European countries with hundreds of miles of coastline. Among the main attractions along the northeast coast of the U.S. are strong, steady winds, shallow waters, low wave height, and a growing regional market for renewable energy. Power produced elsewhere cannot necessarily be transmitted for use in the cities. Although there are good wind regimes along the coast, both over and off water, the wind speed dramatically drops off within just 30 to 60 km inland. For these reasons, developers are considering large wind farms offshore of the northeast U.S.

Offshore wind projects are generally large, often costing $100 to $300 million. To reduce the cost per megawatt of capacity, wind turbines built for offshore applications are larger than those used onshore. The Vestas 2-MW and GE's 3.6-MW turbine are just two examples. Even larger turbines of 7 to 12 MW capacity have been recently developed for offshore installations.

Offshore wind farms fall into two broad categories: near shore (<6 km) and far shore (>6 km). Public opinion favors locations far out to sea even though extra cost is involved. The wind speed and the resulting energy yield improve with distance from the shore as seen in Figure 7.2 for a 3-MW turbine on an 80-m tall tower as an example. With wind coming from the land, the energy yield can be 25 to 30% higher 5 to 6 km from the shore, whereas it is 12 to 15% lower 5 to 6 km inland. Similar gains are seen with wind coming from the sea or along the shore. Thus, the difference between 5 km offshore and 5 km inland can be 30 to 45% in energy yield. Ten kilometers from the shore, the wind speed is typically 1 m/sec (10 to 15%) higher. This gives 30 to 50% higher energy yields. Moreover, wind is less turbulent at sea, which improves the quality of power and extends the life of the blades. On the negative side of being farther out at sea are the deeper seabed, higher construction and operating costs, and higher grid connection cost, all of which partially offset the higher energy yield.

FIGURE 7.1 Vindeby offshore wind farm in Denmark. (From Vestas Wind Systems, Denmark. With permission.)

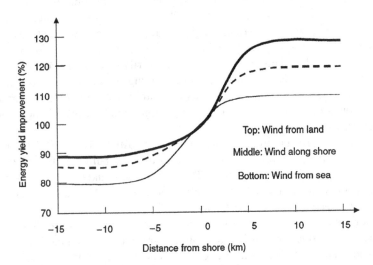

FIGURE 7.2 Energy-yield improvement with distance from the shore in a 3-MW turbine at 80-m hub height. (Adapted from H. Bjerregaard, *Renewable Energy World*, James & James Ltd., London, March–April 2004, p. 102.)

7.1 ENVIRONMENTAL IMPACT

Under the National Environmental Policy Act, an environmental impact assessment of the entire project must be conducted before a private company is allowed to erect any structure on OCS lands. The environmental study must include the Migratory Bird Treaty Act, the Marine Mammal Protection Act, the Endangered Species Act, and the National Environmental Policy Act. It must also include the initial

Offshore Wind Farms

data-gathering structures and work offshore such as localized disturbance of the seabed and possible pollution. Mathematical simulation programs developed by British Maritime Technology to evaluate the environmental impacts of offshore installations and their waste, such as drill cuttings, lubricants, and other chemicals, may be of help here.

As for shipping traffic, the individual turbines could be spaced such that small vessels could sail through the array. However, the proposed site may interfere with shipping routes or large pleasure craft. The wind turbines could lie in the way of fishing areas or even migrating/mating whale habitats. The underground cable may also interfere with fishing areas.

The environmental issues involved in the placement of offshore wind farms include the impact on the avian population and underwater ecosystems. This is especially a factor when birds of known endangered and threatened species reside in or migrate through the area. Avian collisions with turbines resulting in death are a major concern along with the disruption of feeding, nesting, and migrating habits. Though avian deaths by blades had been seen on a large scale at the Altamont Pass, CA, location, it has not been experienced offshore.[2]

For underwater ecosystems, the south shore of Long Island is protected by the state and federal wildlife reserve. The South Shore Estuary Reserve is mostly an inland bay protected from the Atlantic Ocean by island barriers. This provides an environment of mixed freshwater and saltwater where many species of plants and animals flourish. To protect and preserve the underwater sea life here, no site closer than 1.5 mi to the shore may be allowed. However, due to avian migration and nesting in the area, the desirable distance could probably be increased to 2 mi at the minimum.

Sound propagation both in the air and underwater must be considered. The turbines are usually placed at a far enough distance from where the sound propagating through air cannot be heard onshore. Studies have shown that sound propagation through water from the turbine may not affect sea life as long as creatures in that area are used to the sound of passing motorboats, which emit a similar sound. That the shadow of the rotating blades may frighten some fishes is another consideration. Studies have shown that the population of fish around the area of the installed turbines actually increases once installed. This is due to the new area provided on which marine growth can attach and develop.

Although offshore wind farms do not use valuable real estate, they pose a potential visual interference. Waterfront residents, who have paid high prices for oceanview property, do not want to see an offshore wind farm from their windows. The Cape Cod proposal met resistance from many residents fighting against the development. Vacation resorts may be concerned that tourists could see the wind farm from the shore and hear the turbine noise from the beach on a low-wind day.

7.2 OCEAN WATER COMPOSITION

About three-quarters of the earth's surface is covered by oceans. The average ocean depth is 3,800 m, and the maximum depth is 11,524 m in the Mindanao Trench in the Pacific Ocean. In comparison, the average land elevation is 840 m, and the highest elevation is 8,840 m at the top of Mount Everest. Figure 7.3 is the cross-section of the ocean floor, depicting the general terminology of the shore and the continental shelf. About 8% of the ocean floor is shallower than 1,000 m and 5% is shallower than 200 m, where the majority of offshore oil and gas platforms are located. Most offshore wind farms in Europe have been installed in water less than 15 to 20 m in depth.

The ocean shore and shelf material generally consists of sand and gravel coming from land via rivers and blown in by wind. The ocean water density of seawater at atmospheric pressure and 10°C is 1027 kg/m^3. It contains 35 g of salt per kilogram of seawater, which is expressed as the salinity of 35 ppt (parts per thousand). The salinity is measured by measuring the electrical conductivity of seawater, as the two are related. Seawater salt composition by percentage is shown in Figure 7.4: 55% chlorine, 31% sodium, and 14% all others.

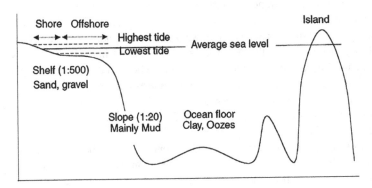

FIGURE 7.3 Ocean-floor terminology.

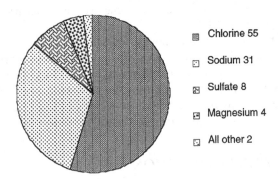

FIGURE 7.4 Salt composition of ocean water (by percentage).

7.3 WAVE ENERGY AND POWER

The primary wave-generating forces are the wind, storms, earthquakes, the moon, and the sun. Figure 7.5 ideally represents a typical ocean wave in shallow offshore water. It is characterized by the following expressions:

$$\frac{2\pi}{d} < 0.25 \text{ and } \frac{d}{gT^2} < 0.0025 \qquad (7.1)$$

where d = water depth and T = wave period.

As shown in Figure 7.6, the water particles under the wave travel in orbits that are circular in deep water, gradually becoming horizontal-elliptical (flat-elliptical) near the surface. The kinetic energy in the wave motion is determined by integrating the incremental energy over the depth and averaging over the wavelength:

$$\text{wave kinetic energy } E_k = \frac{\rho g H^2}{16} \qquad (7.2)$$

The potential energy of the wave is determined by integrating the incremental potential energy in the height of a small column width over one wavelength:

$$\text{wave potential energy } E_p = \frac{\rho g H^2}{16} \qquad (7.3)$$

Note that the wave's potential energy and kinetic energy are equal in magnitude, which is expected in an ideal wave with no energy loss. Therefore, the total energy per unit length of the wave, E_t, is twice that value, i.e.:

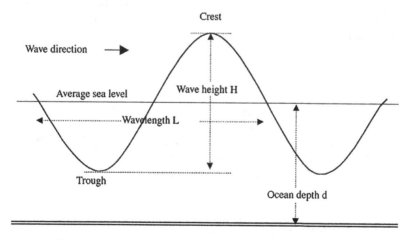

FIGURE 7.5 Ideal representation of ocean waves in shallow offshore water.

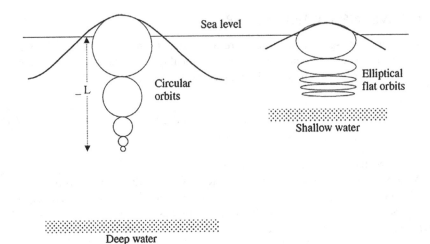

FIGURE 7.6 Orbits of water particles in deep and shallow waters.

$$\text{total energy per unit length } E_1 = E_k + E_p = \frac{\rho g H^2}{8} \tag{7.4}$$

The total energy E in one wavelength L is given by the following:

$$\text{total energy in one wave } E = E_1 L = \frac{\rho g H^2}{8} L \tag{7.5}$$

The mechanical power in the wave is the energy per unit time. It is obtained by multiplying the energy in one wavelength by the frequency (number of waves per second) as follows:

$$\text{wave power } P = \frac{\rho g H^2}{8} Lf \tag{7.6}$$

The energy in a wave thus depends on the frequency, which is a random variable. Ocean wave energy is periodic, with the frequency distribution in ocean surface waves as shown in Figure 7.7. The actual waves may have long waves superimposed on short waves from different directions. A simple linear one-frequency wave theory may be superimposed for an approximate estimate of the total effect.

7.4 OCEAN STRUCTURE DESIGN

Wind-generated waves with a period of 1 to 30 sec are the most important in determining the wave power and force acting on ocean structures.

Offshore Wind Farms

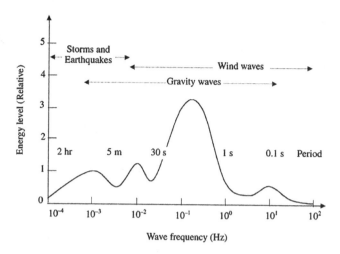

FIGURE 7.7 Approximate energy distribution in ocean surface waves.

7.4.1 FORCES ON OCEAN STRUCTURES

The offshore structure must withstand mechanical forces exerted by the ocean waves, currents, wind, storms, and ice. The wave force is the most dominant of all. The structure must absorb, reflect, and dissipate wave energy without degradation in performance over a long lifetime.

Offshore platforms have been built around the world since the early 1950s to drill for oil and gas, along with undersea pipelines. Offshore mining is also being developed now, and the offshore wind farm is the latest addition to such structures installed to provide a means of producing energy resources and transporting energy to the shore.

The weight and cost of a fixed platform that will withstand wave forces, currents, and wind increase exponentially with the depth of water as seen in Figure 7.8. The offshore cable or pipelines carrying the cable must withstand forces due to inertia, drag, lift, and friction between the floor and the pipe. The water drag and lift forces for an underwater structure in a somewhat streamlined water current can be derived from classical hydrodynamic considerations. They depend on the water velocity near the floor, which follows the one-seventh power law, namely

$$\frac{V}{V_o} = \left(\frac{Y}{Y_o}\right)^{1/7} \qquad (7.7)$$

where V and V_o are the water velocities at heights Y and Y_o, respectively, above the seabed.

The friction coefficient between the pipe surface and the seabed varies with the sediment type. The concrete-coated pipes commonly used offshore have a friction coefficient of 0.3 to 0.6 in clay, about 0.5 in gravel, and 0.5 to 0.7 in sand.

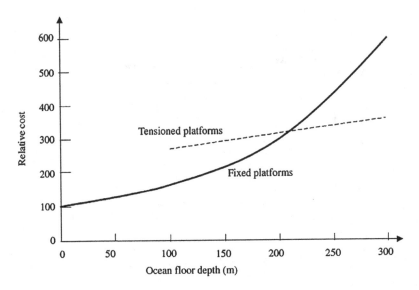

FIGURE 7.8 Cost of offshore structures at various depths.

7.5 CORROSION

Corrosion is caused by the electrochemical reaction in which a metal anode corrodes by oxidation and a cathode undergoes reduction reaction. The seawater works as an electrolyte for the transfer of ions and electrons between the two electrodes. The corrosion rate must be accounted for in the design. Table 7.1 lists the corrosion rates of commonly used metals.

Two types of cathodic protection are widely used for corrosion protection of materials submerged in seawater. The impressed current system gives more permanent protection, but requires electric power. Galvanic protection employs aluminum, magnesium, or zinc anodes attached to the steel structure in seawater. Under the cathodic protection principle, a metal receiving electrons becomes a cathode, which can no longer corrode. The zinc anode is most widely used as a sacrificial material to protect steel hulls of ships. When it has deteriorated, it is replaced for continued protection. Zinc provides about 1000-Ah charge-transfer capacity per kilogram. The sacrificial anode design follows basic electrical circuit principles.[4]

7.6 FOUNDATION

The tower foundation design depends on water depth, wave height, and the seabed type. The foundation for an offshore turbine and its installation is a significant cost component of the overall project, 20 to 30% of the cost of the turbine. The height and consequently the depth of installation are determining factors in choosing the farm location. Turbines installed on land generally have a tower height equal to the

TABLE 7.1
Corrosion Rate of Various Materials in Seawater

Material	Average Corrosion Rate (μm/yr)
Titanium	None
Stainless steel, Nichrome	<2.5
Nickel and nickel-copper alloys	<25
Copper	10–75
Aluminums alloys	25–50
Cast iron	25–75
Carbon steel	100–175

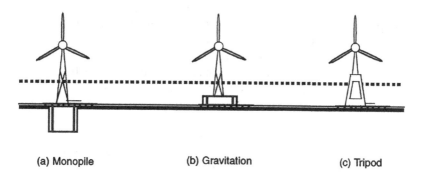

(a) Monopile (b) Gravitation (c) Tripod

FIGURE 7.9 Foundation types for offshore wind turbines.

rotor diameter to overcome wind shear from ground obstacles. However, the offshore tower height can be 70% of the rotor diameter due to the low shear effect of water.

The most favorable water depth for an offshore wind farm is 2 to 30 m. Depths less than 2 m are not accessible by ship, and those greater than 30 m make the foundation too expensive. In this range, three possible foundation designs are shown in Figure 7.9. They are as follows: monopile for 5 to 20 m depths, gravity for 2 to 10 m depths, and tripod or jacket for 15 to 30 m depths.[5]

7.6.1 MONOPILE

The cylindrical steel monopile is the most common and cost-effective foundation in 5 to 20 m deep water. It is driven into the seabed up to a depth of 1.1 times the water depth, depending on the seabed conditions. It does not need seabed preparation, and erosion is not a problem. However, boulders and some layers of bedrock may require drilling or blasting, which would increase the installation cost. The monopile construction is limited to water depths of 20 m due to the stability considerations. A monopile foundation including installation for a 1.5-MW turbine in 15-m deep water could cost about one million dollars.

7.6.2 Gravitation

The gravitation foundation is made of concrete and steel that sits on the seabed. It is often used in bridges and low-depth turbine installations. It is less expensive in shallow water (2 to 10 m deep), but is extremely expensive in water deeper than 15 m.

7.6.3 Tripod

The tripod foundation is suitable for 15–30 m deep water. It utilizes a lightweight, three-legged steel jacket to support the foundation. The jacket is anchored to the seafloor by piles driven into each leg. Each pile is driven up to 20–25 m into the seafloor, depending on seabed conditions. Boulders in the pile area are the only concern, which may be blasted or drilled if necessary. The tripod foundation cannot be used in depths less than 10 m due to the legs' possible interference with vessels. However, in water deeper than 10 m, the tripod is very effective as has been tested in oil rig foundations. Preparation of the seafloor is not necessary, and erosion is not a concern with this installation. For a 1.5-MW turbine utilizing a tripod foundation at a water depth of 15 m, the foundation cost including installation would be about $400,000 in 2005. (Source: Danish Wind Industry Association report, Offshore Foundations: Tripod.)

Other foundations under development for the offshore towers include the following:

- Lightweight foundations guyed for stability
- Floating foundations with the turbine, tower, and foundation in one piece
- Bucket-type foundation

The bucket-type foundation design is used to support Vestas 3-MW turbines. The 135-t structure has the shape of an upturned bucket and is sucked into the seabed through a vacuum process. Early indications are that such a design makes the fabrication, handling, installation, and subsequent removal easier and less expensive. Figure 7.10 depicts a typical wind tower foundation structure.

The most economical way to install turbines in terms of manpower, equipment, and time is to install the turbine and tower as a single unit. The foundation is placed first, and then the turbine, tower, and the rotor are lifted and attached to the foundation in that sequence. A jack-up lift barge is the most likely vessel to be used for installing the foundation as it can carry the foundation and drive the piles in.

Typical foundation cost is about $80 per kW on land and $500 per kW offshore in 10-m deep water, increasing by roughly 2% per meter of additional water depth. These costs vary with the seabed and water conditions.

Approximate costs of foundation installation have been compiled in a chart by the Danish Wind Industry Association.[5] The chart plots the price of the foundation, including installation in water with an average depth of around 25 m. It does not include the cost of installing the turbine tower, nacelle, and rotor.

The issues of remoteness from the onshore facilities and crews to construct and maintain offshore installations are being addressed by developing wind turbines fitted with integrated assembly cranes, helicopter pads, and emergency shelters capable of sustaining crews for several days of harsh weather offshore.

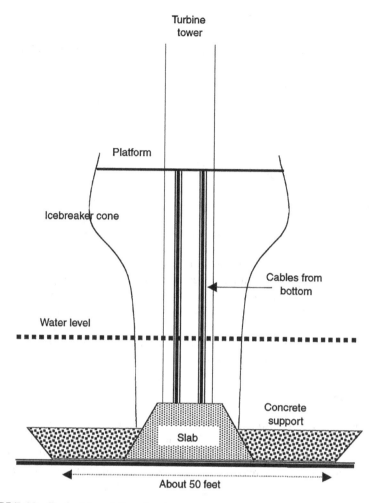

FIGURE 7.10 Typical foundation structure.

7.7 MATERIALS

Offshore equipment is necessarily made to withstand the marine environment. They have marine-grade material, components, seals, and coatings to protect them from the corrosive environment.

Table 7.2 lists materials commonly used in seawater along with their mechanical strengths and relative costs. The marine environment is harsh enough to cause rapid and sometimes unanticipated degradation of equipment. For example, all 81 of the 2-MW turbines at Denmark's Horns Rev—the largest wind farm built to date—were moved back to the shore within a couple of years after installation to repair or modify the generators and transformers at a considerable expense (to remove and reinstall the turbines). It was believed that the equipment design and material would not have withstood the harsh marine environment for a long duration. Similar problems have been encountered before in other offshore farms.

TABLE 7.2
Mechanical Strength of Various Materials Commonly Used in Structures in Seawater

Material	Yield Strength (ksi)	Ultimate Strength (ksi)	Elongation (Percent)	Relative Cost
Mild steels	30–50	50–80	15–20	1
Medium-strength steels	80–180	100–200	15–20	3–5
High-strength steel	200–250	250–300	12–15	10–15
Stainless steels 302 and 316	35–45	80–90	55–60	5–10
Aluminum alloys	30–70	45–75	10–15	7–8
Titanium alloys	100–200	125–225	5–15	20

The coatings on the wind turbine must last over its design life of 20–25 yr. Because the turbines are highly visible on tall towers, they need to be pleasing in appearance for public acceptance. A typical coating system on the onshore wind turbine is a three-coat system. It consists of 50 to 80 μm epoxy-zinc-rich primer, 100–150 μm epoxy midcoat, and 50–80 μm polyurethane topcoat. Protection for offshore wind turbines in highly corrosive marine environments is defined in the EN-ISO-12944 standard. It is necessarily different for parts under water and for the tower parts above water.

7.8 MAINTENANCE

Maintenance of the wind turbine and foundation is important for its longevity and for keeping the energy cost down. Regularly scheduled inspections are absolutely necessary, particularly because offshore towers do not get incidental periodic visits. No federal regulations exist in regard to inspecting the equipment in offshore wind farms. The Naval Facilities Engineering Command outlines procedures for inspecting offshore structures based on good engineering practices. Although not mandatory, their implementation would reduce the maintenance cost in the long run. The inspections must be planned taking into consideration weather conditions and the wave height.

Presently wind farms are built to operate for 20 years with minimal maintenance. Visual inspections of the turbine and tower should be conducted periodically during the plant's lifetime to search for apparent physical signs of deterioration, damage, etc. The procedures for inspecting the tower follow those laid out for inspecting the underwater foundation.

Inspections of the foundation and the tower at the splash line are important as this part of the structure is exposed to corrosion due to galvanic action. A level-one inspection is conducted annually for the underwater structures. A level-two inspection is conducted biannually for the underwater structures, and a level-three inspection is made every 6 years. The procedures and equipment used to carry out these inspections are listed in the Underwater Inspection Criteria published by the Naval Facilities Engineering Service Center.

Generally, visual inspections allow detection and documentation of most forms of deterioration of steel structures. Some types of corrosion may not be detected by visual inspections. For example, inside steel pipe piling, anaerobic bacterial corrosion caused by sulfate-reducing bacteria is especially difficult to detect. Fatigue distress can be recognized by a series of small hairline fractures perpendicular to the line of stress but these are difficult to locate by visual inspection. Cathodic protection systems must be closely monitored both visually and electrically for wear of anodes, disconnected wires, damaged anode suspension systems, and/or low voltage. A nondestructive test plan may be developed and tailored to any specific area of concern at a particular wind farm site.

REFERENCES

1. Long Island Power Authority. *Long Island's Offshore Wind Energy Development Potential: A Preliminary Assessment*. Report for New York State Energy Research and Development Authority by AWS Scientific, Inc., Albany, NY, 2002.
2. Krohn, S., 2000. Offshore Wind Energy: Full Speed Ahead, Danish Wind Industry Association Report, 2002.
3. Ackerman, T., 2000. Transmission Systems for Offshore Wind Farms, *IEEE Power Engineering Review*, December 2000, pp. 23–28.
4. Randall, R.E., 1997. *Elements of Ocean Engineering*, The Society of Naval Architects and Marine Engineers, Jersey City, NJ, 1997.
5. Danish Wind Industry Association, 2003. *Wind Turbine Offshore Foundations*, April 2003.
6. Omid Beik, An HVDC Off-shore Wind Generation Scheme with High Voltage Hybrid Generator, Ph.D. thesis, McMaster University, 2016.
7. Omid Beik, Ahmad S. Al-Adsani, *DC Wind Generation Systems, Design, Analysis, and Multiphase Turbine Technology*, Springer, 2020.
8. Omid Beik, Ahmad S. Al-Adsani, *Wind Energy Systems*, pp. 1–9, Springer, 2020.
9. Omid Beik, Ahmad S. Al-Adsani, *DC Wind Generation System*, pp. 33–69, Springer, 2020.
10. O. Beik and A. S. Al-Adsani, "Parallel Nine-Phase Generator Control in a Medium-Voltage DC Wind System," *IEEE Transactions on Industrial Electronics*, vol. 67, no. 10, pp. 8112–8122, October 2020.
11. O. Beik and N. Schofield, "High-Voltage Hybrid Generator and Conversion System for Wind Turbine Applications," *IEEE Transactions on Industrial Electronics*, vol. 65, no. 4, pp. 3220–3229, April 2018.
12. Beik, O., Schofield, N., '*High voltage generator for wind turbines*', 8th IET International Conference on Power Electronics, Machines and Drives (PEMD 2016), 2016.
13. O. Beik and N. Schofield, "An Offshore Wind Generation Scheme With a High-Voltage Hybrid Generator, HVDC Interconnections, and Transmission," *IEEE Transactions on Power Delivery*, vol. 31, no. 2, pp. 867–877, April 2016.
14. O. Beik and N. Schofield, "*Hybrid generator for wind generation systems*," 2014 IEEE Energy Conversion Congress and Exposition (ECCE), Pittsburgh, PA, 2014, pp. 3886–3893.

8 AC Wind Systems

To achieve a large amount of energy, the wind turbines are grouped in a wind farm and their collective power is sent to the grid. Figure 8.1 shows wind turbines in Walney wind farm, the world's largest wind farm when built. In Walney, the turbine powers are collected using an array grid, sometimes referred to as collector grid, and sent to an off-shore substation using 3-phase AC cable systems. The off-shore substation then sends the total wind power to the on-shore grid. The wind farm that uses AC collector grid and AC transmission system is here referred to as AC wind system.

In majority of existing wind farms, the aggregated power at the off-shore substation is sent to the shore using high-voltage AC (HVAC) system. However, for large wind farms far away from the shore, some high-voltage DC (HVDC) systems are used because DC offers far more efficient power transmission in water from the wind site to the shore. Given the existing technologies for power electronic switches the HVDC systems have lower costs than HVAC transmission above a 'break-even' distance. The greater the distance the more justifiable the HVDC systems. The cost of HVDC substations, which include high-voltage high-power electronics systems, are greater than HVAC substations which include AC transformers and their associated components. Choosing between HVDC and HVAC for transmission system requires a thorough analysis and depends on factors such as power and voltage levels, installation cost, losses, control functions, and the specific components used in the system. However, the cost of HVDC systems will continue to decline as the power electronic systems become more available at lower price.

8.1 OVERVIEW

As an example of AC wind system, Walney wind farm of Dong Energy uses AC collector grid and AC transmission systems. It is located in the Irish Sea about 15 km off the UK coast near Barrow-in-Furness, Cumbria. The total installation consists of two farms Walney 1, Walney 2, and Walney extension. Walney 1, which is studied in detail here, has 51 wind turbines each a Siemens 3.6 MW model SWT-3.6-107. The power from the 51 turbines in Walney is aggregated using an AC 33 kV collector grid with undersea cables, and sent to an off-shore substation. There are two AC transformers on the off-shore substation that step-up the voltage to 132 kV AC and transmit it to the shore via 44 km long 3-phase undersea cables. On the shore, the wind farm power is transferred to the UK national grid at Heysham. Table 8.1 summarizes the main specification of Walney 1 wind farm.

Figure 8.2 shows the turbine conversion system and the connection of off-shore to on-shore substations for the Walney 1 wind farm. The turbines use a geared induction generator (IG) rated at 690V and 3.6 MW. The gearbox is planetary/helical type with three stages and a total ratio of 1:119. The turbine rotor speed varies between 4 and 13 RPM during the wind velocity variations from cut-in to cut-out (3–25 m/s).

TABLE 8.1
Walney 1 Wind Farm General Specifications

Item	Specs
Type of wind turbine	Siemens SWT-3.6-107
Rated power	3.6 MW
Number of turbines	51
Total farm power	183.6 MW
Collector grid	33 kV AC
Transmission system	132 kV AC
Off-shore substation	2 × 120 MVA
Off-shore substation to on-shore distance	44.4 km
Wind farm are	34 km^2
Depth of water	21–26 m

FIGURE 8.1 Walney wind farm. (From Dong Energy)

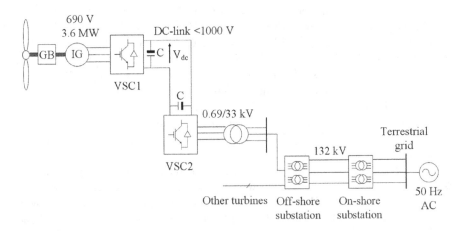

FIGURE 8.2 Walney wind turbine arrangement.

AC Wind Systems

Therefore, the induction generator's speed varies from 476 RPM to 1547 RPM during the entire wind regime. A mechanical brake is provided on the shaft for safety.

The generator is connected to two back-to-back connected voltage source converters (VSC), each fully rated at 3.6 MW. The DC-link voltage is less than 1000 Vdc. Referring to Figure 8.2 the VSC1 acts as a variable speed drive for the induction generator where a vector control is applied to extract a maximum power from the wind at each velocity. VSC2 maintains a fixed DC-link voltage, and controls the output power flow. A transformer steps-up the voltage from 690V to 33 kV at the output of turbine, where it is connected to the collector grid. The two converters VSC1 and VSC2 decouple the wind generator from the grid and allow a variable speed operation of wind turbine. An active blade pitch control is used to adjust pitch angle of the blades based on the strength and direction of wind velocity. This ensures that a desired power is captured from the wind.

In the Walney turbines, the gearbox, induction generator, and VSC1 are placed inside the Nacelle, i.e. at the tower top. The VSC2 and transformer are placed at the turbine base. The 1000 Vdc DC-link extends from the turbine Nacelle to turbine base. There are other emergency and accessory components inside the turbine for measurement, monitoring, and maintenance that are not shown in Figure 8.2. Table 8.2 summarizes the turbine specifications for the Walney 1 wind farm.

The Siemens 3.6 MW wind turbine power is plotted in Figure 8.3. The turbine starts delivering power at a cut-in velocity of 3 m/s. Above the cut-in velocity and up to a rated velocity of 13 m/s the turbine power and rotor speed increase with the wind velocity, and as a function of turbine power coefficient. During this period to maximize wind energy extraction a maximum power point tracking (MPPT) control is followed. From the rated velocity to cut-out velocity of 25 m/s the turbine blades

TABLE 8.2
Siemens Wind Turbine (SWT-3.6-107) Specifications

Item	Specs
Generator	Induction generator
Rated power	3.6 MW
Generator rated voltage	690 V
Generator rated speed	1500 RPM
Generator poles	4
Wind turbine rotor speed range	4–13 RPM
Cut-in velocity	3 m/s
Rated velocity	13 m/s
Cut-out velocity	25 m/s
Converter	2 × VSCs with IGBT switches
Converter voltage	690 V
Converter power	3.6 MW
Transformer ratio	0.69/33
Turbine rotor diameter	107m
Blade length	52m
Hub height	83.5m
Gearbox ratio	1:119
Brake type	Hydraulic disc

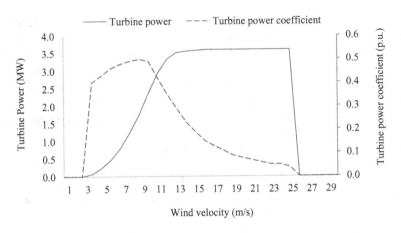

FIGURE 8.3 Siemens 3.6 MW wind turbine power curve.

are pitched and the converter is controlled such that a rated power of 3.6 MW is delivered while the rotor speeds are maintained at its rated value, i.e. around 1500 RPM. Below cut-in velocity and beyond cut-out velocity the turbine does not generate any power, however, the VSC2 always maintains a DC-link voltage such that as soon as wind is reduced below the cut-out or above the cut-in the turbine starts generating power.

8.2 WIND TURBINE AND WIND FARM COMPONENTS

To study the wind farm, its behavior, characteristics, and performance, the components such as induction generator, converters, transformer, and cables are modeled and analyzed. These models are then used in computer simulations to characterize the system. Given Walney wind turbine schematic shown in Figure 8.2 a per-phase equivalent circuit model for the induction generator, turbine transformer, pi model for the cable, and circuit model of back-to-back voltage source converters are presented in Figure 8.4. Each wind turbine in Walney has a 4.5 MVA, 0.69/33 kV tower-mounted transformer manufactured by Siemens. The model of induction generator was previously discussed, however, the transformer model which is similar to that of induction generator shows the series impedance and the parallel magnetizing reactance and loss resistance. Table 8.3 lists the values of parameters for the induction generator and turbine transformer.

As previously discussed, the VSC2 in Figure 8.4 regulates the DC-link voltage to maintain a constant DC-link voltage at all times, i.e. from full-load to no-load. Therefore, all circuits from the DC-link to the point of connection at the on-shore grid are always excited to nominal voltage. A pulse Width Modulation (PWM) vector control of the VSC1 and VSC2 is implemented to maximize wind turbine power conversion. The output of VSC2 is synchronized to the on-shore 50 Hz grid. The VSC1 and VC2 each have three legs with two IGBTs per leg, S1 and S2. Each switch has an anti-parallel diode (D1 to D6), that are freewheeling diodes and allow the

AC Wind Systems

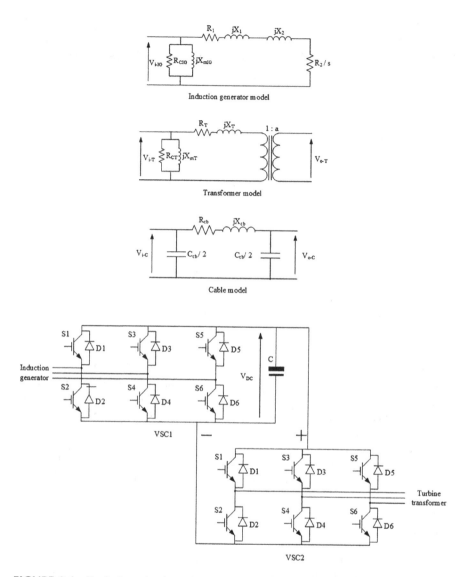

FIGURE 8.4 Equivalent circuit models for wind turbine induction generator, transformer, cable pi model, and voltage source converters.

reverse flow current when appropriate. The IGBT control (PWM) signals are produced by comparing a control signal, i.e. three-phase voltage/current, with a high-frequency triangular signal. In modeling VSCs an average model for dynamic analysis is used while for steady-state analysis the VSCs may be modeled via their input, output, and efficiencies.

Figure 8.5 shows the turbine arrangements and connections of the collector grid for Walney 1 wind farm. There are six groups of turbines each connected to the

TABLE 8.3
Parameters of Walney Wind Farm Component Models (Values Are per Phase)

Induction Generator

Stator resistance	0.0019 ohms
Rotor resistance referred to stator	0.0019 ohms
Loss resistance	4.41 ohms
Magnetizing reactance	1.16 ohms
Stator leakage reactance	0.048 ohms
Rotor leakage reactance referred to stator	0.035 ohms

Turbine Transformer

Power	4.5 MVA
Voltage	0.69/33 kV
No load loss	3.7 kW
Full-load loss	46.5 kW
Impedance	6.4%
Loss resistance	128.68 ohms
Magnetizing reactance	10.62 ohms
Resistance referred to primary	0.001 ohms
Leakage reactance referred to primary	0.00677 ohms

off-shore substation using a 3-phase 3 × 500 mm² undersea cables, while the connections between the turbines are 3-phase 3 × 150 mm² cables. The off-shore to on-shore transmission system cable is however a 3 × 600 mm² with a length of 44.4 km. The model of cables is a PI equivalent circuit shown in Figure 8.4. Table 8.4 lists specifications of the Walney wind farm collector and transmission grid cables.

There are two transformers on the off-shore substation, each rated at 120 MVA, 33/132 kV, and manufactured by ABB. The transformers collect the power from the wind turbine and process it to the shore. Table 8.5 lists parameters of the off-shore substation transformer.

8.3 SYSTEM ANALYSES

To study the performance of the wind generation system, the Walney wind farm is modeled in a simulation environment (Matlab) as an example. Models of the components such as induction generator, transformer, power electronics converters, and cables described in the previous sections are employed to construct the simulation platform. The analyses are conducted with a number of practical assumptions referring to Figure 8.2: (i) the cable impedance and hence voltage drop between induction generator and VSC1, also between VSC2 and turbine transformer are negligible, (ii) the full-load power factor at the VSC2 output is unity, (iii) losses of monitoring, measurement devices and accessory components are negligible, and (iv) the turbine generates no power before cut-in velocity and beyond cut-out velocity.

AC Wind Systems

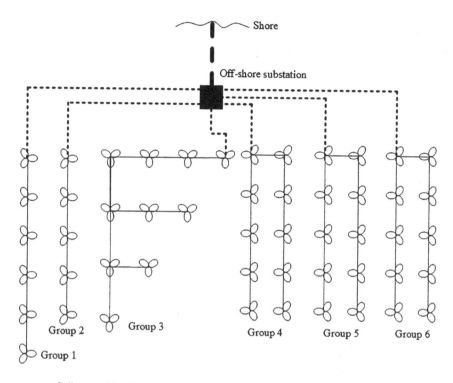

FIGURE 8.5 Walney 1 wind farm turbine arrangements and connection.

TABLE 8.4
Cable Parameters for Walney Collector and Transmission Grids (Values Are per Phase)

	Collector between Turbines	Collector from Group to Off-shore Substation	Transmission between Off-shore to Shore
Cable size	3 × 150 mm²	3 × 500 mm²	3 × 630 mm²
Voltage	33 kV	33 kV	132 kV
Resistance (AC)	0.16 ohms	0.0515 ohms	0.028 ohms
Reactance	0.116 ohms/km	0.0974 ohms/km	0.116 ohms/km
Capacitance	0.191 uF/km	0.3 uF/km	0.21 uF/km

TABLE 8.5
Off-shore Substation Transformer Parameters (Ohm Values Are per Phase)

Power	120 MVA (2x)
Voltage	33/132 kV
No load loss	63.3 kW
Full-load loss	370 kW
Impedance	10.5%
Loss resistance	1741.71 ohms
Magnetizing reactance	213.5 ohms
Resistance referred to primary	0.002 ohms
Leakage reactance referred to primary	0.089 ohms

TABLE 8.6
System Analyses

Wind Turbine	
Induction Generator Line Voltage	690 V Leading PF
Induction generator line current	3076 A
Induction generator output power	3492 kW
Induction generator efficiency	98%
VSC 1 and VSC2 efficiency	98.5%
DC-link voltage	1120 V
DC-link current	3053 A
Turbine transformer output voltage	32.70 kV
Turbine output current	60 A
Transformer efficiency	99%

Off-shore substation, transmission system, and total wind farm	
Output line voltage	130 kV leading PF
Line current	752 A
Collector grid losses	13120 kW
Substation efficiency	98%
Transmission system losses	2110 kW
Wind farm total losses	16500 kW
Wind farm efficiency	91%

The analysis is performed for full-load, i.e., when the turbine delivers its rated power. Therefore, the wind velocity is between 13 m/s and 25 m/s, as depicted in Figure 8.3. Table 8.6 summarizes the system analyses for an AC load flow utilizing the developed simulation platform.

The results in Table 8.6 show that the voltage drop across the collector grid from turbine to the off-shore substation is less than 1%, hence a small voltage difference between turbines. It is seen that the losses of off-shore substation and the transmission cable are a large portion of the total system loss.

AC Wind Systems

The induction generator and turbine output power factors are leading indication that the turbine receives reactive power from the grid and the VSC1 injects reactive power required for the induction generator. In addition, the power factor at the substation input is leading resulting in a reactive power injection from the collector grid to the substation. However, at the point of shore connection the power factor is negative indicating a grid with inductive load behavior at the wind farm output.

8.4 CHALLENGES

The AC systems have several advantages and some disadvantages. The components of the AC systems such as transformers, switch gears, breakers, inductors, and other associated elements are commercially available in high power and high voltages. The AC systems are mature technologies, widely available at competitive prices, have reached a high level of efficiencies, and their risks are well understood and evaluated as they have been in applications for a long time.

However, the impedance in an AC system is a frequency-dependent element, and it contributes to the losses (e.g. cables) in addition to the resistance. The higher the frequency, the higher the impendence and voltage drop, hence the higher the associated losses. A 3-phase AC collector/transmission system requires three cables; and the longer the cables, the increased reactive power management requirements. In contrast, a DC cable system in a steady state is free of the frequency-dependent losses, there are no requirements for reactive power control, and needs only two conductors (lead and return) for each transmission system. The DC systems offer better controllability as they use power electronics converters such as DC/DC and AC/DC converters for stepping up/down the voltage and energy conversion.

Due to these favorable attributes, the DC wind generation systems have recently gained much attention. Therefore, DC wind generation systems, their structure, components, control, and comparison with AC wind systems are discussed in the next chapter.

REFERENCES

1. Walney Off-shore Windfarms. Available: https://orsted.com/.
2. Walney Off-shore Windfarms. Available: https://en.wikipedia.org/wiki/Walney_Wind_Farm.
3. Walney Off-shore Windfarms. Available: https://walneyextension.co.uk/.
4. Siemens wind turbine SWT-3.6-107. Available: http://www.siemens.com.
5. NKT Cables, "Product Catalogue Australia Medium and High Voltage Cables," Available: http://www.nktcables.com.
6. Xian Electric Transformer, "Transformer & Reactor 10kV~500kV," Available: http://www.xianelectric.com.
7. Omid Beik, Ahmad S. Al-Adsani, *Wind Energy Systems*, Springer, 2020, pp. 1–9.
8. Omid Beik, Ahmad S. Al-Adsani, *Wind Turbine Systems*, Springer, 2020, pp. 11–31.

9 DC Wind Systems

To date, both AC and DC wind generation schemes, their associate components such as power electronics interface, and control aspects have been studied and developed for practical applications. In the DC schemes, much focus has been on high-voltage DC (HVDC) transmission systems. An economic and reliability analysis of integration of an HVDC offshore wind power on the Atlantic Coast is discussed in [3]. The policies and regulations around such integration have been discussed, which provide an insight into real-world application and challenges of HVDC systems. The [4] discusses the parallel integration of offshore high-voltage AC (HVAC) and HVDC systems with central and local control where the HVDC is based on diode rectifiers. A single offshore substation collecting power from two wind farms with multilevel VSC HVDC is presented in [5] with its control and operational aspects. The control and power electronics interfaces such as modular VSC's for wind turbines [6], control of multilevel matrix converters [7], and AC/AC cascade converters [8] among other topologies have been studied.

9.1 MAKING A CASE FOR ALL-DC WIND SYSTEM

There have been studies aiming at introducing DC system for both transmission and interconnection/array grid making a wind farm an all-DC system. A control structure for a parallel DC interconnection system is proposed by [9] as an economical system for offshore wind power. The authors in [10] have proposed a distributed series DC system for offshore wind generation systems. The proposed system in [10] eliminates the AC transformers and offshore substation while the turbine conversion system utilizes a permanent magnet (PM) generator coupled to a passive rectifier and a DC/DC converter, i.e. an AC/DC/DC conversion scheme. The authors in [11] discuss fault analysis of a wind scheme with DC collector and HVDC transmission grid while [12] proposes a DC wind power system using series-connected wind turbines. The wind turbine conversion system in [12] includes a permanent-magnet (PM) generator connected to a 3-phase AC-to-AC power electronics converter, a 3-phase transformer, and a passive rectifier. Steady-state and fault analysis of a wind farm with a medium voltage DC (MVDC) array grid and an HVDC transmission is addressed in [13]. A coordinated control approach for a DC wind system is presented in [14] where from the AC grid point of view the wind farm acts as a synchronous generator.

9.2 OVERVIEW

In an all-DC wind system, an example of which is shown in Figure 9.1, both the collector and transmission grids are DC, although at different voltage levels. The wind turbines outputs are DC and directly connected to the DC collector grid. A suitable

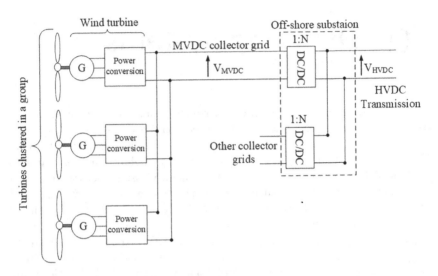

FIGURE 9.1 An all-DC wind system.

voltage for the collector grid may be at a medium voltage DC (MVDC) level. A high-voltage DC (HVDC) could be considered for the collector grid, providing the output of turbine is high voltage.

With the growing interests in the all DC system, the application of DC/DC converters both inside the wind turbines and on the offshore platform has gained much attention. In an all-DC system, the substation steps up the collector grid voltage which is MVDC, to suitable levels for transmission voltage, which is HVDC. A high-voltage dual active bridge DC/DC converter topology using silicon-carbide (SiC) devices is proposed in [15], which offers bidirectional capability, while the authors in [16] propose a three-level DC/DC converter topology. Other high-voltage high-power configurations such as resonant DC/DC converters have been proposed [17]–[19]. The authors in [20] propose a DC/DC converter by combining resonant circuits and switching configurations that results in an increased ratio of step up. A modular DC/DC converter is proposed in [21] that employs a proportional resonant control approach which results in reduced losses and improved efficiency. A switched-capacitor DC/DC converter is presented in [22] and compared with other topologies with improvements in the losses and efficiency.

In such systems, the power electronics systems in terms of voltage source converters (VSC's) decouple the wind generators from a fixed voltage and frequency grid and allow variable speed operation of turbines capturing maximum power from the wind kinetic energy. Currently, most of these conversion systems are rated at low voltages (up to 1000V AC), mainly due to unavailability of commercial switches at high voltage and current. Therefore, to achieve high-voltage, high-power turbine converters, a number of power electronics switches need to be connected in parallel and series, making the converter bulky and costly, hence unfit for a confined space in the wind turbine. Additionally, at high voltages, the large VSC DC-link capacitors are prone to electrolyte failure compromising turbine lifespan and incurring

DC Wind Systems

maintenance costs. To avoid these, wind generator systems and generators are designed for low voltages. A power transformer is then used to increase the voltage at the turbine output, which is suitable for transmission to off-shore or on-shore substation.

To adapt the existing wind turbines for DC array grids, the 3-phase output of wind turbine transformer needs to be rectified as schematically shown in Figure 9.2. In this scheme, the control over the DC output voltage is performed by VSC1 while VSC2 controls the generator for maximum power point tracking (MPPT). The passive rectifier consists of three legs, each with two series diode rated at medium-to-high voltages and currents.

Another approach to achieve a medium-to-high-voltage DC at the output of wind turbine is illustrated in Figure 9.3. In this scheme, the generator outputs are rectified using a passive rectifier whose output is stepped-up with a turbine DC/DC converter. In this case also, the turbine conversion system is rated at low voltages. To achieve a high voltage and power, the DC/DC has a multilevel and modular structure. The multilevel structure facilitates reduction of capacitive smoothing elements at DC/DC converter output. The scheme in Figure 9.2 has greater controllability as there are two converters that control the generator and the output power/voltage. In contrast, in Figure 9.3 scheme, the generator control for maximum power point tracking and the control of output MVDC voltage is performed by the DC/DC converter. In the next sections, the structure of such converters will be discussed in more details.

Figure 9.4 shows an alternative approach that simplifies the wind turbine conversion system for connection to collector DC grids. This scheme uses a hybrid generator and a passive rectifier. The hybrid generator has a 9-phase system with double rotor topology, and the passive rectifier has 9 legs with 18 passive diodes. Using this

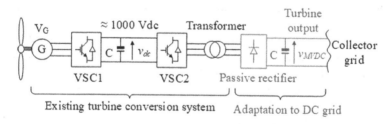

FIGURE 9.2 Adaptation of existing wind turbine conversion system (Walney) to DC collector grid.

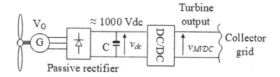

FIGURE 9.3 A wind turbine conversion system with DC turbine output.

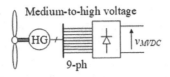

FIGURE 9.4 Another wind turbine conversion system with DC turbine output.

scheme, the active power electronics is removed from the turbine leading to a system amenable to higher voltages. The control is instead shifted to the generator, hence the dual rotor topology.

9.3 ALL-DC SYSTEM COMPONENTS

Compared to AC wind systems, the components count in the DC systems is reduced. Major components used in all-DC wind systems are presented in this section.

9.3.1 DC-DC CONVERTERS

The medium-to-high-voltage DC/DC converters are an essential part of all-DC wind generation systems. Various topologies have been proposed for DC/DC converters that allow for high step-up ratio at high power. Figure 9.5 shows two DC/DC converter topologies using Thyristors. A single active side DC/DC converter uses a Thyristor controlled half bridge in the input and a diode half bridge on the output, while in a dual active side DC/DC converter both input and output half bridges uses Thyristors and provides further controllability. Thyristors are commercially available for high-voltage and high-power ratings, hence provide a readily available solution for the DC/DC converters.

To achieve a higher level of controllability, active power electronics switches in a full-bridge arrangement may be used. Figure 9.6 presents a single-active and a dual-active bridge DC/DC converters. In this topology, the input converter is a 3-phase voltage source converter with IGBT switches that inverts the input DC to a 3-phase controlled medium-to-high frequency AC. A 3-phase transformer then steps-up the voltage. The choice of medium/high frequency is to reduce the transformer size/mass compared to grid frequency (50/60 Hz) transformer. In single-active bridge, a passive rectifier is used on the output to rectify the high-frequency AC voltage to a DC voltage. In dual-active bridge, a voltage source converter is used on the output. The configuration in Figure 9.6 may also be realized using single-phase converters and a transformer. However, the higher number of phases improve the power processing capability and the DC power quality.

Other topologies for high-voltage high-power DC/DC converter have been studied such as resonant based converters [17]–[19]. However, the topologies presented in Figures 9.5 and 9.6 are the most common high-voltage high-power DC/DC converters.

DC Wind Systems

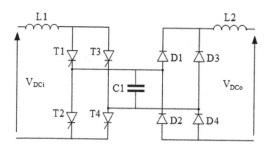

Single active side

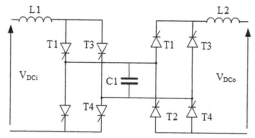

Dual active side

FIGURE 9.5 DC/DC converter using thyristors.

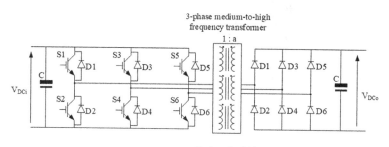

Single active bridge

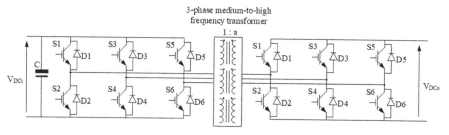

Dual active bridge

FIGURE 9.6 DC/DC converter using active switches.

9.3.2 Generator System

The wind turbine conversion for DC systems are presented in Figures 9.2–9.4. The hybrid generator in the scheme in Figure 9.4 uses an unconventional generator systems, referred to as hybrid generator, which is discussed below.

Figure 9.7 shows the schematic of hybrid generator which has a dual rotor topology and its application in a wind turbine. The permanent magnet (PM) machines have a fixed magnetic field that induces a fixed stator voltage at each speed. Hence, a power electronic converter is required to control the machine performance. If the power electronics converter is eliminated and replaced with a passive rectifier, as shown in Figure 9.4, the control needs to be implemented within the generator with an additional winding that is capable of regulating the magnetic field. The hybrid generator therefore utilizes two rotor excitation, a permanent magnet (PM) and a wound field (WF) rotor, and the name hybrid refers to the combination of PM and WF rotors. Both the PM and WF rotors are assemble on a shared shaft and share the same stator as shown in Figure 9.7.

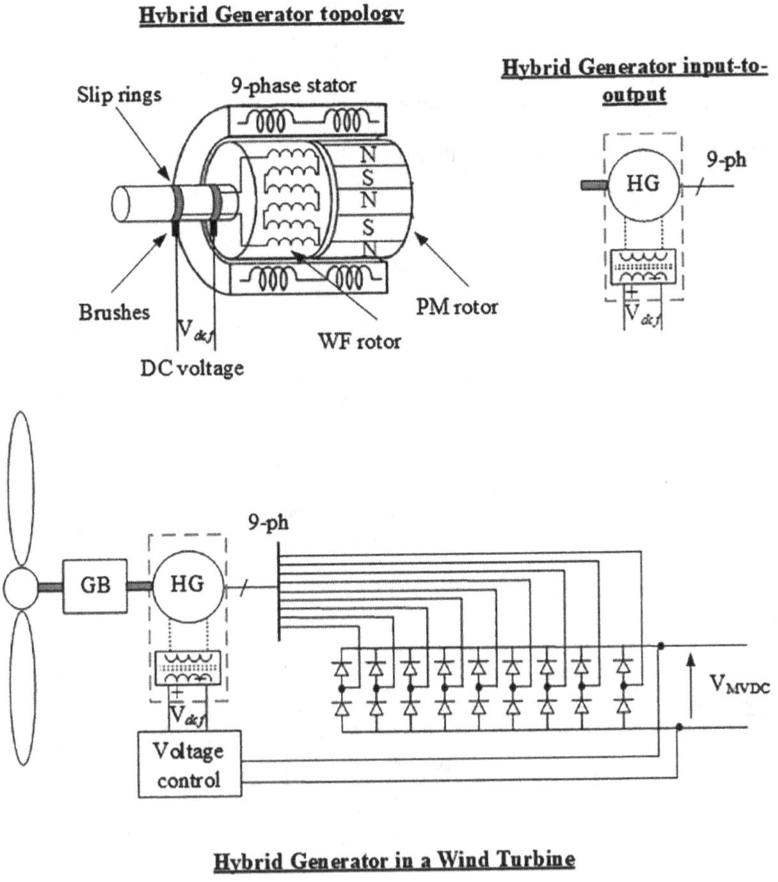

FIGURE 9.7 The hybrid generator and 9-phase rectifier.

DC Wind Systems

Therefore, the hybrid generator combines voltages from fixed field PM rotor and variable field WF rotor. The total hybrid generator output voltage is a vector sum of these two voltages. The WF is supplied a DC current and voltage which are provided by brushes and slip rings or using a brushless exciter system. The brushless excitation systems suitable for hybrid generators are the same as in conventional synchronous generators.

The hybrid generator has a 9-phase stator winding as shown in Figure 9.7. Compared to 3-phase counterparts, the 9-phase stator winding results in improved power density, improved rectified output power and voltage quality, and hence reduced capacitive filter requirements on the DC side.

9.3.3 MULTILEG RECTIFIER

In the scheme presented in Figure 9.7 the hybrid generator is interfaced to a multileg passive rectifier. There are nine legs each having two series diodes the midpoint of which is connected to a phase lead. A 9-phase terminal voltage may be expressed as follows:

$$\begin{aligned} v_1 &= v_p \sin(\omega t) \\ v_2 &= v_p \sin(\omega t - 40) \\ v_3 &= v_p \sin(\omega t - 80) \\ v_4 &= v_p \sin(\omega t - 120) \\ v_5 &= v_p \sin(\omega t - 160) \\ v_6 &= v_p \sin(\omega t - 200) \\ v_7 &= v_p \sin(\omega t - 240) \\ v_8 &= v_p \sin(\omega t - 280) \\ v_9 &= v_p \sin(\omega t - 320) \end{aligned} \tag{9.1}$$

where v_p is peak phase voltage magnitude. Sinusoidal voltage waveform at the hybrid generator terminals is assumed to facilitate the analysis.

Figure 9.8 plots 9-phase voltages. At each moment in time, there are two rectifier legs that conduct the phase current. Therefore, the voltage difference between midpoints of the two legs, i.e., line-to-line voltage, appears on the DC side during commutation of the legs. However, as discussed in Chapter 5, there are four different line-to-line voltages in a 9-phase system, viz:

$$\begin{aligned} v_{LL1} &= v_1 - v_2 = v_p \left(\sin(\omega t) - \sin(\omega t - 40) \right) \\ v_{LL2} &= v_1 - v_3 = v_p \left(\sin(\omega t) - \sin(\omega t - 80) \right) \\ v_{LL3} &= v_1 - v_4 = v_p \left(\sin(\omega t) - \sin(\omega t - 120) \right) \\ v_{LL4} &= v_1 - v_5 = v_p \left(\sin(\omega t) - \sin(\omega t - 160) \right) \end{aligned} \tag{9.2}$$

The four different line-to-line voltages are shown in Figure 9.8. As seen in the figure, the highest line-to-line voltage is between phase 1 and phase 5 with a normalized amplitude of 2.0 p.u. compared to 1.0 p.u. for the phase voltage. Therefore, the highest line-to-line voltage appears on the rectified DC side, as this drives turn-on and turn-off of the diodes. Figure 9.9 shows rectified DC voltage that consists of pieces of line-to-line voltage where applicable. There are 18 line-to-line voltages, hence 18 pieces for the rectified voltage with DC ripple frequency 18 times that of generator terminal voltage.

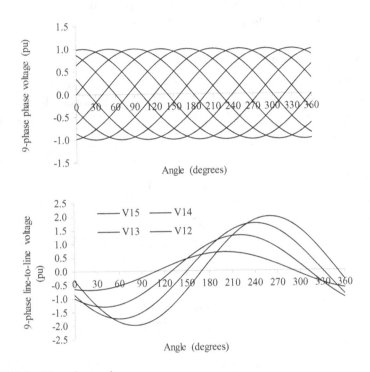

FIGURE 9.8 Nine-phase voltages.

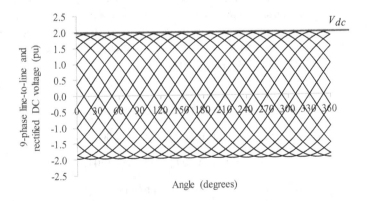

FIGURE 9.9 Rectified voltages from 9-phase.

DC Wind Systems 133

9.4 SYSTEM ANALYSES

The analyses of an all-DC system are presented in this section. To conduct comparable analyses, the AC site layout from the Walney wind farm (discussed in Chapter 8) is adapted for the DC wind farm, and shown in Figure 9.10. A simulation model (in Matlab) is set-up using the components models and equivalent circuits that were discussed previously, and the system performance is analyzed. A number of assumptions are made:

- The distance from the hybrid generator terminals to the multileg passive rectifier is negligible. Hence, the drop voltage in terminal leads is not considered.
- The DC/DC converters adjust their step-up ratio based on the wind velocity to achieve desirable voltages.
- Turbine generator nominal power is 3.6 MW, i.e. the same as Walney wind turbines.
- A phase terminal voltage of 38 kV is considered for the hybrid generator, i.e., a high-voltage conversion system achieved by elimination of active power electronics in the turbine Nacelle. This is compared to low-voltage induction generator (690V) for Walney wind turbines.

Given a phase voltage of 38 kV RMS for the 9-phase hybrid generator, the rectified DC voltage is calculated using Equation 9.3 (which was discussed in Chapter 5) as

$$V_{dc} = v_p \frac{2n}{\pi} \sin\left(\frac{\pi}{n}\right) = 38\sqrt{2} \frac{2 \times 9}{\pi} \sin\left(\frac{\pi}{9}\right) = 105 \text{ kV} \quad (9.3)$$

Therefore, the nominal DC collector grid voltage is 105 kV DC. In the Walney wind farm the off-shore substation transformers have a step-up ratio of 1:4. Given the same step-up ratio is adapted for the DC/DC converters on the off-shore substation in all-DC grid, the HVDC transmission grid nominal voltage is 420 kV DC.

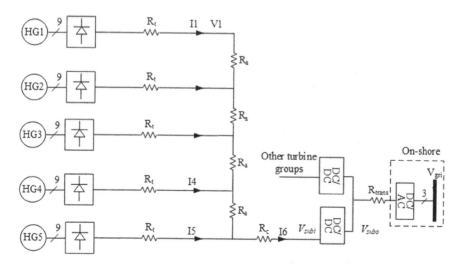

FIGURE 9.10 All-DC wind farm layout.

To have a comparable thermal performance for the cable systems, the current densities for the AC cables in Walney system are used for the DC cables in the all-DC system. Therefore, cable conductor cross-section area and resistance are calculated as follows:

$$A_{dc} = \frac{I_{dc}}{J_{dc}} \qquad (9.4)$$

$$R_{dc} = \frac{\rho_{cu} l_{dc}}{A_{dc}} \qquad (9.5)$$

where A_{dc} is DC cable cross-section area, I_{dc} is DC current flowing in the cable, J_{dc} is DC cable current density, R_{dc} is DC cable resistance, and l_{dc} is DC cable length. The loss and voltage drop across a DC cable are calculated as follows:

$$P_{dc} = 2 R_{dc} I_{dc}^2 \qquad (9.6)$$

$$\Delta v_{dc} = 2 R_{dc} I_{dc} \qquad (9.7)$$

The factor 2 in above equations is for the return DC cable that needs to be taken into account. Note that in AC configuration the 2-phase system is modeled using a per-phase equivalent circuit. Table 9.1 lists the main specification of the all-DC system.

Table 9.2 shows the results of system analysis at full-load when the system operates at steady state. The voltage drop across the collector grid cable connecting the first turbine to the offshore substation is 0.1% compared to 0.4% for the Walney wind farm. As discussed, the all-DC system is analyzed and compared to Walney wind farm keeping the geometrical topology of the wind farm, i.e. the distances, site layout, turbine arrangements, etc. unchanged. The loss analysis shows that the total wind farm losses in the all-DC system are 60% of that in AC Walney wind farm. Majority of losses are from the collector grid where it contributes 65% of the all-DC wind farm loss, compared to 25% from the DC/DC converters on off-shore

TABLE 9.1
All-DC System Specifications

Item	Specs
Hybrid generator nominal power	3.6 MW
Hybrid generator phase voltage	38 kV RMS
DC collector system nominal voltage	105 kV DC
Transmission system nominal voltage	420 kV DC
DC/DC step-up ratio (off-shore substation)	1:4
DC cables current density	2.0 A/mm²
Copper resistivity	0.017 μΩm
Collector grid cables cross-section area	87 mm²
Transmission grid cables cross-section area	363 mm²
Collector grid DC cable resistance	0.5 ohms
Transmission grid DC cable resistance	5 ohms

DC Wind Systems

TABLE 9.2
System Analyses Results

Turbine	
Hybrid generator output power	3492 kW
Hybrid generator phase voltage	38.11 kV RMS
Hybrid generator efficiency	97%
Rectifier efficiency	99%
Tower cable losses	0.15 kW
Turbine rectified voltage	105.60 kV DC
Turbine output DC current	33 A
Wind Farm	
DC voltage at substation input	105.50 kV DC
DC current at substation input	165 A
DC voltage at substation output	421 kV DC
DC current at substation output	414 A
Wind farm voltage at the point of connection to grid	420.20 kV DC
Wind farm current at the point of connection to grid	414 A
Off-shore substation input power	177100 kW
Off-shore substation efficiency	98%
Collector grid total loss	6520 kW
HVDC transmission losses	717 kW
Wind farm total output power	173700 kW
Wind farm efficiency at the point of connection to the grid	95%

substation and the rest of 10% from HVDC transmission and wind turbines. Total wind farm efficiency for the All-DC system is 95% compared to 91% for the Walney wind farm.

A comparison of the All-DC system mass with Walney system is presented in Table 9.3. The all-DC system results in 15% lower mass for the turbine conversion system and a 70% reduction in the total cable system. The turbine conversion system in Walney includes a gearbox, an induction generator, two voltage source converters, and a transformer (5 series-connected components), while the all-DC system the turbine conversion system includes a Hybrid Generator (HG), and a rectifier (2 series-connected components). The large reduction in mass in the all-DC system is due to higher DC voltage for the collector grid (105 kV DC for all-DC compared to 33 kV AC for Walney), and the reduction in the number of cables (2 DC cables per circuit for all-DC compared to 3 AC cables per circuit for Walney).

Mass of a cable is calculated as

$$m_{mass} = \delta_{cu} V_{vol} \tag{9.8}$$

where δ_{cu} is mass density of copper (8960 kg/m^3), and the cable volume is calculated using its length and cross-section as

$$V_{vol} = A_{cable} \, l_{cable} \tag{9.9}$$

TABLE 9.3
Comparison of All-DC and AC Walney Wind Systems Mass

	All-DC Wind Farm (Tonnes)	All-AC Walney Wind Farm (Tonnes)
Hub mass	38.5	38.5
Hybrid generator mass	52	–
Induction generator and gearbox mass	–	52
Two voltage source converter mass	–	2.4
Rectifier mass	0.25	–
Tower mass	260	260
Turbine cables	0.03	4.78
Collector grid cables	9.5	225
Transmission system cables	91	752

9.5 VARIABLE VOLTAGE DC COLLECTOR GRID

As the wind velocity varies from cut-in 3 m/s to cut-out 25 m/s, the turbine rotor speed and hence the hybrid generator shaft speed (RPM) vary. In the previous section, it was assumed that the hybrid generator rectified voltage is maintained at its nominal voltage (105 kV) as the wind velocity varies. This was made possible by controlling the wound field (WF) of the hybrid generator. Therefore, in the analyses of all-DC presented in the previous section the collector grid voltage (105 kV DC) is assumed fixed as the wind velocity varies. In addition, in the previous section, it was assumed that the DC/DC converters on the off-shore substations use a fixed ratio of 1:4 to step-up the voltage, hence the HVDC transmission grid voltage in the analyses were assumed to be fixed at 420 kV DC.

A variable voltage generator and collector grid voltage is investigated in this section. However, the transmission grid voltage is assumed to be fixed, therefore the DC/DC step-up ratio is variable. Figure 9.11 shows turbine characteristics. From cut-in (3 m/s) to rated velocity (13 m/s), i.e. MPPT region, the rotor speed and hybrid generator voltage change with the wind velocity. In this example, a hybrid generator with a rated 11 kV phase voltage is utilized. The WF therefore controls the hybrid generator phase voltage from zero to 11 kV in the MPPT region as seen from Figure 9.11. The turbine rotor speed in MPPT region changes from zero to 13 RPM, which is controlled by the turbine blade pitch system.

Beyond the rated and up to cut-out velocity the hybrid generator is controlled (by the wound field) to maintain a fixed phase terminal voltage at 11 kV, a fixed turbine rotor speed controlled by the pitch system to 13 RPM, and a rated power of 3.6 MW. Note that there will be slight perturbations around the nominal values due to transients. In the Walney wind farm, the turbines are equipped with a gearbox with 1:119 ratio and the turbine rotor speed is increased to 476–1547 RPM for the induction generator. However, in this section, a scaled-down gearbox with 1:46 ratio is used that increases the hybrid generator speed to 184–600 RPM.

Below cut-in velocity, the turbine output power is zero due to insufficient wind; in this region the turbine rotor speed increases from zero to cut-in while turbine is

DC Wind Systems

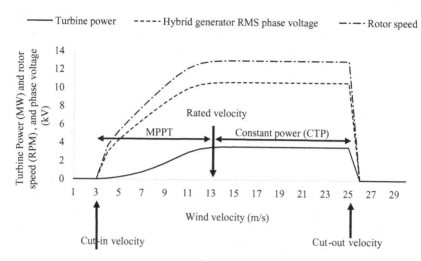

FIGURE 9.11 Wind turbine characteristics.

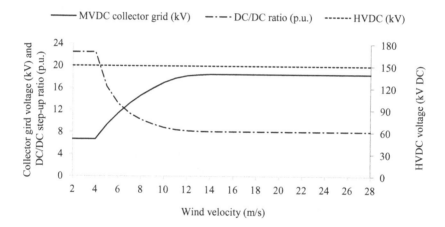

FIGURE 9.12 Collector grid and HVDC voltage, and DC/DC step-up ratio.

initializing without delivering power. The WF current up to cut-in speed is maintained at zero, hence given there is no turbine rotor speed the HG output voltage is zero as seen from Figure 9.11. Below cut-in velocity, a control system that supervises the wind farm operation adjusts the DC/DC converter step-up ratio such that the collector grid voltage matches the hybrid generator rectified DC voltage at cut-in velocity. Therefore, as soon as the wind velocity increases beyond the cut-in velocity the turbine starts delivering power. The DC/DC step-up ratio in this region is 22.44, while the collector grid voltage is maintained at 6.5 kV DC as shown in Figure 9.12.

In the MPPT region, the collector grid voltage increases with wind velocity while the control system reduces DC/DC converter step-up ratio to maintain a fixed HVDC voltage, as shown in Figure 9.12. In the constant power (CTP) region where the wind velocity varies from 13 to 25 m/s, the collector grid voltage is maintained at 18.50 kV

DC and the DC/DC step-up ratio is maintained at 8.11 p.u. as shown in Figure 9.12. Beyond cut-out velocity (25 m/s), the turbine is stopped while blades are feathered by pitch control and depending on the wind strength the Nacelle may be controlled out of the wind using yaw control system. Beyond the cut-out velocity, the power and speed are gradually reduced to zero at a rate that depends on the dynamics of pitch and yaw control systems, as seen from Figure 9.11. In this region, the collector grid voltage is maintained at its nominal value (18.5 kV DC), hence as soon as the wind velocity is returned below the cut-out velocity the power starts flowing to the offshore substation. The HVDC voltage in the entire wind regime is fixed at 150 kV DC using DC/AC converter on the shore.

Figure 9.13 shows an algorithm for the wind farm supervisory control system. The input to the control system are the measurements of the wind velocity at each wind turbine. The control system decides on the step-up ratio of the DC/DC converters on the off-shore substation based on the region of operation. There are four operational regions as depicted in Figure 9.11, below cut-in, MPPT, Constant Power (CTP), and the region beyond cut-out velocity. Once the supervisory control system sets up the DC/DC ratio, the collector grid voltage is set. The WF control system then adjusts the current for each individual wind turbine hybrid generator.

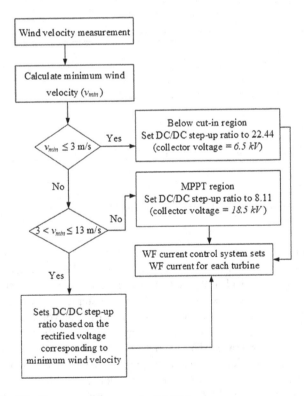

FIGURE 9.13 Windfarm supervisory control system.

REFERENCES

1. Omid Beik, An HVDC Off-shore Wind Generation Scheme with High Voltage Hybrid Generator, Ph.D. thesis, McMaster University, 2016.
2. Department of Defense, The Military Handbook for Reliability Prediction of Electronic Equipment, Tech. Rep. MIL-HDBK-217F-2, 1995.
3. J. Lin, Integrating the First HVDC-Based Offshore Wind Power into PJM System—A Real Project Case Study *IEEE Transactions on Industry Applications*, 2016, 52, (3), pp. 1970–1978.
4. Yu, L., Li, R., Xu, L., Parallel Operation of Diode Rectifier based HVDC Link and HVAC Link for Offshore Wind Power Transmission, *IET Journal of Engineering*, 2019, pp. 4713–4717.
5. Raza, A., Dianguo, X., Xunwen, S., Weixing L., Williams, B.W., A Novel Multiterminal VSC-HVdc Transmission Topology for Offshore Wind Farms, *IEEE Transactions on Industry Applications*, 2017, 53, (2), pp. 1316–1325.
6. Beik, O., Dekka A., Narimani, M., *A New Modular Neutral Point Clamped Converter with Space Vector Modulation Control*,' Proceedings IEEE International Conference on Industrial Technology, Lyon, France, 2018, pp. 591–595.
7. Diaz, M., Cardenas, R., Espinoza, M., Rojas, F., Mora, A., Clare, J. C., Wheeler, P., Control of Wind Energy Conversion Systems Based on the Modular Multilevel Matrix Converter, *IEEE Transactions on Industrial Electronics*, 2017, 64, (11), pp. 8799–8810.
8. Thitichaiworakorn, N., Hagiwara M., Akagi, H., A Medium-Voltage Large Wind Turbine Generation System Using an AC/AC Modular Multilevel Cascade Converter. *IEEE Journal of Emerging and Selected Topics in Power Electronics*, 2016. 4, (2), pp. 534–546.
9. Meyer, C., Hoing, M., Peterson, A., De Doncker, R. W., Control and Design of DC Grids for Offshore Wind Farms, *IEEE Transactions on Industry Applications*, 2007, 43, (6), pp. 1475–1482.
10. Veilleux E., Lehn, P. W., Interconnection of Direct-drive Wind Turbines using a Series-connected DC Grid, *IEEE Transactions on Sustainable Energy*, 5, (1), pp. 139–147.
11. Deng, F., Chen, Z., Operation and Control of a DC-Grid Offshore Wind Farm under DC Transmission System Faults, *IEEE Transactions on Power Delivery*, 2013, 28, (3), pp. 1356–1363.
12. Holtsmark, N., Bahirat, H. J., Molinas, M., Mork B. A., Hoidalen, H. K., An All-DC Offshore Wind Farm With Series-Connected Turbines: An Alternative to the Classical Parallel AC Model?, *IEEE Transactions on Industrial Electronics*, 2013, 60, (6), pp. 2420–2428.
13. Shi, T., Shi, L., Yao, L., Detailed Modelling and Simulations of an All-DC PMSG-based Offshore Wind Farm, *IET Journal of Engineering*, 2019, (16), pp. 845–849.
14. Yang, R., Shi, G., Cai X., Zhang, X., Voltage Source Control of Offshore All-DC Wind Farm, *IET Journal of Engineering*, 2019, (18), pp. 4718–4722.
15. Akagi, H., Kinouchi S., Miyazaki, Y., Bidirectional Isolated Dual-active-bridge (DAB) DC-DC Converters using 1.2-kV 400-A SiC-MOSFET Dual Modules, *CPSS Transactions on Power Electronics and Applications*, 2016, 1, (1), pp. 33–40.
16. Filba-Martinez, A., Busquets-Monge, S., Nicolas-Apruzzese J., Bordonau, J., Operating Principle and Performance Optimization of a Three-Level NPC Dual-Active-Bridge DC–DC Converter, *IEEE Transactions on Industrial Electronics*, 2016, 63, (2), pp. 678–690.
17. Wu, J., Li, Y., Sun, X., Liu, F., A New Dual-Bridge Series Resonant DC–DC Converter with Dual Tank, *IEEE Transactions on Power Electronics*, 2018, 33, (5), pp. 3884–3897.

18. Jovcic D., Zhang, L., LCL DC/DC converter for DC grids, *IEEE Transactions on Power Delivery*, 2013, 28, (4), pp. 2071–2079.
19. Zhang, X., Green, T. C., The Modular Multilevel Converter for High Step-Up Ratio DC–DC Conversion, *IEEE Transactions on Industrial Electronics*, 2015, 62, (8), pp. 4925–4936.
20. Abbasi M., Lam, J., A Step-Up Transformerless, ZV–ZCS High-Gain DC/DC Converter with Output Voltage Regulation Using Modular Step-Up Resonant Cells for DC Grid in Wind Systems, *IEEE Journal of Emerging and Selected Topics in Power Electronics*, 2017, 5, (3), pp. 1102–1121.
21. Liu, H., Dahidah, M. S. A., Yu, J., Naayagi R. T., Armstrong, M., Design and Control of Unidirectional DC–DC Modular Multilevel Converter for Offshore DC Collection Point: Theoretical Analysis and Experimental Validation, *IEEE Transactions on Power Electronics*, 2019, 34, (6), pp. 5191–5208.
22. Chen, W., Huang, A. Q., Li, C., Wang, G., Gu, W., Analysis and Comparison of Medium Voltage High Power DC/DC Converters for Offshore Wind Energy Systems, *IEEE Transactions on Power Electronics*, 2013, 28, (4), pp.2014–2023.

Part B

Photovoltaic Power Systems

10 Photovoltaic Power

Photovoltaic (PV) power technology uses semiconductor cells (wafers), generally several square centimeters in size. From the solid-state physics point of view, the cell is basically a large-area p-n diode with the junction positioned close to the top surface. The cell converts sunlight into DC electricity. Numerous cells are assembled in a module to generate the required power (Figure 10.1). Unlike the dynamic wind turbine, the PV installation is static, does not need strong tall towers, produces no vibration or noise, and needs no active cooling.

The PV-cell-manufacturing process is energy intensive. In the past, every square centimeter of cell area consumed more than a kWh before it faced the sun and produced the first kWh of energy. However, the energy consumption during manufacturing is steadily declining with the continuous implementation of new production processes.

Major advantages of PV power are as follows:

- Short lead times to design, install, and start up a new plant
- Highly modular; hence, the plant economy is not strongly dependent on size
- Power output matches well with peak-load demands
- Static structure, no moving parts; hence, no noise
- High power capability per unit of weight
- Longer life with little maintenance because of no moving parts
- Highly mobile and portable because of lightweight

At present, PV power is extensively used in utility-scale installations, on residential and commercial buildings, and for stand-alone power in remote villages around the world, particularly in hybrid systems with diesel power generators. It is expected that PV power will continue to find expanding markets in many countries. The driving force is the energy need in developing countries and the environmental concerns in developed countries.

The phenomenal growth of solar power installations in both the utility-scale solar parks and on residential roof-tops is attributed to its rapidly declining module prices as shown in Figure 10.2. The PV module prices per watt of generation capacity have fallen to less than $0.50 per watt at present at an average rate of 12% price reduction per year over the last four decades. In the USA, Japan, Korea, and Australia, the present installed solar power system (PV modules + d.c. to a.c. inverters + grid interface connection) prices are below $1.50 per watt capacity in utility-scale installations and below $3 per watt on residential roof-top installations. In Germany, China, and India, they are about $1 per watt and $1.20 per watt, respectively.

Very large solar PV power plants are being built around the world at present. For example, the 580 MW Solar Star plant (Units I and II) in California was the world's largest PV power station when completed in 2015. The 850 MW capacity Longyangxia

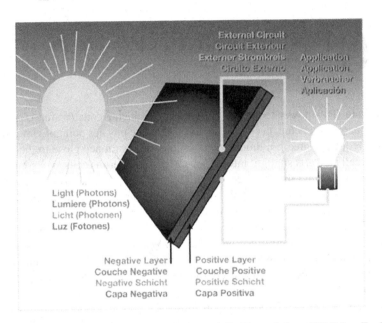

FIGURE 10.1 PV module in sunlight generates DC. (From Solarex/BP Solar, Frederick, MD. With permission.)

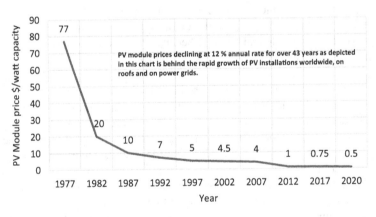

FIGURE 10.2 PV module prices declining at 12% per year rate for four decades. (from various industry sources)

Dam Solar Park in China became operational as the largest solar farm in the world in 2017. This was passed in 2019 with the completion of the Pavagada Solar Park in Karnataka, India, with a capacity of 2050 MW. The force of solar power competing with traditional electrical power sources is visible in the 750 MW capacity Rewa Ultra Mega Solar power plant in MP, India, commissioned in 2020. It is spread over

Photovoltaic Power

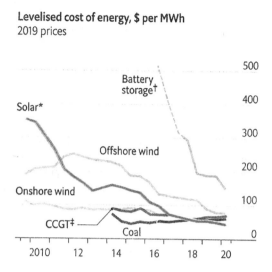

FIGURE 10.3 Cost of energy from various sources, $/MWh (2019 prices). (Vaclav Smil, BP Statistical Review of World Energy, *The Economist*, June 2020) (CCGT = Combine cycle gas turbines, + = estimated,* = average of fixed and sun-tracking solar arrays)

an area of 1,590 acres (6.4 km^2), with its first-year kWh energy production cost was $0.04 versus $0.06 from the coal-based power plants in India. Figure 10.3 shows the worldwide energy cost of various electrical energy sources, showing the solar energy cost now lower than all other sources. With such heads-on competition with all other power sources, solar power is now here to stay and grow.

The total installed cumulative PV power generation capacity globally has grown from mere 0.4 gigawatts (GW, billion watts) to 600 GW, amounting to a 30% annual growth rate over the past three decades, relentlessly doubling every three years for almost three decades. Worldwide at present, the total electrical power generating capacity from all sources is about 5500 GW, which annually produces about 30 Trillion kWh electrical energy globally, including 4.5 Trillion kWh in the USA.

Globally in 2020, PV power is about 10% in cumulative installed MW generation capacity and about 3% in kWh energy production. The percentage in the solar kWh energy production is less than 1/3rd of that in the MW power generation capacity because the effective sunlight for electrical power generation varies seasonally between 6 and 8 hours in a 24-hour day. The fossil fuel and nuclear power plants, on the other hand, produce energy around the clock.

10.1 BUILDING-INTEGRATED PV SYSTEM

The near-term potentially large application of PV technology is for cladding buildings to power air-conditioning and lighting loads. One of the attractive features of the

PV system is that the power output matches very well with the peak-load demand. It produces more power on a sunny summer day when the air-conditioning load strains the grid lines (Figure 10.4). Figure 10.5 shows a PV-clad building in Germany with an integrated grid-connected power system.

With the U.S. Department of Energy's 5-year cost-sharing program in the past, Solarex (now BP Solar) in Maryland developed low-cost, easy to install, pre-engineered building-integrated photovoltaic (BIPV) modules. Such modules made in shingles and panels can replace traditional roofs and walls. The building owners have to pay only the incremental cost of these components. The land is paid for, the support structure is already installed, the building is already wired, and developers may finance the BIPV modules as part of their overall project. The major advantage of the BIPV system is that it produces distributed power generation at the point of

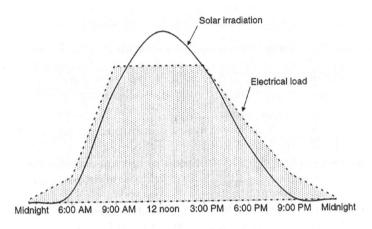

FIGURE 10.4 Power usage in a commercial building on a typical summer day.

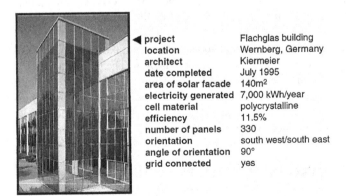

project	Flachglas building
location	Wernberg, Germany
architect	Kiermeier
date completed	July 1995
area of solar facade	140m²
electricity generated	7,000 kWh/year
cell material	polycrystalline
efficiency	11.5%
number of panels	330
orientation	south west/south east
angle of orientation	90°
grid connected	yes

FIGURE 10.5 Building-integrated PV systems in Germany. (From the Institution of Mechanical Engineers, U.K. With permission.)

consumption. The BIPV system, therefore, offers the first potentially widespread commercial implementation of PV technology in the industrialized countries. On the consumer side, the DoE had also in the same time frame launched the Million Solar Roofs Initiative to place one million solar power systems on homes and buildings across the U.S.

10.2 PV CELL TECHNOLOGIES

In comparing alternative power generation technologies, the most important measure is the energy cost per kWh delivered. In PV power, this cost primarily depends on two parameters: the PV energy conversion efficiency, and the capital cost per watt capacity. Together, these two parameters indicate the economic competitiveness of the PV electricity.

The conversion efficiency of the PV cell is defined as follows:

$$\eta = \frac{\text{Electrical power output}}{\text{Solar power impinging on the cell}}$$

The primary goals of PV cell research and development are to improve the conversion efficiency and other performance parameters to reduce the cost of commercial solar cells and modules. The secondary goal is to significantly improve manufacturing yields while reducing the energy consumption and manufacturing costs, and reducing the impurities and defects. This is achieved by improving our fundamental understanding of the basic physics of PV cells. The continuing development efforts to produce more efficient low-cost cells have resulted in various types of PV technologies available in the market today in terms of the conversion efficiency and the module cost. The major types are discussed in the following subsections.[1]

10.2.1 SINGLE-CRYSTALLINE SILICON

Single-crystal silicon is the most widely available cell material and has been the workhorse of the industry. Its energy conversion efficiency ranges from 15 to 20%. In the most common method of producing it, the silicon raw material is first melted and purified in a crucible. A seed crystal is then placed in the liquid silicon and drawn at a slow constant rate. This results in a solid, single-crystal cylindrical ingot (Figure 10.6). The manufacturing process is slow and energy intensive, resulting in high raw material costs. The ingot is sliced using a diamond saw into 200- to 400-µm (0.005 to 0.010 in.)-thick wafers. The wafers are further cut into rectangular cells to maximize the number of cells that can be mounted together on a rectangular panel. Unfortunately, almost half of the expensive silicon ingot is wasted in slicing the ingot and forming square cells. The material waste can be minimized by making full-sized round cells from round ingots (Figure 10.7). Using such cells would be economical when panel space is not at a premium. Another way to minimize waste is to grow crystals on ribbons. Some U.S. companies have set up plants to draw PV ribbons, which are then cut by laser beam to reduce waste.

148 Wind and Solar Power Systems

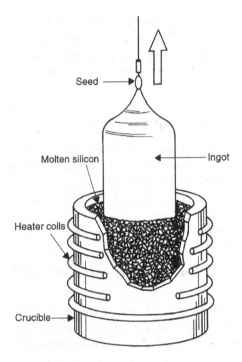

FIGURE 10.6 Single-crystal ingot making by Czochralski process. (From Photovoltaic Fundamentals, DOE/NREL Report DE91015001, February 1995.)

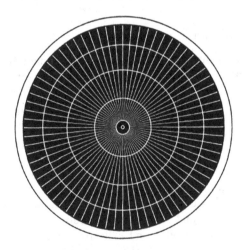

FIGURE 10.7 Round-shaped PV cell reduces material waste typically found in rectangular cells. (Depiction based on cell used by Applied Solar Energy Corporation.)

Photovoltaic Power

10.2.2 Polycrystalline and Semicrystalline Silicon

This is a relatively fast and low-cost process to manufacture crystalline cells. Instead of drawing single crystals using seeds, the molten silicon is cast into ingots. In the process, it forms multiple crystals. The conversion efficiency is lower, but the cost is much lower, giving a low cost per watt of power. Because the crystal structure is somewhat random (imperfect) to begin with, it cannot degrade further with imperfections in the manufacturing process or in operation. It comes in both thick- and thin-film cells and is overtaking the cell market in commercial applications.

10.2.3 Thin-Film Cell

These are new types of PV cells that have entered the market. Copper indium diselenide ($CuInSe_2$ or CIS), cadmium telluride (CdTe), and gallium arsenide (GaAs) are all thin-film materials, typically a few micrometers or less in thickness, directly deposited on a glass, plastic, stainless steel, ceramic, or other compatible substrate material. In this manufacturing process, layers of different PV materials are applied sequentially to a substrate. This technology uses much less material per square area of the cell, and hence, is less expensive per watt of power generated.

Researchers at NREL have also developed efficient and low-cost CIS cells with a focus on processes that are capable of being inexpensive while maintaining high performance, achieving a thin-film CIS solar cell efficiency of over 25%. The DOE-funded research at NREL is also directed at developing reproducible processes for making high-efficiency thin-film CdS/CdTe cells, as well as developing alternative processes that can improve cell performance, reproducibility, and manufacturability through improvements in five areas: transparent conducting oxide layers, CdS/window layers, absorbers and junctions, back contacts, and nanoparticle devices.

10.2.4 Amorphous Silicon

In this technology, a 2-µm-thick amorphous silicon vapor film is deposited on a glass or stainless steel roll, typically 2000 ft long and 13 in. wide. Compared to crystalline silicon, this technology uses only about 1% of the material. Its efficiency is about half that of crystalline silicon technology at present, but the cost per watt is significantly lower. On this premise, two large plants to manufacture amorphous silicon panels had started in the U.S. since 2000.

The DoE-funded research concentrates on stabilizing a-Si module efficiency through four tasks:

- Improving thin-film deposition through promising new techniques
- Developing materials with lower bandgaps to better balance photon utilization in multijunction devices
- Understanding light-induced metastability to deposit more stable a-Si material
- Developing low-temperature epitaxy

Table 10.1 compares amorphous silicon and crystalline silicon technologies.

TABLE 10.1
Comparison of Crystalline and Amorphous Silicon Technologies

	Crystalline Silicon	Amorphous Silicon
Present Status	Workhorse of terrestrial and space applications	New low-cost, low-efficiency technology
Thickness	200–400 µm (0.004–0.008 in.)	2 µm (less than 1% of that in crystalline silicon)
Raw Material	High	About 3% of that in crystalline silicon
Conversion Efficiency	20–25%	10–12%

10.2.5 SPHERAL CELL

This is yet another technology that is being explored in the laboratories. The raw material is low-grade spherical silicon crystalline beads, which are applied on typically 4-in squares of thin perforated aluminum foil. In the process, the impurities are pushed to the surface, where they are etched away. Because each sphere works independently, the individual sphere failure has a negligible impact on the average performance of the bulk surface. The Southern California Edison Company had once estimated that a 100-ft^2 spheral panel can generate 2000 kWh per year in an average Southern California climate.

10.2.6 CONCENTRATOR CELL

In an attempt to improve conversion efficiency, sunlight is concentrated tens or hundreds of times the normal intensity by focusing on a small area using low-cost lenses (Figure 10.8). A primary advantage of this is that such a cell requires a small fraction of area compared to the standard cells, thus significantly reducing the PV material requirement. However, the total sunlight collection area remains approximately the same for a given power output. Besides increasing the power and reducing the size or number of cells, the concentrator cell has the additional advantage that the cell efficiency increases under concentrated light up to a point. Another advantage is its small active cell area. It is easier to produce a high-efficiency cell of small area than to produce large-area cells with comparable efficiency. An efficiency over 40% has been achieved in a cell designed for terrestrial applications, which is a modified version of the triple-junction cell that Spectrolab developed for space applications. On the other hand, the major disadvantage of the concentrator cell is that it requires focusing optics, which adds to the cost. Concentrator PV cells have seen a significant interest in Australia and Spain in the past.

10.2.7 MULTIJUNCTION CELL

The single-junction n-on-p silicon cell converts only red and infrared light into electricity, but not blue and ultraviolet. The PV cell converts light into electricity most efficiently when the light's energy matches the semiconductor's energy level, known

Photovoltaic Power

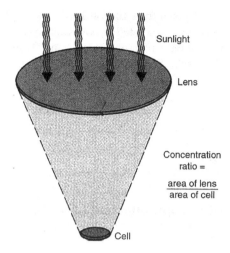

FIGURE 10.8 Lens concentrating the sunlight on a small area reduces the need for active cell material. (From Photovoltaic Fundamentals, DOE/NREL Report DE91015001)

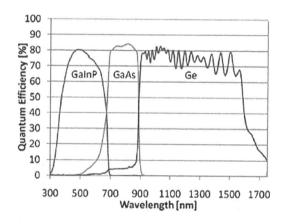

FIGURE 10.9 Improving conversion efficiency by capturing sun energy over a wide spectrum by cascading multi-junctions.

as its bandgap. The layering of multiple semiconductors with a wide range of band gaps converts more energy levels (wavelengths) of light into electricity. The multi-junction cell uses multiple layers of semiconductor materials to convert a broader spectrum of sunlight into electricity, thus improving the efficiency. The lights which escape the first junction without being converted into electricity is captured by the second junction that converts additional energy into electricity, and so does the 3rd junction (Figure 10.9). It is like a layer cake that captures different tastes. Moreover, concentrated sunlight by few hundred times also improves the photoconversion efficiency. With the gallium indium phosphide/gallium arsenide/germanium (GaInP/GaAs/Ge) triple-junction cell, NREL and Spectrolab have reported over 40% efficiency under concentrated sunlight. The cell captures infrared photons as well. Under

NASA and NREL funding, Tecstar (now Emcore) and Spectrolab (now Boeing) currently make the GaInP/GaAs/Ge for satellite power systems, but may soon trickle down to terrestrial applications. Figure 10.10 is the Technical data Sheet of Spectrolab's triple-junction solar cell that offers 28.3% efficiency in commercially available cells, widely used in space applications at present.

SPECTROLAB   A BOEING COMPANY

28.3% Ultra Triple Junction (UTJ) Solar Cells

Typical Electrical Parameters
(AM0 (135.3 mW/cm²) 28°C, Bare Cell)

Parameters	< 32 cm²	> 50 cm²
Jsc	17.05 mA/cm²	17.05 mA/cm²
Jmp	16.30 mA/cm²	16.30 mA/cm²
Jload min avg	16.40 mA/cm²	16.40 mA/cm²
Voc	2.660 V	2.660 V
Vmp	2.350 V	2.300 V
Vload	2.310 V	2.270 V
Cff	0.85	0.83
Effload	28.0%	27.5%
Effmp	28.3%	27.6%

Radiation Degradation
(Fluence 1MeV Electrons/cm²)

Parameters	1x10¹⁴	5x10¹⁴	1x10¹⁵
Imp/Imp₀	0.99	0.98	0.96
Vmp/Vmp₀	0.94	0.91	0.89
Pmp/Pmp₀	0.93	0.89	0.86

Thermal Properties

Solar Absorptance = 0.92 (5 mil CMG/AR, 0.90 for bare cells)
Emittance (Normal) = 0.85 (Ceria Doped Microsheet)

Weight

84 mg/cm² (Bare) @ 140 μm (5.5 mil) Ge wafer thickness

Temperature Coefficients (15°C - 80°C)
(Fluence 1MeV Electrons/cm²)

Parameters	BOL	5x10¹⁴	1x10¹⁵
Jmp (μA/cm²/°C)	1.2	5.3	6.9
Jsc (μA/cm²/°C)	5.3	6.5	6.9
Vmp (mV/°C)	-6.5	-6.7	-6.8
Voc (mV/°C)	-5.9	-6.3	-6.5

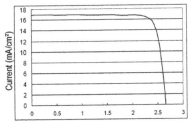

Typical IV Characteristic
AM0 (135.3 mW/cm²) 28°C, Bare Cell

Voltage (V)

*A/R: Anti-Reflective Coating

The information contained on this sheet is for reference only. Specifications subject to change without notice.
Revised 10/5/2010

© 2010 Spectrolab, Inc All Rights Reserved

FIGURE 10.10 Triple junctions PV cell datasheet from Spectrolab (Boeing Aerospace) for widely used in space applications. (Source: Spectrolab's product catalog, Wikipedia)

10.2.8 Inverted Metamorphic Multijunction (IMM) Cell

The IMM cell was invented at NREL to combine the efficiency improvements from the multi-junction cascading and also from concentrated sunlight. It was further funded by DoE for solar power installations on earth. The IMM cells are made with vapor deposited in thin films on substrates, and the cells with cover glass can be mounted on flexible cloth. The IMM cell yields photoconversion efficiency of 28.3% at normal sun, and 42.6% under concentrated sunlight 327 times the normal on earth. The space qualified cells made by Spectrolab (CA) and Emcore (NM) come in 3, 4, and 6 junctions with even higher efficiency. It is estimated that 6-junctison cell under 400 times concentrated sun can have efficiency exceeding 50% in near future. That is more than we can convert from coal or gas in utility-scale power plants.

REFERENCES

1. Carlson, D. E., 1995. *Recent Advances in Photovoltaics, Proceedings of the Intersociety Engineering Conference on Energy Conversion*, 1995, pp. 621–626.
2. Omid Beik, Ahmad S. Al-Adsani, *DC Wind Generation Systems, Design, Analysis, and Multiphase Turbine Technology*, Springer, 2020.
3. D. Schumacher, O. Beik and A. Emadi, "Standalone Integrated Power Electronics System: Applications for Off-Grid Rural Locations," *IEEE Electrification Magazine*, vol. 6, no. 4, pp. 73–82, December 2018.

11 Photovoltaic Power Systems

The photovoltaic (PV) effect is the electrical potential developed between two dissimilar materials when their common junction is illuminated with radiation of photons. The PV cell, thus, converts light directly into electricity. A French physicist, Becquerel, discovered the PV effect in 1831. It was limited to the laboratory until 1954, when Bell Laboratories produced the first silicon cell. It soon found application in U.S. space programs for its high power-generating capacity per unit weight. Since then, it has been extensively used to convert sunlight into electricity for earth-orbiting satellites. Having matured in space applications, PV technology is now spreading into terrestrial applications ranging from powering remote sites to feeding utility grids around the world.

11.1 PV CELL

The physics of the PV cell is very similar to that of the classical diode with a p-n junction (Figure 11.1). When the junction absorbs light, the energy of absorbed photons is transferred to the electron–proton system of the material, creating charge carriers that are separated at the junction. The charge carriers may be electron–ion pairs in a liquid electrolyte or electron–hole pairs in a solid semiconducting material. The charge carriers in the junction region create a potential gradient, get accelerated under the electric field, and circulate as current through an external circuit. The square of the current multiplied by the resistance of the circuit is the power converted into electrical form. The remaining power of the photon elevates the temperature of the cell and dissipates into the surroundings.

The origin of the PV potential is the difference in the chemical potential, called the *Fermi level*, of the electrons in the two isolated materials. When they are joined, the junction approaches a new thermodynamic equilibrium. Such equilibrium can be achieved only when the Fermi level is equal in the two materials. This occurs by the flow of electrons from one material to the other until a voltage difference is established between them, which has a potential just equal to the initial difference of the Fermi level. This potential drives the photocurrent in the PV circuit.

Figure 11.2 shows the basic cell construction.[1] Metallic contacts are provided on both sides of the junction to collect electrical current induced by the impinging photons. A thin conducting mesh of silver fibers on the top (illuminated) surface collects the current and lets the light through. The spacing of the conducting fibers in the mesh is a matter of compromise between maximizing the electrical conductance and minimizing the blockage of the light. Conducting-foil (solder) contact is provided over the bottom (dark) surface and on one edge of the top surface. In addition to these basic elements, several enhancement features are also included in the construction.

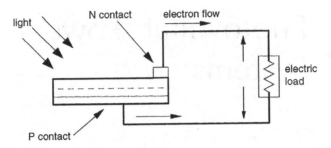

FIGURE 11.1 PV effect converts the photon energy into voltage across the p-n junction.

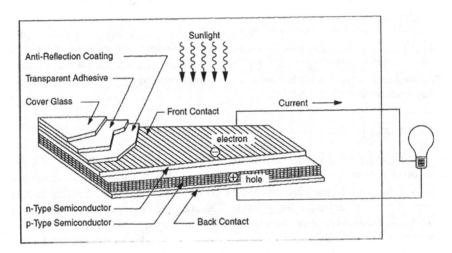

FIGURE 11.2 Basic construction of PV cell with performance-enhancing features (current-collecting silver mesh, antireflective coating, and cover-glass protection).

For example, the front face of the cell has an antireflective coating to absorb as much light as possible by minimizing the reflection. The mechanical protection is provided by a cover glass applied with a transparent adhesive.

11.2 MODULE AND ARRAY

The solar cell described in the preceding subsection is the basic building block of the PV power system. Typically, it is a few square inches in size and produces about 1 W of power. To obtain high power, numerous such cells are connected in series and parallel circuits on a panel (module) area of several square feet (Figure 11.3). The solar array or panel is defined as a group of several modules electrically connected in a series–parallel combination to generate the required current and voltage. Figure 11.4 shows the actual construction of a module in a frame that can be mounted on a structure. The average size of solar panels (modules) used in a rooftop solar installation is approximately 1.67 m × 1 m (5.4 feet by 3.25 feet or 65 inches × 39 inches).

Photovoltaic Power Systems

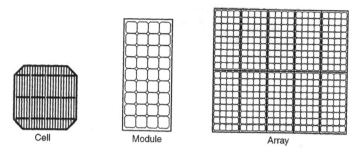

FIGURE 11.3 Several PV cells make a module, and several modules make an array.

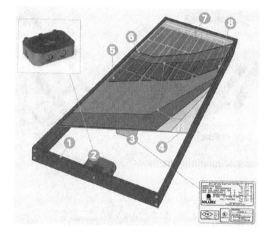

FIGURE 11.4 Construction of PV module: (1) frame, (2) weatherproof junction box, (3) rating plate, (4) weather protection for 30-year life, (5) PV cell, (6) tempered high-transmissivity cover glass, (7) outside electrical bus, (8) frame clearance. (From Solarex/BP Solar, Frederick, MD. With permission.)

Mounting of the modules can be in various configurations as seen in Figure 11.5. In roof mounting, the modules are in a form that can be laid directly on the roof. In the newly developed amorphous technology, the PV sheets are made in shingles that can replace the traditional roof shingles on a one-to-one basis, providing better economy in regard to building material and labor.

11.3 EQUIVALENT ELECTRICAL CIRCUIT

The complex physics of the PV cell can be represented by the equivalent electrical circuit shown in Figure 11.6. The circuit parameters are as follows. The current I delivered to the external load at the output terminals is equal to the light-generated current I_L, less the diode current I_D and the shunt-leakage current I_{sh}. The series resistance R_s represents the internal resistance to the current flow, and depends on the p-n

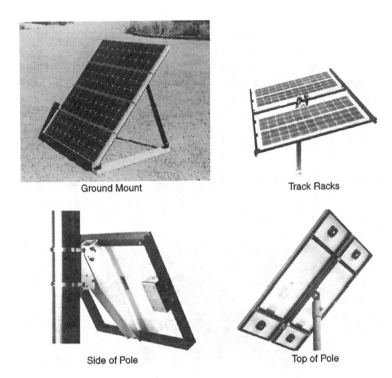

FIGURE 11.5 PV module mounting methods.

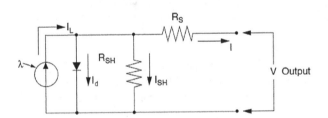

FIGURE 11.6 Equivalent circuit of PV module showing the diode and ground leakage currents.

junction depth, impurities, and contact resistance. The shunt resistance R_{sh} is inversely related to the leakage current to ground. In an ideal PV cell, $R_s = 0$ (no series loss), and $R_{sh} = \infty$ (no leakage to ground). In a typical high-quality 1 square inch silicon cell, R_s varies from 0.05 to 0.10 Ω and R_{sh} from 200 to 300 Ω. The PV conversion efficiency is sensitive to small variations in R_s, but is insensitive to variations in R_{sh}. A small increase in R_s can decrease the PV output significantly.

The open-circuit voltage V_{oc} of the cell is obtained when the load current is zero, i.e., when $I = 0$, and is given by the following:

$$V_{oc} = V + IR_{sh} \tag{11.1}$$

Photovoltaic Power Systems

The actual diode current is given by the classical diode current expression:

$$I_d = I_D \left[e^{\frac{QV_{oc}}{AkT}} - 1 \right] \quad (11.2)$$

where
I_D = the saturation current of the diode
Q = electron charge = 1.6×10^{-19} C
A = curve-fitting constant
k = Boltzmann constant = 1.38×10^{-23} J/°K
T = temperature on absolute scale °K

The load current is therefore given by the expression:

$$I = I_L - I_D \left[e^{\frac{QV_{oc}}{AkT}} - 1 \right] - \frac{V_{oc}}{R_{sh}} \quad (11.3)$$

The last term is the leakage current to the ground. In practical cells, it is negligible compared to I_L and I_D and is generally ignored. The diode-saturation current can therefore be determined experimentally by applying a voltage V_{oc} to the cell in the dark and measuring the current going into the cell. This current is often called the *dark current* or the *reverse diode-saturation current*.

11.4 OPEN-CIRCUIT VOLTAGE AND SHORT-CIRCUIT CURRENT

The two most important parameters widely used for describing cell electrical performance are the open-circuit voltage V_{oc} and the short-circuit current I_{sc} under full illumination. The short-circuit current is measured by shorting the output terminals and measuring the terminal current. Ignoring the small diode and ground leakage currents under zero terminal voltage, the short-circuit current under this condition is the photocurrent I_L.

The maximum photovoltage is produced under the open-circuit voltage. Again, by ignoring the ground leakage current, Equation 11.3 with $I = 0$ gives the open-circuit voltage as follows:

$$V_{oc} = \frac{AkT}{Q} = \log_n \left(\frac{I_L}{I_D} + 1 \right) \quad (11.4)$$

The term kT/Q is expressed in voltage (0.026 V at 300°K). In practical photocells, the photocurrent is several orders of magnitude greater than the reverse saturation current. Therefore, the open-circuit voltage is many times the kT/Q value. Under conditions of constant illumination, I_L/I_D is a sufficiently strong function of the cell temperature, and the solar cell ordinarily shows a negative temperature coefficient of the open-circuit voltage.

11.5 I-V AND P-V CURVES

The electrical characteristic of the PV cell is generally represented by the current vs. voltage (I-V) curve. Figure 11.7 shows the I-V characteristic of a PV module under two conditions, in sunlight and in the dark. In the first quadrant, the top left of the I-V curve at zero voltage is called the short-circuit current. This is the current we would measure with output terminals shorted (zero voltage). The bottom right of the curve at zero current is called the *open-circuit voltage*. This is the voltage we would measure with output terminals open (zero current). In the left-shaded region, the cell works as a constant current source, generating a voltage to match with the load resistance. In the shaded region on the right, the current drops rapidly with a small rise in the voltage. In this region, the cell works like a constant voltage source with an internal resistance. Somewhere in the middle of the two shaded regions, the curve has a knee point.

If a voltage is externally applied in the reverse direction, for instance, during a system fault transient, the cell current remains flat, and the power is absorbed by the cell with a negative voltage and positive current. However, beyond a certain negative voltage, the junction breaks down as in a diode, and the current rises to a high value. In the dark, the current is zero for any voltage up to the breakdown voltage, which is the same as in the illuminated condition.

The power output of the panel is the product of the voltage and current outputs. In Figure 11.8, the power is plotted against the voltage. Note that the cell produces no power at zero voltage or zero current, and produces the maximum power at the voltage corresponding to the knee point of the I-V curve. This is why the PV power circuit is always designed to operate close to the knee point with a slight slant on the left-hand side. For this reason, the PV circuit is modeled approximately as a constant current source in the electrical analysis of the system.

Figure 11.9 is the I-V characteristic of a 22-W panel at two solar illumination intensities, 1000 W/m^2 and 500 W/m^2. These curves are at AM1.5 (air mass 1.5). The AM0 (air mass zero) represents the outer-space conditions (vacuum), in which the

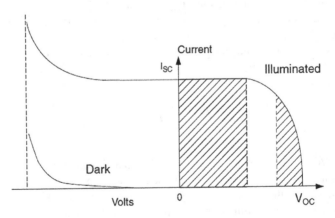

FIGURE 11.7 Current vs. voltage (I-V) characteristic of the PV module in sunlight and in the dark.

Photovoltaic Power Systems

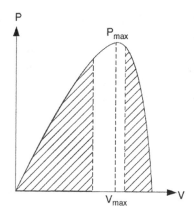

FIGURE 11.8 Power vs. voltage (P-V) characteristic of the PV module in sunlight.

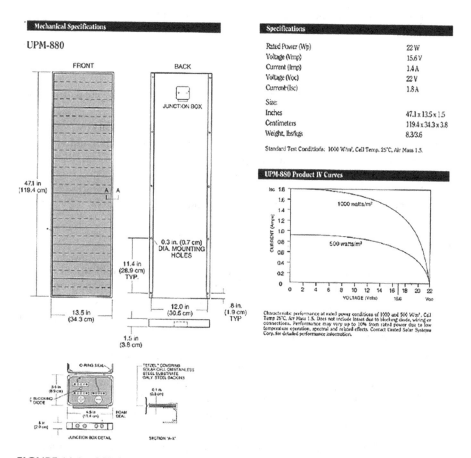

FIGURE 11.9 I-V characteristic of a 22-W PV module at full and half sun intensities. (From United Solar Systems, San Diego, CA. With permission.)

solar radiation is 1350 W/m². AM1 represents conditions normal to the sun in a clear unpolluted atmosphere of the earth on a dry noon. Thus, AM1 represents ideal conditions in pure air when sunlight experiences the least resistance to reach the earth. AM1.5 represents average quality air with average humidity and pollution at an average inclination. AM1.5 is, therefore, taken as a reference value for terrestrial PV designs. In northern altitudes with the sun at 15° from the horizon, the AM index can be as high as 4 when sunlight has a high resistance to cut through before reaching the earth surface.

Solar power impinging on a normal surface on a bright day with AM1.5 is about 1000 W/m², and it would be low on a cloudy day. The 500 W/m² solar intensity is another reference condition the industry uses to report I-V curves.

The photoconversion efficiency of the PV cell is defined by the following ratio:

$$\eta = \frac{\text{Electrical power output}}{\text{Solar power impinging the cell}} \quad (11.5)$$

Obviously, the higher the efficiency, the higher the output power we get under a given illumination.

11.6 ARRAY DESIGN

The major factors influencing the electrical design of the solar array are as follows:

- The sun intensity (also called illumination or solar flux intensity)
- The sun angle deviating from normal sun (zero angle gives maximum power)
- The load matching for maximum power
- The operating temperature of the PV cells

These factors are discussed in the following subsections.

11.6.1 SUN INTENSITY

The magnitude of the photocurrent is maximum under a full bright sun (1.0 sun). On a partially sunny day, the photocurrent diminishes in direct proportion to the sun intensity. At a lower sun intensity, the I-V characteristic shifts downward as shown in Figure 11.10. On a cloudy day, therefore, the short-circuit current decreases significantly. The reduction in the open-circuit voltage, however, is small.

The photoconversion efficiency of the cell is insensitive to the solar radiation in the practical working range. For example, Figure 11.11 shows that the efficiency is practically the same at 500 W/m² and at 1000 W/m². This means that the conversion efficiency is practically the same on a bright sunny day as on a cloudy day. We get a lower power output on a cloudy day not because the photoconversion efficiency is lower, but because of the lower solar energy impinging on the cell.

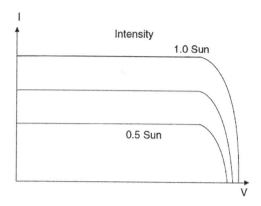

FIGURE 11.10 I-V characteristic of PV module shifts down at lower sun intensity, with small reduction in voltage.

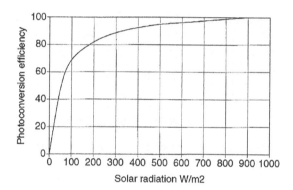

FIGURE 11.11 Photoconversion efficiency relative to its maximum at full sun vs. actual solar radiation (practically insensitive over a wide range of radiation from 300 to 1000 W/m2).

11.6.2 SUN ANGLE

The cell output current is given by $I = I_o \cos\theta$, where I_o is the current with normal sun (reference), and θ is the angle of the sun line measured from the normal. This cosine law holds well for sun angles ranging from 0 to about 50°. Beyond 50°, the electrical output deviates significantly from the cosine law, and the cell generates no power beyond 85°, although the mathematical cosine law predicts 7.5% power generation (Table 11.1). The actual power-angle curve of the PV cell, called the *Kelly cosine*, is shown in Figure 11.12. This is an empirically derived curve.

11.6.3 SHADOW EFFECT

The array may consist of many parallel strings of series-connected cells. Two such strings are shown in Figure 11.13. A large array may get partially shadowed due to

TABLE 11.1
Kelly Cosine Values of the Photocurrent in Silicon Cells

Sun Angle Degrees	Mathematical Cosine Value	Kelly Cosine Value
30	0.866	0.866
50	0.643	0.635
60	0.500	0.450
80c	0.174	0.100
85	0.087	0

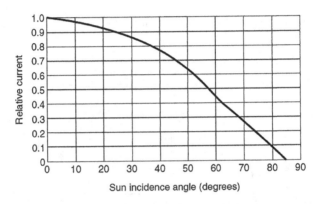

FIGURE 11.12 Kelly cosine curve for PV cell at sun angles from 0 to 90°.

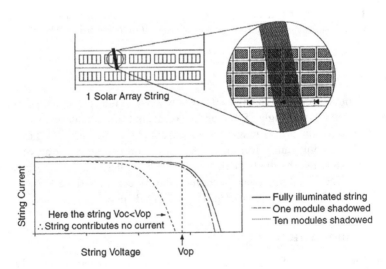

FIGURE 11.13 Shadow effect on one long series string of an array (power degradation is small until shadow exceeds the critical limit).

Photovoltaic Power Systems

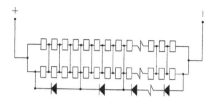

FIGURE 11.14 Bypass diode in PV string minimizes power loss under heavy shadow.

a structure interfering with the sun line. If a cell in a long series string gets completely shadowed, it loses the photovoltage but still must carry the string current by virtue of its being in series with all other cells operating in full sunlight. Without internally generated voltage, the shadowed cell cannot produce power. Instead, it acts as a load, producing local I^2R loss and heat. The remaining cells in the string must work at higher voltage to make up the loss of the shadowed cell voltage. A higher voltage in healthy cells means a lower string current as per the I-V characteristic of the string. This is shown in the bottom left of Figure 11.13. The current loss is not proportional to the shadowed area, and may go unnoticed for a mild shadow on a small area. However, if more cells are shadowed beyond the critical limit, the I-V curve goes below the operating voltage of the string, making the string current fall to zero, losing all the power of the string. This causes loss of one whole string from the array.

The commonly used method to eliminate loss of string power due to a possible shadow is to subdivide the circuit length in several segments with bypass diodes (Figure 11.14). The diode across the shadowed segment bypasses only that segment of the string. This causes a proportionate loss of the string voltage and current, without losing the whole-string power. Some modern PV modules come with such internally embedded bypass diodes.

11.6.4 Temperature Effects

With increasing temperature, the short-circuit current of the cell increases, whereas the open-circuit voltage decreases (Figure 11.15). The effect of temperature on PV power is quantitatively evaluated by examining the effects on the current and the voltage separately. Suppose I_o and V_o are the short-circuit current and the open-circuit voltage at the reference temperature T, and α and β are their respective temperature coefficients. If the operating temperature is increased by ΔT, then the new current and voltage are given by the following:

$$I_{sc} = I_o\left(1 + \alpha \times \Delta T\right) \text{ and } V_{oc} = V_o\left(1 - \beta \times \Delta T\right) \tag{11.6}$$

Because the operating current and the voltage change approximately in the same proportion as the short-circuit current and open-circuit voltage, respectively, the new power is as follows:

$$P = VI = I_o\left(1 + \alpha \times \Delta T\right)V_o\left(1 - \beta \times \Delta T\right) \tag{11.7}$$

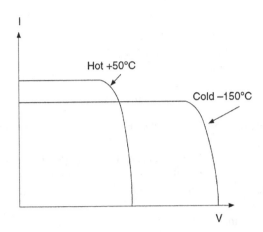

FIGURE 11.15 Effect of temperature on I-V characteristic (cell produces less current but greater voltage, with net gain in the power output at cold temperatures).

This can be simplified in the following expression by ignoring a small term:

$$P = P_o\left[1 + (\alpha - \beta)\Delta T\right] \quad (11.8)$$

For a typical single-crystal silicon cell, α is about 20 μu/°C and β is about 5 mu/°C, where u stands for unit. The power is, therefore, given by the following:

$$P = P_o\left[1 + \left(20 \times 10^{-6} - 5 \times 10^{-3}\right)\Delta T\right] \text{ or } P = P_o\left[1 - 0.005 \times \Delta T\right] \quad (11.9)$$

This expression indicates that for every degree centigrade rise in the operating temperature above the reference temperature, the silicon cell power output decreases by about 0.5%. Because the increase in current is much less than the decrease in voltage, the net effect is a decrease in power at a higher operating temperature.

The effect of temperature on the power output is shown in the power vs. voltage characteristics at two operating temperatures in Figure 11.16. The figure shows that the maximum power available at a lower temperature is higher than that at a higher temperature. Thus, a cold day is actually better for the PV cell, as it generates more power. However, the two P_{max} points are not at the same voltage. To extract maximum power at all temperatures, the PV system must be designed such that the module output voltage can increase to V_2 for capturing P_{max2} at a lower temperature and can decrease to V_1 for capturing P_{max1} at a higher temperature. This is done by a power electronics control box called peak power tracker (PPT) that adds some system design complexity.

11.6.5 Effect of Climate

On a partly cloudy day, the PV module can produce up to 80% of its full sun power. It can produce about 30% power even with heavy clouds on an extremely overcast

Photovoltaic Power Systems

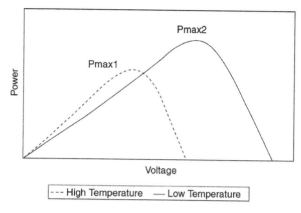

FIGURE 11.16 Effect of temperature on P-V characteristic (cell produces more power at cold temperatures).

day. Snow does not usually collect on the module, because it is angled to catch the sun. If snow does collect, it quickly melts. Mechanically, the module is designed to withstand golf-ball-size hail.

11.6.6 ELECTRICAL LOAD MATCHING

The operating point of any power system is the intersection of the source line and the load line. If the PV source having the I-V and P-V characteristics shown in Figure 11.17(a) is supplying power to a resistive load R_1, it will operate at point A_1. If the load resistance increases to R_2 or R_3, the operating point moves to A_2 or A_3, respectively. The maximum power is extracted from the module when the load resistance is R_2 (Figure 11.17b). Such a load that matches with the source is always necessary for the maximum power extraction from a PV source.

The operation with a constant-power load is shown in Figure 11.17(c) and Figure 11.17(d). The constant-power load line has two points of intersection with the source line, denoted by B_1 and B_2. Only the point B_2 is stable, as any perturbation from it generates a restoring power to take the operation back to B_2, and the system continues to operate at B_2 with an inherent stability.

Therefore, the necessary condition for the electrical operating stability of the solar array is as follows:

$$\left[\frac{dP}{dV}\right]_{load} > \left[\frac{dP}{dV}\right]_{source} \quad (11.10)$$

Some loads such as heaters have constant resistances, which absorb power that varies with the square of the voltage. Other loads such as induction motors behave more like constant-power loads. They draw more current at lower voltage and *vice versa*. In most large systems with mixed loads, the power varies approximately in a linear proportion with voltage.

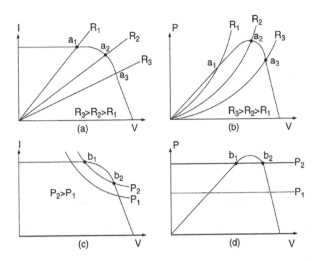

FIGURE 11.17 Operating stability and electrical load matching with constant-resistive load and constant-power load.

11.6.7 SUN TRACKING

More energy is collected by the end of the day if the PV module is installed on a tracker with an actuator that follows the sun like the sunflower. There are two types of sun trackers:

- One-axis tracker, which follows the sun from east to west during the day.
- Two-axis tracker, which follows the sun from east to west during the day, and from north to south during the seasons of the year (Figure 11.18).

A sun-tracking design can increase the energy yield up to 40% over the year compared to the fixed-array design. Dual-axis tracking is done by two linear actuator motors, which follow the sun within one degree of accuracy (Figure 11.19). During the day, it tracks the sun east to west. At night it turns east to position itself for the next morning's sun. Old trackers did this after sunset using a small nickel-cadmium battery. The new design eliminates the battery requirement by doing the turning in the weak light of the dusk and/or dawn. The Kelly cosine presented in Table 11.1 is useful in assessing accurately the power available in sunlight incident at extreme angles in the morning or evening.

When a dark cloud obscures the sun, the tracker may aim at the next brightest object, which is generally the edge of a cloud. When the cloud is gone, the tracker aims at the sun once again, and so on and so forth. Such sun hunting is eliminated in newer sun trackers.

One method of designing the sun tracker is to use two PV cells mounted on two 45° wedges (Figure 11.20), and connecting them differentially in series through an actuator motor. When the sun is normal, the currents on both cells are equal to $I_o \cos 45°$. As they are connected in series opposition, the net current in the motor is

Photovoltaic Power Systems

FIGURE 11.18 Dual-axis sun tracker follows the sun throughout the year. (From American Sun Company, Blue Hill, Maine. With permission.)

zero, and the array stays put. On the other hand, if the array is not normal to the sun, the sun angles on the two cells are different, giving two different currents as follows:

$$I_1 = I_o \cos(45+\delta) \text{ and } I_2 = I_o \cos(45-\delta) \qquad (11.11)$$

The motor current is therefore:

$$I_m = I_1 - I_2 = I_o \cos(45+\delta) - I_o \cos(45-\delta) \qquad (11.12)$$

Using Taylor series expansion:

$$f(x+h) = f(x) + hf'(x) + \frac{h^2}{2!}f''(x) + \cdots$$

we can express the two currents as follows:

FIGURE 11.19 Actual motor of the sun tracker. (From American Sun Company, Blue Hill, Maine. With permission.)

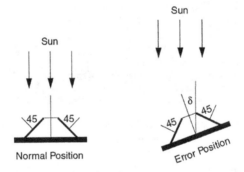

FIGURE 11.20 Sun-tracking actuator principle. (Two differentially connected sensors at 45° generate signals proportional to the pointing error.)

$$I_1 = I_o \cos 45 - I_o \delta \sin 45 \text{ and } I_2 = I_o \cos 45 + I_o \delta \sin 45 \qquad (11.13)$$

The motor current is then

$$I_m = I_1 - I_2 = 2I_o \delta \sin 45° = \sqrt{2} I_o \delta \text{ if } \delta \text{ is in radians} \qquad (11.14)$$

A small pole-mounted panel can use one single-axis or dual-axis sun tracker. A large array, on the other hand, is divided into small modules, each mounted on its own sun tracker. This simplifies the structure and eliminates the problems related to a large movement in a large panel.

Photovoltaic Power Systems 171

11.7 PEAK-POWER OPERATION

The sun tracker drives the module mechanically to face the sun to collect the maximum solar radiation. However, that in itself does not guarantee the maximum power output from the module. As was seen in Figure 11.16, the module must operate electrically at a certain voltage that corresponds to the peak power point under a given operating condition. First, we examine the electrical principle of peak-power operation.

If the array is operating at any point at voltage V and current I on the I-V curve, the power generation is $P = VI$ watts. If the operation moves away from the preceding point such that the current is now $I + \Delta I$, and the voltage is $V + \Delta V$, then the new power is as follows:

$$P + \Delta P = (V + \Delta V)(I + \Delta I) \qquad (11.15)$$

which, after ignoring a small term, simplifies to the following:

$$\Delta P = \Delta V \times I + \Delta I \times V \qquad (11.16)$$

ΔP would be zero if the array were operating at the peak power point, which necessarily lies on a locally flat neighborhood. Therefore, at the peak power point, the preceding expression in the limit becomes:

$$\frac{dV}{dI} = -\frac{V}{I} \qquad (11.17)$$

We note here that dV/dI is the dynamic impedance of the source, and V/I the static impedance. Thus, at the peak power point, the following relation holds:

$$\text{Dynamic impedance } Z_d = -\text{static impedance } Z_s \qquad (11.18)$$

There are three electrical methods of extracting the peak power from a PV source, as described in the following text:

1. In the first method, a small signal current is periodically injected into the array bus, and the dynamic bus impedance ($Z_d = dV/dI$) and the static bus impedance ($Z_s = V/I$) are measured. The operating voltage is then increased or decreased until Z_d equals $-Z_s$. At this point, the maximum power is extracted from the source.
2. In another method, the operating voltage is increased as long as dP/dV is positive. That is, the voltage is increased as long as we get more power. If dP/dV is sensed negative, the operating voltage is decreased. The voltage stays the same if dP/dV is near zero within a preset deadband.
3. The third method makes use of the fact that for most PV cells, the ratio of the voltage at the maximum power point to the open-circuit voltage (i.e., V_{mp}/V_{oc}) is approximately constant, say K. For example, for high-quality crystalline

silicon cells, K = 0.72. An unloaded cell is installed on the array and kept in the same environment as the power-producing cells, and its open-circuit voltage is continuously measured. The operating voltage of the power-producing array is then set at $K \cdot V_{oc}$, which will produce the maximum power.

11.8 SYSTEM COMPONENTS OF STAND-ALONE SYSTEM

The array by itself does not constitute the PV power system. We may also need a structure to mount it, a sun tracker to point the array to the sun, various sensors to monitor system performance, and power electronic components that accept the DC power produced by the array, charge the battery, and condition the remaining power in a form that is usable by the load. If the load is AC, the system needs an inverter to convert the DC power into AC at 50 or 60 Hz.

Figure 11.21 shows the necessary components of a stand-alone PV power system. The peak-power tracking controller senses the voltage and current outputs of the array and continuously adjusts the operating point of the PWM switching regulator to extract the maximum power under varying climatic conditions. The output of the array goes to the inverter, which converts the DC into AC. The array output in excess of the load requirement is used to charge the battery. The battery charger is usually a DC–DC buck converter. If excess power is still available after fully charging the battery, it is shunted in dump heaters, which may be a room or water heater in a stand-alone system. When the sun is not available, the battery discharges to the inverter to power the load. The battery discharge diode Db is to prevent the battery from being charged when the charger is opened after a full charge or for other reasons. The array diode Da is to isolate the array from the battery, thus keeping the array from acting as the load on the battery at night. The diode Da also isolates any string with internal short from drawing current from healthy solar strings. The charge current controller (also called the mode controller) collects system signals, such as the array and the battery currents and voltages, and keeps track of the battery state of charge by bookkeeping the charge/discharge ampere-hours. It uses this information

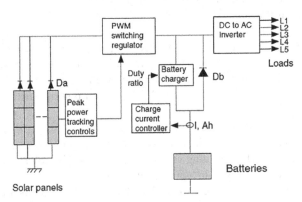

FIGURE 11.21 PV power system with peak-power-tracking control system.

Photovoltaic Power Systems

to turn on or off the battery charger, discharge converter, and dump loads as needed. Thus, the mode controller is the central controller of the entire system.

In the grid-connected system, dump heaters are not required, as all excess power is always fed to the grid lines. The battery is also eliminated, except for a few small critical loads, such as the start-up controller and the computer. DC power is first converted into AC by the inverter, ripples are filtered, and only then is the filtered power fed into the grid lines.

In the PV system, the inverter is a critical component, which converts the array DC power into AC for supplying the loads or interfacing with the grid. A new product line recently introduced into the market is the AC PV module, which integrates an inverter directly into module design. It is presently available in a few hundred watts capacity. It provides utility-grade 60-Hz power directly from the module junction box. This greatly simplifies PV system design.

The constant current delivered by the panel depends on the cell area and the solar flux (radiation in watts/m^2). The panel's generated voltage depends on the band-gap of the p-n junction. The cell output voltage is independent of the panel area, but linearly depends on the load resistance in the operating range. Since the power delivered by the panel depends on its output voltage as shown in Figure 11.8, the maximum (peak) power P_{max} can be extracted from the panel only if the load resistance is matched such that the operating voltage is at the knee point of the cell's I-V characteristic, which is denoted by V_{max}. All practical solar power installations are designed to operate at this peak power point at voltage V_{max} with the help of power electronics voltage converter, the peak power tracker (a d.c.-d.c. power electronic converter), connected between the load and the solar panel. With proper switching duty ratio control of the peak power tracker, the panel output voltage is always maintained at V_{max} (the maximum power point) of the panel, and the load side voltage is always maintained at the specified load voltage. When operating in such a mode, the maximum power output of the solar panel facing the sun at off-normal angle δ from the sun line is:

$$\text{PV panel power output} = \text{Panel Area} \times \text{Solar flux} \times \cos\delta \times \text{Conversion efficiency} \quad (11.19)$$

Equation 11.19 can be used to estimate the panel area required for a desired solar power. From a fixed panel, the maximum kWh energy is captured over the year with daily and seasonal variations in the sun angle, the panel is installed to face perpendicular to the sun at local noon on an equinox day. This is simply done by tilting the panel from the horizontal approximately equal to the latitude of the location. From a panel installed this way, a rough estimate of the yearly average energy capture on a bright day is approximately given by:

$$\text{Energy capture in kWh on Bright day} = \text{Panel area} \times 1\,\text{kW}/\text{m}^2 \times \text{PV efficiency} \times 6\,\text{hours} \quad (11.20)$$

These are rough estimates of the power and energy output of the solar array. These estimates need to be adjusted for the known solar radiation, temperature, and altitude at the installation site.

REFERENCES

1. Cook, G., Billman, L., and Adcock R., Photovoltaic Fundamentals, DOE/Solar Energy Research Institute Report No. DE91015001, February 1995.
2. Omid Beik, Ahmad S. Al-Adsani, *DC Wind Generation Systems, Design, Analysis, and Multiphase Turbine Technology*, Springer, 2020.
3. D. Schumacher, O. Beik and A. Emadi, "Standalone Integrated Power Electronics System: Applications for Off-Grid Rural Locations," *IEEE Electrification Magazine*, vol. 6, no. 4, pp. 73–82, December 2018.

12 Solar Power Conversion Systems

12.1 OVERVIEW

The number of people without electricity is expected to increase in the next decade in some parts of the world [1]. For instance, in India there were 289 million people without electrical power in 2009. As shown in Table 12.1 the number of people without electricity in Africa will increase by 12% to 646 million by the year 2030. This is primarily in the Africa's rural areas where the access to the power grid is limited, and the existing aging grids are out of service. The cost of developing power plants, long distances for extending electric power grids, and scattered population, are some of the reasons behind limited access to electricity in rural areas.

Electrification of rural areas has begun and continues to grow in different parts of the world, mainly in developing countries. Rural electrifications take different forms and are traditionally powered by distributed energy resources such as wind, solar, or hydro. Alternative sources such as biofuels like oil, wood, and diesel are also popular, with the diesel being the most common option due to its availability, cost, and accessibility. Since diesel is a fossil fuel that contributes to climate change, renewable energy sources (wind and solar) are the ideal choice due to their sustainability and abundance. As the renewable energy systems grow, their cost continues to decline and their supporting technologies, such as power electronics converters and energy storage systems, become more efficient. These improvements facilitate worldwide use of distributed and renewable energy systems for both on-grid and off-grid applications, with the off-grid being more suitable for small rural areas. The low cost of infrastructure including road and tower building and short lead time for expanding power grid make the distributed energy sources more attractive for small and distant rural areas. In addition, when the load is far away from the source in extended grid, the reliability of the system is adversely impacted with higher risk of power outages.

Figure 12.1 shows an economic comparison of diesel and solar in the African continent. The blue (dark) shade identifies locations where the diesel is more economical, while the yellow (light) shade shows the area where the PV is more economic. Based on the comparison shown in Figure 12.1, solar power is a suitable source of energy in vast areas of Africa. It is expected that the solar locations continue to grow due to availability of technology as its cost continues to decline rapidly.

12.2 SOLAR POWETR ELECTRONICS SYSTEMS

The previous section discussed the opportunities for off-grid renewable energy systems, and presented solar systems as a suitable alternative to diesel for rural areas in African continent. In this section solar power electronics systems,

TABLE 12.1
Population without Immediate Access to Electrical Energy

Country/Region	Million People (2009)	Estimated Million People (2030)	Percentage Change from 2009 to 2030 (Two Decades)
China	8	0	−100%
India	289	154	−50%
Other parts of Asia	379	221	−42%
Africa	577	646	12%
Latin America	30	10	−67%
Middle East	21	5	−76%

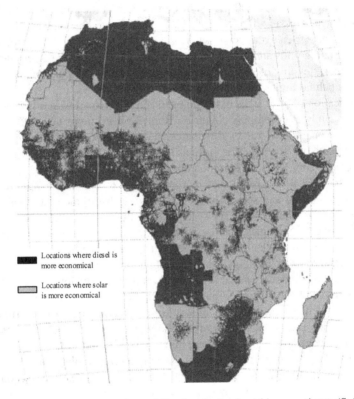

FIGURE 12.1 Economic comparison of diesel and solar for African continent. (S. Szabó, K. Bódis, T. Huld, and M. Moner-Girona, "Energy solutions in rural Africa: Mapping electrification costs of distributed solar and diesel generation versus grid extension," *Environmental Research Letters*, vol. 6, no. 3, pp. 34002, 2011.)

specifically for off-grid applications are addressed. Off-grid solar systems are independent of a power grid and may be formed as an isolated micro grid supplying power to a local community, manufacturing facilities, or an isolated area. The off-grid solar systems may also come as portable systems for smaller applications and individual use.

Solar Power Conversion Systems

12.2.1 Solar Conversion Architecture

A solar conversion system with a main AC bus is shown in Figure 12.2. The scheme includes solar input(s), energy storage systems (ESSs), and AC and DC loads. As shown in Figure 12.2 the DC output of solar array is controlled using a DC/DC converter and then converted to AC and connected to the main AC bus. An energy storage system (ESS) is connected to the AC bus using a bidirectional DC/DC and AC/DC converter. The bidirectional converters are controlled to store the energy in the ESS or feed the loads when necessary. The AC loads are directly fed off the AC bus, while for the DC loads an AC/DC converter is used. Note that a DC/DC converter to control the DC voltage/current may be used for the DC loads.

A solar conversion system with a main DC bus is shown in Figure 12.3. Compared to solar system with AC bus the scheme in Figure 12.3 is more simplified with fewer power electronics converters, hence reduced complexity and facilitated control. The ESS is connected to the DC bus using a bidirectional DC/DC converter as it requires to be charged or discharged depending on the status of the control system. Note that

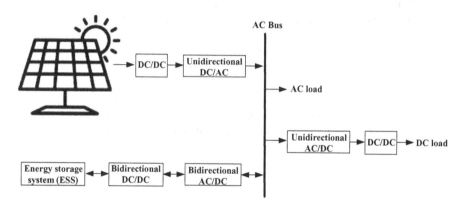

FIGURE 12.2 A solar conversion system with a main AC bus.

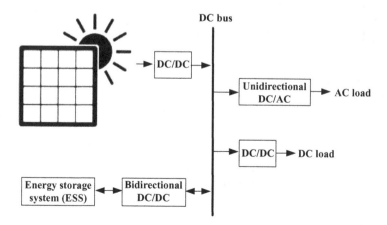

FIGURE 12.3 A solar conversion system with a main DC bus.

since the schemes presented here are off-grid systems, i.e. operating in standalone mode, the ESS is charged solely by the solar source.

The choice between the two off-grid solar schemes, Figure 12.2 with the main AC bus and Figure 12.3 with the main DC bus, depends on the application requirements. Depending on the load the DC/DC converters used in these schemes may incorporate isolated or non-isolated topology. In general, when a low-voltage DC connects to a high-voltage DC, a topology with galvanic isolation is employed. For both off-grid solar schemes, the DC loads are considered to be at low voltages, i.e. 5V–60V while the AC loads are considered to be at 120V–240V.

12.2.2 AN OFF-GRID SOLUTION

As an example, a schematic of a 4.5 kW off-grid solar power conversion system is illustrated in Figure 12.4. In this scheme three loads, a single-phase AC, a 3-phase AC, and a DC load are supplied from a solar system. A DC ESS is used to store the energy when applicable.

A DC boost converter increases the DC input from the solar array. The solar input nominal voltage is 40V; however, due to inherent characteristics of solar array its voltage varies from 24V-48V depending on the operating conditions. The boost converter steps-up this voltage to a fixed 60V low-voltage DC-link. A circuit model for the boost converter is shown in Figure 12.5, where the DC/DC boost converter has a 3-phase topology, i.e. it consists of three inductors, three diodes, and three MOSFET switches. The higher phase numbers result in reduced current, i.e. a 1/3rd of the current for three phases compared to conventional boost converter and hence reduced losses, reduced voltage ripple, and a significant reduction in current ripple. These factors contribute to the performance improvement of the system.

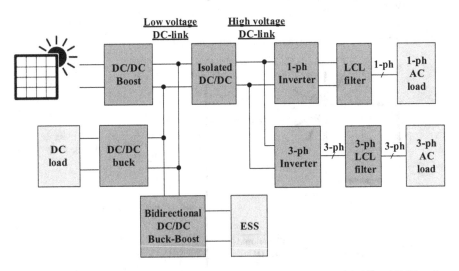

FIGURE 12.4 Schematic of an Off-grid Power Conversion System, with AC and DC Loads, and a DC ESS.

Solar Power Conversion Systems 179

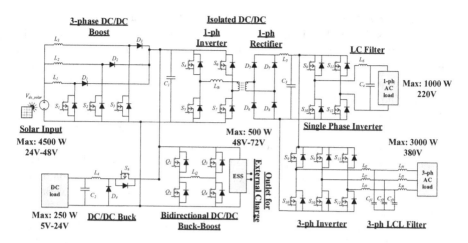

FIGURE 12.5 Circuit model for the solar power conversion scheme under study.

The DC load is fed from the low-voltage DC-link via a conventional DC/DC buck converter. The buck converter consists of a MOSFET, a parallel diode, and an inductor that stores the energy. The DC load could range from 5V ports used as a cellphone charger to other portable electronics that use low voltages to operate or get charged. Most of the DC loads as electronically powered devices have a voltage range of 5–24V, and a power of up to 250W. The buck converter for the DC load is selected as a non-isolated topology, due to simplicity and cost effectiveness.

A 500W ESS is connected to the low-voltage DC-link through a bidirectional buck-boost converter. This allows the ESS to have either a voltage lower (48V) or higher (72V) than the low-voltage DC-link. The bidirectional capability is required such that the ESS can be charged and discharged depending on the system requirements and status of the control system. As shown in Figure 12.5 the buck-boost converter consists of a full-bridge converter with four MOSFETS, and an inductor that connects middle of one leg to the other leg. Note that all of the MOSFETs used in any of the converters shown in Figure 12.5 have an anti-parallel diode where it acts as a freewheeling diode creating a path for the reverse current where applicable.

To power the AC loads, which are at higher power than the DC loads, an isolated DC-DC converter is used to step up the low-voltage DC-link to a high-voltage DC-link with a nominal voltage of 400V, as shown in Figure 12.5. Therefore, the step-up ratio is around 7 in this case, which is implemented using a medium frequency transformer. The isolated DC/DC converter consist of a single-phase inverter, that converts the low-voltage DC-link to a single-phase AC voltage where it is stepped-up by a transformer and rectified using a single-phase inverter. The isolated topology is used to provide a galvanic isolation as a maximum power of 4500 W is processed at high voltages. The isolated DC/DC converter may take other topologies such as fly-back, push-pull, half-bridge, full-bridge, or forward converter. The half-bridge and full-bridge converters employ two quadrants of the transformer B-H curve, while the fly-back converter employs a single quadrant. Therefore, the fly-back converter rated for the same power will have double the size of transformer compared to other topologies, while its switching devices have higher current and voltage ratings. The push-pull converter, however, needs extra care when it comes to the flux as there are higher risks of an imbalanced flux degrading the performance. The full-bridge converter offers lower losses as the currents are half of the half-bridge converter. However, it incurs higher costs due to higher number of active switches. The isolated converter used in the solar conversion system shown in Figure 12.5 offers a phase shift capability, and facilitated zero voltage and current switching. Therefore, these led to lower overall losses with greater controllability compared to the other possible isolated topologies. Compared to the grid frequency transformers (50 or 60 Hz), the medium frequency transformer in the isolated DC/DC converter topology results in reduced mass and size. This is an important consideration for the off-grid portable solar systems as it makes them more applicable in a wide variety of applications.

A 1000 W single-phase load is fed from the high-voltage DC-link using a single-phase inverter. The AC load has a voltage of 120V and a frequency of 50 Hz. An LC filter is used at the output of the inverter to filter high-frequency PWM harmonics and deliver a sinusoidal waveform to the load. The high-voltage DC-link also feeds a 3-phase load with a rated power of 3000 W, and a line-to-line voltage of 380V at a 50 Hz frequency. The three-phase inverter consists of three legs each with two MOSFETS. A 3-phase LCL filter is used to eliminate PWM harmonics.

Given the maximum power of the loads, (DC load 250 W, single-phase AC load 1000 W, 3-phase AC load 3000 W), the solar array is rated at 4500 W, while the ESS has a capacity of 500 W. The ESS is dedicated to supply the power to the DC loads, such as cellphones, laptops, and other electronics when the solar array is not providing power. The ESS may be used to provide partial power for the AC loads, or for starting-up the AC loads if the solar array is not capable of providing the full power.

All of the converters in the solar conversion schemes presented in Figure 12.5 use a current control scheme with PWM modulation. There is no feedback from the load, hence an open-loop system, however, the control on the current and voltage is performed by the relevant converter. The system also includes over-current and over-voltage protection, emergency shut down, and an outlet that allows the ESS to be externally charged. Therefore, the user can charge the ESS when the solar array is not operational or the sun is not present.

12.2.3 System Characteristics

To characterize the off-grid solar power conversion system discussed previously, a simulation platform in MATLAB is set-up, with a minimum solar input voltage, i.e. 24V. Figure 12.6 shows the low- and high-voltage DC-link voltages. The average voltage for the low-voltage DC-link is 60V, however, the voltage waveform has a ripple of 17%. The high-voltage DC-link ripple is 2% around an average voltage of 400V. The DC-link capacitors provide smoothing functions for the voltage waveforms. For the high-voltage DC-link the voltage is passed through capacitor C_1 and C_3 in Figure 12.5, hence lower voltage ripple and smoother waveform with less high-frequency harmonics.

Solar Power Conversion Systems

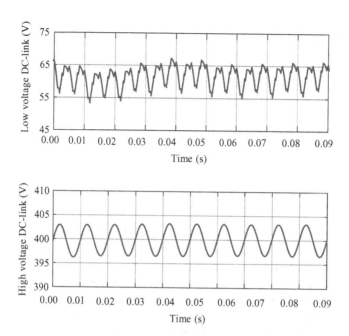

FIGURE 12.6 Low-voltage and high-voltage DC-link waveforms.

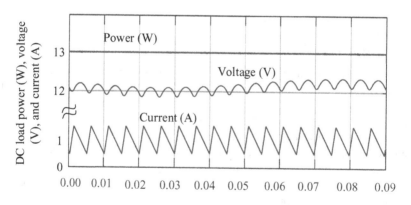

FIGURE 12.7 DC load power, voltage, and current.

Figure 12.7 shows the results at the output of the buck converter supplying the load (Figure 12.5). The average DC voltage is 12V with a ripple of 1.5%, while the supplied power is 13W. The DC load is an electronic device with a battery that is being charged by the buck converter. The load current is also shown in Figure 12.7, and as seen there it has a triangular waveform typical of buck converters. The current has a ripple of 50%, due to a small inductor used for the buck converter.

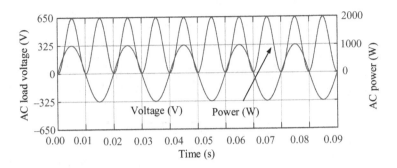

FIGURE 12.8 AC load voltage and power.

In the simulations, a single-phase AC load of 1000 W is also fed from the solar conversion system. Figure 12.8 shows AC power and voltage. The phase voltage has an amplitude of 325 V, and an RMS value of 230 V, at a frequency of 50 Hz. The LC filter at the output of the inverter in Figure 12.5 filters the higher order harmonics, hence the results is a sinusoidal waveform. The frequency of the AC power is twice the voltage frequency, i.e. 100 Hz, as expected. The average power is 1000 W, however, the power fluctuates from a minimum of 0 to a peak 2000 W. The single-phase power, which is product of the phase voltage and line current, results in large fluctuations, as expected. Note that the fluctuations for the 3-phase power are significantly less, due to the phase shift between the phases that results in canceling out the positive and negative peaks.

12.3 CHALLENGES

The previous sections discuss application of solar power conversion system and their off-grid applications. Although the era of electrification has gained momentum and it is spreading across the globe, there are challenges facing off-grid electrification, where these challenges vary from region to region with a variety of reasons.

One of the main challenges for the off-grid systems is the structural difference from the grid-supplied power. The grids are built based on large generation of electricity, transmission of power, and then distributing it to the paying customers with varying demand. The power generation plants and grids are usually designed and operated around a concept of central energy supply. For the off-grid systems, the regulations, policies, and more importantly the incentives are designed like a grid connection, with the idea of central power generation. Therefore, the discussion is shifted to grid connection, stability of the grid, grid-scale energy storage, and the regulations and programs around it. An off-grid electrification then becomes a solution to assist the grid as a temporary solution. The testament to this is the fast pacing development of renewable energy systems as wind farms, solar parks, hydro plants, as centrally controlled power supply. Individual and off-grid systems largely remain secondary solutions.

The prices of off-shore solar systems are another hindering subject to their widespread development. The initial cost, if not subsidized or incentivized, is one of the reasons that individuals prefer to stay on gasoline-fueled electric systems, or to stay connected to the grid. Additionally, currently the off-grid systems come with limited storage capacity. Therefore, being independent from the power grid has risks of being without power when the off-grid system does not supply sufficient power.

In some cases, lack of proper planning leads to the off-grid solution being unattractive, or poor implementation results in failures. There are areas of the world with abundance of sun, water and wind, where the local authorities are willing and ready to implement the off-grid solutions. However, lack of properly selected technology and capability to handle the management of the off-grid technology prevents such solutions. In some cases, the technology gap is the main reason that prevents developing off-grid solutions. For example, in some areas of the world where locally a technology is at its developing stage, not fully understood to place in use, while the mature technology is either too expensive or inaccessible.

12.4 TREND AND FUTURE

As per the World Bank estimate, the off-grid solar power industry has expanded into a $1.75 billion annual market adding 420 million users over the past decade and continues to grow [2]. To achieve universal access to electricity by 2030, the off-grid solar sector would need to serve as many as 132 million households, which in turn would require between $6.6 billion to $11 billion in additional financing.

Canada Energy Regulator (CER) [3] reported that there are over 280 communities in Canada, most of which remote (over 200,000 people), are not connected to the power grid. These off-grid communities rely on diesel-fueled generation, or regional grids based on hydropower or transported liquefied natural gas (LNG) [4]–[8]. Approximately, a third of the off-grid communities are located in the northern territories of Canada. Figure 12.9 shows the method of electricity generation for each province in Canada.

Diesel is stored easily, relatively affordable, easy to install, and can be rather scaled-up easily. However, diesel has high operating costs, and it generates emissions, which contributes to the climate change. The diesel poses challenges for the Nunavut (sparsely populated territory of northern Canada forming most of the Canadian Arctic islands), who face some of the highest energy costs in Canada.

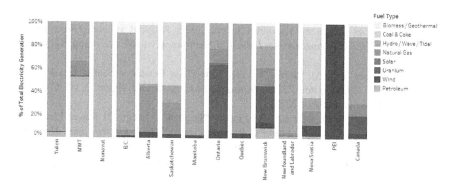

FIGURE 12.9 Canada's electricity generation by province.

Canada's territories have each developed an energy strategy that involves a move towards renewable energy and reducing GHG emissions. Yukon has a high percentage of electricity from renewable generation. Yukon's Energy Strategy prioritizes replacing diesel with renewable energy and a move towards electrification in remote communities. In the Nunavut region of Canada, solar photovoltaic (PV) projects are currently underway due to its strong solar potentials. A hybrid solar/diesel/battery system began operation in 2015, and within its first year reduced over 37000 liters of diesel fuel consumption. The southern part of Nunavut especially has a stronger potential for solar power, as it receives up to 20 hours of sunlight in the summer and 5 hours of sunlight in the winter. With the growing off-grid technologies, especially the solar PV systems, the diesel use will be replaced faster than once thought possible. A few years back, the Government of Nunavut unveiled a net metering program that provides residents with a credit on their energy bill for excess power that is generated using renewable sources and delivered to the community grid.

REFERENCES

1. IRENA, *"Africa 2030: Roadmap for a Renewable Energy Future,"* International Renewable Energy Agency, 2017.
2. The World Bank, "Market Snapshot: Overcoming the Challenges of Powering Canada's Off-grid Communities," https://www.cer-rec.gc.ca/nrg/ntgrtd/mrkt/snpsht/2018/10-01-lcndffgrdcmmnts-eng.html, 2018.
3. Canada Energy Regulator, "Market Snapshot: Overcoming the Challenges of Powering Canada's Off-grid Communities," https://www.cer-rec.gc.ca/nrg/ntgrtd/mrkt/snpsht/2018/10-01-lcndffgrdcmmnts-eng.html, 2018.
4. Omid Beik, An HVDC Off-shore Wind Generation Scheme with High Voltage Hybrid Generator, Ph.D. thesis, McMaster University, 2016.
5. Omid Beik, Ahmad S. Al-Adsani, *DC Wind Generation Systems, Design, Analysis, and Multiphase Turbine Technology*, Springer, 2020.
6. Omid Beik, Ahmad S. Al-Adsani, *Wind Energy Systems*, pp. 1–9, Springer, 2020.
7. Omid Beik, Ahmad S. Al-Adsani, *DC Wind Generation System*, pp. 33–69, Springer, 2020.
8. Beik, O., Dekka A., Narimani, M., *A New Modular Neutral Point Clamped Converter with Space Vector Modulation Control,' Proc. IEEE Int. Conf. Ind. Tech*, Lyon, France, 2018, pp. 591–595.
9. A. Dekka, O. Beik and M. Narimani, "Modulation and Voltage Balancing of a Five-Level Series-Connected Multilevel Inverter With Reduced Isolated Direct Current Sources," *IEEE Transactions on Industrial Electronics*, vol. 67, no. 10, pp. 8219–8230, October 2020.
10. O. Beik and A. S. Al-Adsani, "Parallel Nine-Phase Generator Control in a Medium-Voltage DC Wind System," *IEEE Transactions on Industrial Electronics*, vol. 67, no. 10, pp. 8112–8122, October 2020.
11. D. Schumacher, O. Beik and A. Emadi, "Standalone Integrated Power Electronics System: Applications for Off-Grid Rural Locations," *IEEE Electrification Magazine*, vol. 6, no. 4, pp. 73–82, December 2018.
12. O. Beik and N. Schofield, "High-Voltage Hybrid Generator and Conversion System for Wind Turbine Applications," *IEEE Transactions on Industrial Electronics*, vol. 65, no. 4, pp. 3220–3229, April 2018.

Solar Power Conversion Systems

13. Beik, O., Schofield, N., *'High Voltage Generator for Wind Turbines'*, *8th IET International Conference on Power Electronics, Machines and Drives (PEMD 2016)*, 2016.
14. O. Beik and N. Schofield, "An Offshore Wind Generation Scheme With a High-Voltage Hybrid Generator, HVDC Interconnections, and Transmission," *IEEE Transactions on Power Delivery*, vol. 31, no. 2, pp. 867–877, April 2016.
15. O. Beik and N. Schofield, *"Hybrid Generator for Wind Generation Systems,"* *2014 IEEE Energy Conversion Congress and Exposition (ECCE)*, Pittsburgh, PA, 2014, pp. 3886–3893.
16. M. R. Aghamohammadi, and A. Beik-Khormizi, *"Small Signal Stability Constrained Rescheduling using Sensitivities Analysis by Neural Network as a Preventive Tool,"* *Proc. 2010 IEEE PES Transmission and Distribution Conference and Exposition*, pp. 1–5.
17. A. B. Khormizi and A. S. Nia, *"Damping of Power System Oscillations in Multi-machine Power Systems using Coordinate Design of PSS and TCSC,"* *2011 10th International Conference on Environment and Electrical Engineering (EEEIC)*, Rome. pp. 1–4

Part C

System Integration

13 Energy Storage

Electricity is more versatile in use than other types of power, because it is a highly ordered form of energy that can be converted efficiently into other forms. For example, it can be converted into mechanical form with efficiency approaching 100% or into heat with 100% efficiency. Heat energy, on the other hand, cannot be converted into electricity with such high efficiency, because it is a disordered form of energy in atoms. For this reason, the overall thermal-to-electrical conversion efficiency of a typical fossil thermal power plant is less than 50%.

A disadvantage of electricity is that it cannot be easily stored on a large scale. Almost all electric energy used today is consumed as it is generated. This poses no hardship in conventional power plants, in which fuel consumption is continuously varied with the load requirement. Wind and photovoltaics (PVs), both being intermittent sources of power, cannot meet the load demand at all times, 24 h a day, 365 d a year. Energy storage, therefore, is a desired feature to incorporate with such power systems, particularly in stand-alone plants. It can significantly improve the load availability, a key requirement for any power system.

The present and future energy storage technologies that may be considered for stand-alone wind or PV power systems fall into the following broad categories:

- Electrochemical battery
- Flywheel
- Compressed air
- Superconducting coil

13.1 BATTERY

The battery stores energy in an electrochemical form and is the most widely used device for energy storage in a variety of applications. The electrochemical energy is in a semi-ordered form, which is in between the electrical and thermal forms. It has a one-way conversion efficiency of 85 to 90%.

There are two basic types of electrochemical batteries:

The *primary battery*, which converts chemical energy into electric energy. The electrochemical reaction in a primary battery is nonreversible, and the battery is discarded after a full discharge. For this reason, it finds applications where a high energy density for one-time use is required.

The *secondary battery*, which is also known as the *rechargeable battery*. The electrochemical reaction in the secondary battery is reversible. After a discharge, it can be recharged by injecting a direct current from an external source. This type of battery converts chemical energy into electric energy in the discharge mode. In the charge mode, it converts the electric energy into chemical energy. In both modes, a small fraction of energy is converted into heat, which is dissipated to the surrounding medium. The round-trip conversion efficiency is between 70 and 80%.

The internal construction of a typical electrochemical cell is shown in Figure 13.1. It has positive and negative electrode plates with insulating separators and a chemical electrolyte in between. The two groups of electrode plates are connected to two external terminals mounted on the casing. The cell stores electrochemical energy at a low electrical potential, typically a few volts. The cell capacity, denoted by C, is measured in ampere-hours (Ah), meaning it can deliver C A for one hour or C/n A for n hours.

The battery is made of numerous electrochemical cells connected in a series–parallel combination to obtain the desired battery voltage and current. The higher the battery voltage, the higher the number of cells required in series. The battery rating is stated in terms of the average voltage during discharge and the ampere-hour capacity it can deliver before the voltage drops below the specified limit. The product of the voltage and ampere-hour forms the watthour (Wh) energy rating the battery can deliver to a load from the fully charged condition. The battery charge and discharge rates are stated in units of its capacity in Ah. For example, charging a 100-Ah battery at $C/10$ rate means charging at $100/10 = 10$ A. Discharging that battery at $C/2$ rate means drawing $100/2 = 50$ A, at which rate the battery will be fully discharged in 2 h. The state of charge (SOC) of the battery at any time is defined as the following:

$$SOC = \frac{\text{Ah capacity remaning in the battery}}{\text{Rated Ah capacity}}$$

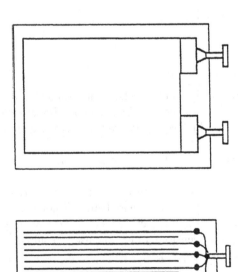

FIGURE 13.1 Electrochemical energy storage cell construction.

Energy Storage

13.2 TYPES OF BATTERY

There are at least six major rechargeable electrochemistries available today. They are as follows:

- Lead-acid (Pb-acid)
- Nickel-cadmium (NiCd)
- Nickel-metal hydride (NiMH)
- Lithium-ion (Li-ion)
- Lithium-polymer (Li-poly)
- Zinc-air

New electrochemistries are being developed by the United States Advanced Battery Consortium for a variety of applications, such as electric vehicles, spacecraft, utility load leveling and, of course, for renewable power systems.[1]

The average voltage during discharge depends on the electrochemistry, as shown in Table 13.1. The energy densities of various batteries, as measured by the Wh capacity per unit mass and unit volume, are compared in Figure 13.2. The selection

TABLE 13.1
Average Cell Voltage during Discharge in Various Rechargeable Batteries

Electrochemistry	Cell Volts	Remark
Lead-acid	2.0	Least-cost technology
Nickel-cadmium	1.2	Exhibits memory effect
Nickel-metal hydride	1.2	Temperature sensitive
Lithium-ion	3.6	Safe, contains no metallic lithium
Lithium-polymer	3.0	Contains metallic lithium
Zinc-air	1.2	Requires good air management to limit self-discharge rate

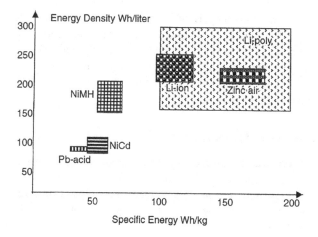

FIGURE 13.2 Specific energy and energy density of various electrochemistries.

of the electrochemistry for a given application is a matter of performance and cost optimization.

Some construction and operating features of these electrochemistries are presented in the following sections.

13.2.1 LEAD-ACID

This is the most common type of rechargeable battery used in the past and even today in some applications because of its maturity and high performance-over-cost ratio, even though it has the least energy density by weight and volume. In a Pb-acid battery under discharge, water and lead sulfate are formed, the water dilutes the sulfuric acid electrolyte, and the specific gravity of the electrolyte decreases with the decreasing *SOC*. Recharging reverses the reaction, in which the lead and lead dioxide are formed at the negative and positive plates, respectively, restoring the battery into its originally charged state.

The Pb-acid battery comes in various versions. The shallow-cycle version is used in automobiles, in which a short burst of energy is drawn from the battery to start the engine. The deep-cycle version, on the other hand, is suitable for repeated full charge and discharge cycles. Most energy storage applications require deep-cycle batteries. The Pb-acid battery is also available in a sealed "gel-cell" version with additives, which turns the electrolyte into nonspillable gel. The gel-cell battery, therefore, can be mounted sideways or upside down. The high cost, however, limits its use in military avionics.

13.2.2 NICKEL-CADMIUM

The NiCd is a matured electrochemistry, in which the positive electrode is made of cadmium and the negative electrode of nickel hydroxide. The two electrodes are separated by Nylon™ separators and placed in potassium hydroxide electrolyte in a stainless steel casing. With a sealed cell and half the weight of the conventional Pb-acid, the NiCd battery has been used to power most rechargeable consumer applications. It has a longer deep-cycle life and is more temperature tolerant than the Pb-acid battery. However, this electrochemistry has a memory effect (explained later), which degrades the capacity if not used for a long time. Moreover, cadmium has recently come under environmental regulatory scrutiny. For these reasons, NiCd has been replaced by Li-ion batteries in laptop computers and other similar high-priced consumer electronics.

13.2.3 NICKEL-METAL HYDRIDE

The NiMH battery was the result of joint development programs by automobile industry targeted for large-scale application in electric vehicles. It is an extension of the NiCd technology that offers an improvement in energy density over that in NiCd. Another performance improvement is that it has no memory effect. However, compared to NiCd, the NiMH is less capable of delivering high peak power, has a high self-discharge rate, and is susceptible to damage due to overcharging, and is

Energy Storage

expensive. The major construction difference is that the anode is made of a metal hydride. This eliminates the environmental concerns of cadmium. For its high energy density and low environmental concern, NiMH has been used widely in electrical vehicles in the recent past until it was replaced by Li-ion battery.

13.2.4 LITHIUM-ION

The Li-ion technology is a new development, which offers three times the energy density over that of Pb-acid. Such a large improvement in energy density comes from lithium's low atomic weight of 6.9 vs. 207 for lead. Moreover, Li-ion has a higher cell voltage, 3.5 V vs. 2.0 V for Pb-acid and 1.2 V for other electrochemistries. This requires fewer cells in series for a given battery voltage, thus reducing the manufacturing cost.

On the negative side, the lithium electrode reacts with any liquid electrolyte, creating a sort of passivation film. Every time the cell is discharged and then charged, the lithium is stripped away, a free metal surface is exposed to the electrolyte, and a new film is formed. This is compensated for by using thick electrodes or else the battery life would be shortened. For this reason, Li-ion is more expensive at present, but its price has been rapidly declining.

In operation, the Li-ion electrochemistry is vulnerable to damage from overcharging or other shortcomings in battery management. Therefore, it requires more elaborate charging circuitry with adequate protection against overcharging.

13.2.5 LITHIUM-POLYMER

This is a lithium battery with solid polymer electrolytes. It is constructed with a film of metallic lithium bonded to a thin layer of solid polymer electrolyte. The solid polymer enhances the cell's specific energy by acting as both the electrolyte and the separator. Moreover, the metal in solid electrolyte reacts less than it does with a liquid electrolyte. With no liquid to spill, it finds applications in air force planes that needs to move rapidly in any reorientation.

13.2.6 ZINC-AIR

The zinc-air battery has a zinc negative electrode, a potassium hydroxide electrolyte, and a carbon positive electrode, which is exposed to the air. During discharge, oxygen from the air is reduced at the carbon electrode (the so-called air cathode), and the zinc electrode is oxidized. During discharge, it absorbs oxygen from the air and converts it into oxygen ions for transport to the zinc anode. During charge, it evolves oxygen. Good air management is essential for the performance of the zinc-air battery.

13.3 EQUIVALENT ELECTRICAL CIRCUIT

For steady-state electrical performance calculations, the battery is represented by an equivalent electrical circuit shown in Figure 13.3. In its simplest form, the battery

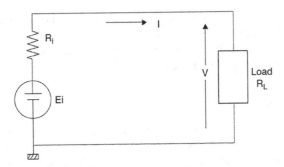

FIGURE 13.3 Equivalent electrical circuit of the battery showing internal voltage and resistance.

works as a constant voltage source with a small internal resistance. The open-circuit (or electrochemical) voltage E_i of the battery decreases linearly with the Ah discharged (Q_d), and the internal resistance R_i increases linearly with Q_d. That is, the battery open-circuit voltage is lower, and the internal resistance is higher in a partially discharged state as compared to the E_0 and R_0 values in a fully charged state. These parameters are expressed quantitatively as follows:

$$E_i = E_0 - K_1 Q_d$$
$$R_i = R_0 - K_2 Q_d \tag{13.1}$$

where K_1 and K_2 are constants found by curve-fitting the test data.

The terminal voltage drops with increasing load as shown by the V_b line in Figure 13.4, in which the operating point is the intersection of the source line and the load line (point P). The power delivered to the external load resistance is $I^2 R_L$.

In a fast-discharge application, such as for starting a heavily loaded motor, the battery may be required to deliver the maximum possible power for a short time. The peak power it can deliver is derived using the maximum power transfer theorem in electrical circuits. It states that the maximum power can be transferred from the source to the load when the internal impedance of the source equals the conjugate of the load impedance. The battery can deliver the maximum power to a DC load when $R_L = R_i$. This gives the following:

$$P_{max} = \frac{E_i^2}{4 R_i} \tag{13.2}$$

Because E_i and R_i vary with the SOC, the P_{max} also varies accordingly. The internal loss is $I^2 R_i$. The efficiency at any SOC is therefore:

$$\eta = \frac{R_L}{R_L + R_i} \tag{13.3}$$

Energy Storage

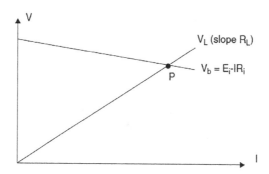

FIGURE 13.4 Battery source line intersecting with load line at the operating point.

The efficiency decreases as the battery is discharged, thus generating more heat at a low *SOC*.

13.4 PERFORMANCE CHARACTERISTICS

The basic performance characteristics, which influence the battery design, are as follows:

- Charge/discharge (C/D) voltages
- C/D ratio
- Round-trip energy efficiency
- Charge efficiency
- Internal impedance
- Temperature rise
- Life in number of C/D cycles

13.4.1 C/D VOLTAGES

The cell voltage variation during a typical C/D cycle is shown in Figure 13.5 for a cell with nominal voltage of 1.2 V, such as NiCd and NiMH, as an example. The voltage is maximum when the cell is fully charged (SOC = 1.0 or Ah discharged = 0). As the cell is discharged, the cell voltage (V_c) drops quickly to a plateau value of 1.2 V, which holds for a long time before dropping to 1.0 at the end of capacity (SOC = 0). In the reverse, when the cell is recharged, the voltage quickly rises to a plateau value of 1.45 V and then reaches a maximum value of 1.55 V. The C/D characteristic also depends on how fast the battery is charged and discharged (Figure 13.6). Other electrochemistries undergo a similar pattern with different numbers.

13.4.2 C/D RATIO

After discharging a certain Ah to load, the battery requires more Ah of charge to restore the full *SOC*. The C/D ratio is defined as the Ah input over the Ah output with

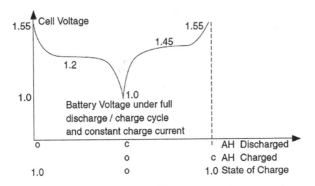

FIGURE 13.5 Voltage variation during C/D cycle of NiCd cell with nominal voltage of 1.2 V.

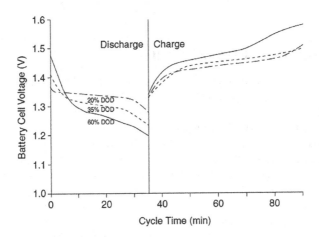

FIGURE 13.6 Cell voltage curves at different C/D rates.

no net change in the *SOC*. This ratio depends on the charge and discharge rates and also on temperature, as shown in Figure 13.7. At 20°C, for example, the C/D ratio is 1.1, meaning the battery needs 10% more Ah charge than that which was discharged for restoring to its fully charged state.

13.4.3 Energy Efficiency

The energy efficiency over a round trip of a full charge and discharge cycle is defined as the ratio of the energy output over the energy input at the electrical terminals of the battery. For a typical battery of capacity C with an average discharge voltage of 1.2 V, average charge voltage of 1.45 V, and C/D ratio of 1.1, the efficiency is calculated as follows:

The energy output over the full discharge = $1.2 \times C$
The energy input required to restore full charge = $1.45 \times 1.1 \times C$

Energy Storage

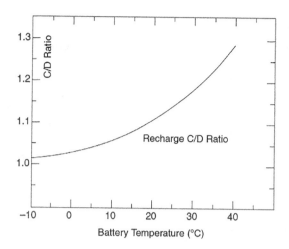

FIGURE 13.7 Temperature effect on C/D ratio.

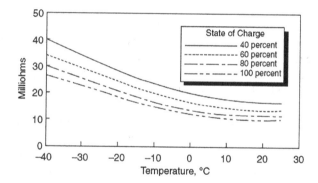

FIGURE 13.8 Temperature effect on internal resistance in 25-Ah NiCd cell.

Therefore, the round-trip energy efficiency is as follows:

$$\eta_{energy} = \frac{1.2 \times C}{1.45 \times 1.1 \times C} = 0.75 \text{ or } 75\%$$

13.4.4 Internal Resistance

The efficiency calculations in the preceding text indicate that 25% of the energy is lost per C/D cycle, which is converted into heat. This characteristic of the battery can be seen as having an internal resistance R_i. The value of R_i is a function of the battery capacity, operating temperature, and the *SOC*. The higher the cell capacity, the larger the electrodes and the lower the internal resistance. R_i varies with *SOC* as per Equation 13.1. It also varies with temperature as shown in Figure 13.8, which is for a high-quality 25-Ah NiCd cell.

13.4.5 CHARGE EFFICIENCY

Charge efficiency is defined as the ratio of the Ah being deposited internally between the plates over that delivered to the external terminals during the charging process. It is different from energy efficiency. The charge efficiency is almost 100% when the cell is empty of charge, the condition in which it converts all Ah received into useful electrochemical energy. As the *SOC* approaches one, the charge efficiency tapers down to zero. The knee point at which the charge efficiency starts tapering off depends on the charge rate (Figure 13.9). For example, at *C*/2 charge rate, the charge efficiency is 100% up to about 75% *SOC*. At a slow charge rate of *C*/40, on the other hand, the charge efficiency at 60% *SOC* is only 50%.

13.4.6 SELF-DISCHARGE AND TRICKLE-CHARGE

The battery slowly self-discharges even with no load on its terminals (open circuit). To maintain full *SOC*, it is continuously trickle-charged to counter the self-discharge rate. This rate is usually less than 1% per day for most electrochemistries in normal working conditions.

After the battery is fully charged, the charge efficiency drops to zero. Any additional charge will be converted into heat. If overcharged at a higher rate than the self-discharge rate for an extended period of time, the battery would overheat, posing a safety hazard of potential explosion. Excessive overcharging produces excessive gassing, which scrubs the electrode plates. Continuous scrubbing at high rate produces excessive heat and wears out electrodes, leading to shortened life. For this reason, the battery charger should have a regulator to cut back the charge rate to the trickle rate after the battery is fully charged. Trickle charging produces a controlled amount of internal gassing. It causes mixing action of the battery electrolyte, keeping it ready to deliver the full charge.

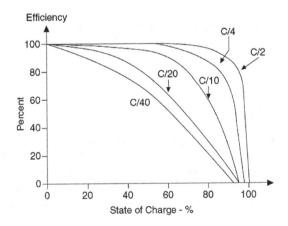

FIGURE 13.9 Charge efficiency vs. SOC at various charge rates.

Energy Storage

13.4.7 Memory Effect

One major disadvantage of the NiCd battery is the memory effect. It is the tendency of the battery to remember the depth at which it has delivered most of its capacity in the past. For example, if the NiCd battery is repeatedly charged and discharged 25% of its capacity to point M in Figure 13.10, it will remember point M. Subsequently, if the battery is discharged beyond point M, the cell voltage will drop much below its original normal value shown by the dotted line in Figure 13.10. The end result is the loss of full capacity after repeatedly using many shallow discharge cycles. The phenomenon is like losing a muscle due to lack of use over a long time. A remedy for restoring the full capacity is "reconditioning," in which the battery is fully discharged to almost zero voltage once every few months and then fully charged to about 1.55 V per cell. Other types of batteries do not have such memory effect.

13.4.8 Effects of Temperature

As seen in the preceding sections, the operating temperature significantly influences the battery performance as follows:

- The capacity and charge efficiency decrease with increasing temperature.
- The capacity drops at temperatures above or below a certain range, and drops sharply at temperatures below freezing.
- The self-discharge rate increases with temperature.
- The internal resistance increases with decreasing temperature.

Table 13.2 shows the influence of temperature on the charge efficiency, discharge efficiency, and self-discharge rate in the NiCd battery. The process of determining the optimum operating temperature is also indicated in the table. It is seen that different attributes have different desirable operating temperature ranges shown by the bold-faced numbers. With all attributes jointly considered, the most optimum operating

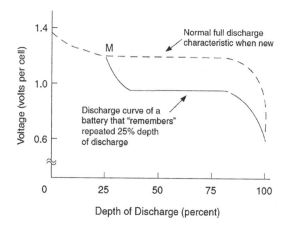

FIGURE 13.10 Memory effect degrades discharge voltage in NiCd cell.

TABLE 13.2
Optimum Working-Temperature Range for NiCd Battery

Operating Temperature (°C)	Charge Efficiency (%)	Discharge Efficiency (%)	Self-Discharge Rate (% Capacity/Day)
−40	0	72	0.1
−35	0	80	0.1
−30	15	85	0.1
−25	40	90	0.2
−20	75	95	0.2
−15	85	97	0.2
−10	**90**	**100**	**0.2**
−5	**92**	**100**	**0.2**
0	**93**	**100**	**0.2**
5	**94**	**100**	**0.2**
10	**94**	**100**	**0.2**
15	**94**	**100**	**0.3**
20	**93**	**100**	**0.4**
25	**92**	**100**	**0.6**
30	91	100	1.0
35	90	100	1.4
40	88	100	2.0
45	85	100	2.7
50	82	100	3.6
55	79	100	5.1
60	75	100	8.0
65	70	100	12
70	60	100	20

temperature is the intersection of all the desirable ranges. For example, if we wish to limit the self-discharge rate below 1%, discharge efficiency at 100%, and charge efficiency at 90% or higher, Table 13.2 indicates that the optimum working-temperature range is between 10°C and 25°C, which is the common belt through the boldfaced parts of the three columns.

13.4.9 INTERNAL LOSS AND TEMPERATURE RISE

The battery temperature varies over the C/D cycle. Taking NiCd as an example, the heat generated in one such cycle with 1.2 h of discharge and 20.8 h of charge every day is shown in Figure 13.11. Note that the heat generation increases with the depth of discharge (DoD) because of the increased internal resistance at higher DoD. When the battery is put to charge, the heat generation is negative for a while, meaning that the electrochemical reaction during the initial charging period is endothermic (absorbing heat), as opposed to the exothermic reaction during other periods with a positive heat generation. The temperature rise during the cycle depends on the cooling method used to dissipate the heat by conduction, convection, and radiation.

Energy Storage

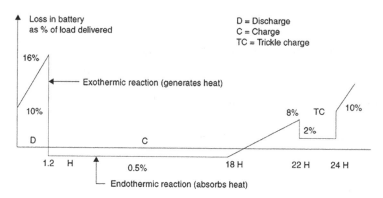

FIGURE 13.11 Internal energy loss in battery during C/D cycle showing endothermic and exothermic periods.

Different electrochemistries, however, generate internal heat at different rates. The heat generation of various batteries can be meaningfully compared in terms of the adiabatic temperature rise during discharge, which is given by the following relation:

$$\Delta T = \frac{WH_d}{MC_p}\left[1-\eta_v+\frac{E_d}{E_o}\right] \quad (13.4)$$

where
ΔT = adiabatic temperature rise of the battery, °C
WH_d = watthour energy discharged, Wh
M = mass of the battery, kg
C_p = battery-specific heat, Wh/kg·C
η_v = voltage efficiency factor on discharge
E_d = average cell entropy energy per coulomb during discharge, i.e., average power loss per ampere of discharge, W/A
E_o = average cell open-circuit voltage, V

For full discharge, the WH_d/M ratio in Equation 13.4 becomes the specific energy. This indicates that a higher specific energy cell would also tend to have a higher temperature rise during discharge, requiring an enhanced cooling design. Various battery characteristics affecting the thermal design are listed in Table 13.3. Figure 13.12 depicts the adiabatic temperature rise ΔT for various electrochemistries after a full discharge in short bursts.

13.4.10 RANDOM FAILURE

The battery fails when at least one cell in a series fails. *Cell failure* is theoretically defined as the condition in which the cell voltage drops below a certain value before discharging the rated capacity at room temperature. The value is generally taken as 1.0 V in cells with nominal voltage of 1.2 V. This is a very conservative definition of battery failure. In practice, if one cell shows less than 1.0 V, other cells can make up

TABLE 13.3
Battery Characteristics Affecting Thermal Design

Electrochemistry	Operating Temperature Range (°C)	Overcharge Tolerance	Heat Capacity (Wh/kg-K)	Mass Density (kg/l)	Entropic Heating on Discharge W/A
Lead-acid	–10 to 50	High	0.35	2.1	–0.06
Nickel-cadmium	–20 to 50	Medium	0.35	1.7	0.12
Nickel-metal hydride	–10 to 50	Low	0.35	2.3	0.07
Lithium-ion	10 to 45	Very low	0.38	1.35	0
Lithium-polymer	50 to 70	Very low	0.40	1.3	0

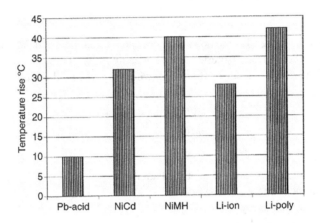

FIGURE 13.12 Adiabatic temperature rise for various electrochemistries.

the difference without detecting the failure at the battery level. Even if all cells show stable voltage below 1.0 V at full load, the load can be reduced to maintain the desired voltage for some time until the voltage degrades further.

The cell can fail in open, short, or be in some intermediate state (a soft short). A short that starts soft eventually develops into a hard short. In a low-voltage battery, any attempt to charge with a shorted cell may result in physical damage to the battery or the charge regulator. On the other hand, the shorted cell in a high-voltage battery with numerous series-connected cells may work forever. It, however, loses the voltage and ampere-hour capacity, and hence, would work as a load on the healthy cells. An open cell, on the other hand, disables the entire battery of series-connected cells.

In a system having two parallel batteries (a common design practice), if one cell in one battery gets shorted, the two batteries would have different terminal characteristics. Charging or discharging such batteries as a group can result in highly uneven current sharing, subsequently overheating one of the batteries. Two remedies are available to avoid this. One is to charge and discharge both batteries with individual current controls such that they both draw their rated share of the load. The other is to replace the failed cell immediately, which can sometimes be impractical. In general,

Energy Storage

an individual C/D control for each battery is the best strategy. It may also allow replacement of any one battery with a different electrochemistry or different age, which would have different load-sharing characteristics. Batteries are usually replaced several times during the economic life of a plant.

13.4.11 WEAR-OUT FAILURE

In addition to a random failure, the battery cell eventually wears out and fails. This is associated with the electrode wear due to repeated C/D cycles. The number of times the battery can be discharged and recharged before the electrodes wear out depends on the electrochemistry. The battery life is measured by the number of C/D cycles it can deliver before a wear-out failure. The life depends strongly on the depth of discharge and the temperature as shown in Figure 13.13, which is for a high-quality NiCd battery. The life also depends, to a lesser degree, on the electrolyte concentration and the electrode porosity. The first two factors are application related, whereas the others are construction related.

It is noteworthy from Figure 13.13 that the life at a given temperature is an inverse function of the depth of discharge. At 20°C, the life is 10,000 cycles at 30% DoD and about 6,000 cycles at 50% DoD. This makes the product of the number of cycles until failure and the DoD remain approximately constant. This product decreases with increasing temperature. This is true for most batteries. This means that the battery at a given temperature can deliver the same number of equivalent full cycles of energy regardless of the depth of discharge. The total Wh energy the battery can deliver over its life is approximately constant. Such observation is useful in comparing the costs of various batteries for a given application.

The life consideration is a dominant design parameter in battery sizing. Even when the load may be met with a smaller capacity, the battery is oversized to meet the life requirement as measured in number of C/D cycles. For example, with the same Wh load, the battery that must charge or discharge twice as many cycles over its life needs approximately double the capacity to have the same calendar life.

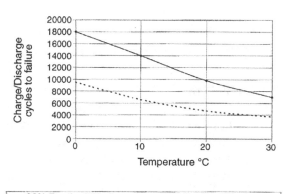

FIGURE 13.13 C/D cycle life of sealed NiCd battery vs. temperature and DoD.

13.4.12 Battery Types Compared

The performance characteristics and properties of various electrochemistries presented in the preceding sections are summarized and compared in Tables 13.4 and 13.5. Note that the overall cost of the Pb-acid battery is low compared to NiCd, NiMH, and Li-ion batteries. Because of its least cost per Wh delivered over the life, the Pb-acid battery in cost-sensitive applications has been the workhorse of industry.[2]

13.5 MORE ON LEAD-ACID BATTERY

The Pb-acid battery is available in small to large capacities in various terminal voltages, such as 6 V, 12 V, and 24 V. As in other batteries, the ampere-hour capacity of the Pb-acid battery is sensitive to temperature. Figure 13.14 shows the capacity variations with temperature for deep-cycle Pb-acid batteries. At 20°F, for example, the high-rate battery capacity is about 20% of its capacity at 100°F. The car is hard to start in winter for this reason. On the other hand, the self-discharge rate decreases significantly at cold temperatures, as seen in Figure 13.15.

TABLE 13.4
Specific Energy and Energy Density of Various Batteries

Electrochemistry	Specific Energy (Wh/kg)	Energy Density (Wh/ liter)	Specific Power (W/kg)	Power Density (W/liter)
Lead-acid	30–40	70–75	~200	~400
Nickel-cadmium	40–60	70–100	150–200	220–350
Nickel-metal hydride	50–65	140–200	~200	450–500
Lithium-ion	100–150	300–400	>500	500–600
Lithium-polymer	100–200	150–300	>200	>350
Zinc-air	140–180	200–200	~150	~200

TABLE 13.5
Comparison of Various Batteries

Electrochemistry	Cycle Life in Full Discharge Cycles	Calendar Life in Years	Self-Discharge (%/month at 25°C)
Lead-acid	500–1000	5–8	3–5
Nickel-cadmium	1000–2000	10–15	20–30
Nickel-metal hydride	1000–2000	8–10	20–30
Lithium-ion	1500–3500	—	5–10
Lithium-polymer	1000–1500	—	1–2
Zinc-air	200–300	—	4–6

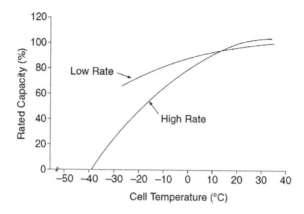

FIGURE 13.14 Pb-acid battery capacity variations with temperature.

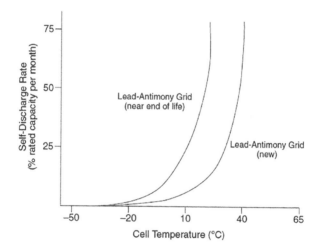

FIGURE 13.15 Pb-acid battery self-discharge rate vs. temperature.

Table 13.6 shows the effect of *SOC* on the voltage, specific gravity, and freezing point of the Pb-acid battery. The electrolyte in a fully charged battery has a high specific gravity and freezes at 65°F. On the other hand, a fully discharged battery freezes at +15°F. The table shows the importance of keeping the battery fully charged on cold days.

The cycle life vs. DoD for the Pb-acid battery is depicted in Figure 13.16, again showing the "half-life at double the DoD" rule of thumb.

The discharge rate influences the Pb-acid battery capacity, as shown in Figure 13.17. The shorter the discharge time (i.e., higher the discharge rate), the lower the ampere-hour capacity the battery can deliver.

The Pb-acid cell voltage is 2.0 V nominal, and the internal resistance is around 1 mΩ per cell. The cycle life is 500 to 1000 full C/D cycles for medium-rated batteries. The ranges of operating temperatures are between −20 and 50°C and the survival temperature between −55 and 60°C.

TABLE 13.6
Effects of SOC on Specific Gravity and Freezing Point of Lead-Acid Battery

State of Charge	Specific Gravity	Freezing Point	120-V Battery Voltage
1 (Fully Charged)	1.27	−65°F	128
75%	1.23	−40°F	124
50%	1.19	−10°F	122
25%	1.15	+5°F	120
0 (Fully Discharged)	1.12	+15°F	118

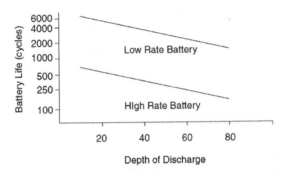

FIGURE 13.16 Pb-acid battery life in cycles to failure vs. DoD.

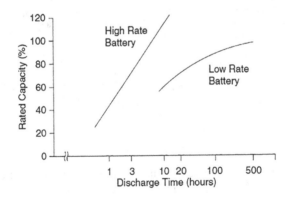

FIGURE 13.17 Pb-acid battery capacity with discharge time.

13.6 BATTERY DESIGN

The battery design for a given application depends on the following system requirements:

- Voltage and current
- C/D and duration
- Operating temperature during C/D
- Life in number of C/D cycles
- Cost, size, and weight constraints

Energy Storage

Once these system-level design parameters are identified, the battery design proceeds in the following steps:

1. Select the electrochemistry suitable for the overall system requirements.
2. Determine the number of series cells required to meet the voltage requirement.
3. Determine the ampere-hour discharge required to meet the load demand.
4. For the required number of C/D cycles, determine the maximum allowable DoD.
5. Ampere-hour capacity of the battery is then determined by dividing the ampere-hour discharge required by the allowable DoD calculated earlier.
6. Determine the number of battery packs required in parallel for the total ampere-hour capacity.
7. Determine the temperature rise and thermal controls required.
8. Provide the C/D rate controls as needed.

Each cell in the battery pack is electrically insulated from the others and from the ground. The electrical insulation must be a good conductor of heat to maintain a low temperature gradient between the cells and also to the ground.

The battery performs better under slow C/D rates. It accepts less energy when charged at a faster rate. Also, the faster the discharge rate, the faster the voltage degradation and lower the available capacity to the load. For these reasons, high-C/D-rate applications require different design considerations from the low-rate applications.

Because the battery design is highly modular, built from numerous cells, there is no fundamental technological limitation on the size of the energy storage system that can be designed and operated using electrochemical batteries. The world's largest 40-MW peak-power battery was commissioned in 2003 at a cost of $30 million. The system used 14,000 sealed NiCd cells manufactured from recycled cadmium by Saft Corporation at a total cell cost of $10 million. The cells will be recycled again after their 20-year life. The battery system is operated by Golden Valley Electric Association in Fairbanks for an Alaskan utility company. The spinning energy reserve of the battery provides continuous voltage support and cuts down on blackout possibilities.

13.7 BATTERY CHARGING

During battery charging, the energy management software monitors the SOC, the overall health, and safe termination criteria. The continuously monitored operating parameters are the battery voltage, current, and temperature. The charging timer is started after all initial checks are successfully completed. Charging may be suspended (but not reset) if it detects any violation of critical safety criteria. The timer stops charging if the defect persists beyond a certain time limit.

Normal charging has the following three phases:

- Bulk (fast) charge, which deposits 80 to 90% of the drained capacity
- Taper charge, in which the charge rate is gradually cut back to top off the remaining capacity
- Trickle (float) charge after the battery is fully charged to counter the selfdischarge rate

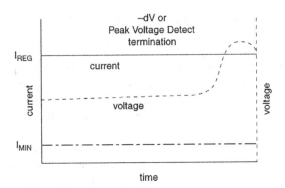

FIGURE 13.18 Constant current charging of NiCd and NiMH batteries.

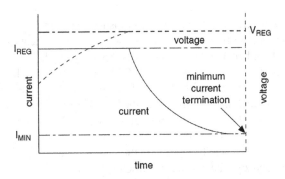

FIGURE 13.19 Constant voltage charging of Li-ion battery.

The bulk-charge and the taper-charge termination criteria are preloaded in the battery management software to match the battery electrochemistry and system-design parameters. For example, the NiCd and NiMH batteries are generally charged at a constant current (Figure 13.18), terminating the charging when the continuously monitored ΔV is detected negative. On the other hand, the Li-ion battery, being sensitive to overcharging, is charged at a constant voltage, tapering off the charge current as needed (Figure 13.19).

13.8 CHARGE REGULATORS

For safety reasons, it is extremely important that excessive charging of the battery be avoided at all times. Overcharging causes internal gassing, which causes loss of water in the Pb-acid battery and premature aging. The charge regulator allows the maximum rate of charging until the gassing starts. Then the charge current is tapered off to the trickle-charge rate so that the full charge is approached gently.

The batteries are charged in the following three different manners:

Energy Storage

13.8.1 MULTIPLE CHARGE RATES

This is the best method, in which the battery is charged gently in multiple steps. First, the battery is charged at a full charge rate until 80 to 90% of the capacity is achieved. The charge current is then cut back in steps until the battery is fully charged. At this time, the charge current is further reduced to a trickle-charge rate, keeping it fully charged until the next load demand comes on the battery. This method, therefore, needs at least three charge rates in the charge regulator design.

13.8.2 SINGLE-CHARGE RATE

This method uses a simple low-cost regulator, which is either on or off. The regulator is designed for only one charge rate. When the battery is fully charged, as measured by its terminal voltage, the charger is turned off by a relay. When the battery voltage drops below a preset value, the charger is again connected in full force. Because the charging is not gentle in this method, full charge is difficult to achieve and maintain. An alternate version of this charging method is the multiple pulse charging. Full current charges the battery up to a high preset voltage just below the gassing threshold. At this time, the charger is shut off for a short time to allow the battery chemicals to mix and the voltage to fall. When the voltage falls below a low preset threshold, the charger is reconnected, again passing full current to the battery.

13.8.3 UNREGULATED CHARGING

This least-cost method can be used in PV power systems. It uses no charge regulator. The battery is charged directly from a solar module dedicated just for charging. The charging module is properly designed for safe operation with a given number of cells in the battery. For example, in a 12-V Pb-acid battery, the maximum PV module voltage is kept below 15 V, making it impossible to overcharge the battery. When the battery is fully charged, the array is fully shunted to ground by a shorting switch (transistor). The shunt transistor switch is open when the battery voltage drops below a certain value. The isolation diode blocks the battery that powers the array or shunt at night, as discussed in Section 9.8.

13.9 BATTERY MANAGEMENT

Drawing electric power from the battery when needed, and charging it back when access power is available, requires a well-controlled charge and discharge process. Otherwise, the battery performance could suffer, life could be shortened, and maintenance would increase. Some common performance problems are as follows:

- Low charge efficiency, resulting in low *SOC*
- Loss of capacity, disabling the battery to hold the rated ampere-hour charge
- Excessive gassing and heating, leading to a short life
- Unpredictable premature failure, leading to loss of load availability
- Positive plate corrosion, shortening the life
- Stratification and sulfation, degrading the performance

The following features incorporated into battery management can avoid the problems given in the preceding text:

- Controlled voltage charging, preferably at a constant voltage
- Temperature-compensated charging, in which the charge termination occurs earlier if the battery temperature is higher than the reference temperature
- Individual charge control if two or more batteries are charged in parallel
- Accurate set points to start and stop the charge and discharge modes

13.9.1 Monitoring and Controls

The batteries in modern power systems are managed by dedicated computer software. The software monitors and controls the following performance parameters:

- Voltage and current
- Temperature and pressure (if applicable)
- Ampere-hour in and out of the battery
- *SOC* and discharge
- Rate of charge and discharge
- Depth of discharge
- Number of charge and discharge cycles

An ampere-hour-integrating meter is commercially available, which keeps track of the ampere-hour in and out of the battery and sends the required signals to the mode controller.

The temperature-compensated maximum battery voltage and the *SOC* can improve battery management, particularly in extreme cold temperatures. It can allow an additional charging during cold periods when the battery can accept more charge. The low-voltage alarm is a good feature to have, as discharging below the threshold low voltage can cause a cell voltage to reverse (become negative). The negative voltage of the cell makes it a load, leading to overheating and a premature failure. The alarm can be used to shed noncritical loads from the battery to avoid potential damage.

Figure 13.20 depicts a commercially available battery management system incorporating a dedicated microprocessor with software.

13.9.2 Safety Considerations

Not to Overcharge: The battery operation requires certain safety considerations. The most important is not to overcharge the battery. Any overcharge above the trickle-charge rate is converted into heat, which, if beyond a certain limit, can cause the battery to explode. This is particularly critical when the battery is charged directly from a PV module without a charge regulator. In such a case, the array is sized below a certain safe limit. As a rule of thumb, the PV array rating is kept below the continuous overcharge current that can be tolerated by the battery. This is typically below $C/15$ A for Pb-acid batteries.

Energy Storage

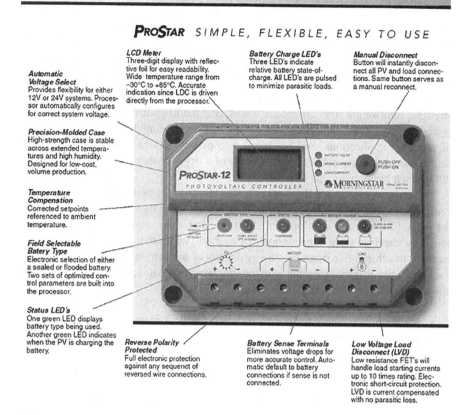

FIGURE 13.20 Battery management microprocessor for PV power system. (From Morningstar Corporation, Newtown, PA. With permission.)

Isolation Diodes: Not all battery cells degrade at the same rate due to manufacturing tolerances and difference in operating temperatures. One or more cells making one string of the battery may age at a higher rate and reach to a lower voltage than other strings connected in parallel. In such a case, the current sharing between parallel-connected strings could be uneven, or highly uneven to the point that the low-voltage string could draw current inward as a load instead of sharing the load. Such situation is highly undesirable and unsafe, as it may cause the battery to overheat. The remedy to avoid such an unsafe situation is to use isolation diode at the top of each battery string to block any reverse current.

Air Circulation: Battery when charging, discharging, or under trickle charge generates gases internally, which pressurizes the battery casing. If the battery is not sealed, like some lead-acid batteries (some old Navy submarines may still have such batteries in operation), the battery room may collect such leaked gas to an unsafe concentration level for personnel safety or fire safety. To avoid such unsafe situation, it is necessary to circulate air in the battery room at certain cubic meter per hour rate required by the standards applicable to the battery chemistry and its location.

13.10 FLYWHEEL

The flywheel stores kinetic energy in a rotating inertia. This energy can be converted from and to electricity with high efficiency. The flywheel energy storage is an old concept, which has now become commercially viable due to advances made in high-strength, lightweight fiber composite rotors, and the magnetic bearings that operate at high speeds. The flywheel energy storage system is being developed for a variety of potential applications, and is expected to make significant inroads in the near future. The round-trip conversion efficiency of a large flywheel system can approach 90%, much higher than that of a battery.

The energy storage in a flywheel is limited by the mechanical stresses due to the centrifugal forces at high speeds. Small- to medium-sized flywheels have been in use for years. Considerable development efforts are underway around the world for high-speed flywheels to store large amounts of energy. The present goal of these developments is to achieve five times the energy density of the currently available secondary batteries. This goal is achievable with the following enabling technologies, which are already in place in their component forms:

- High-strength fibers having an ultimate tensile strength of over one million pounds per square inch
- Advances made in designing and manufacturing fiber-epoxy composites
- High-speed magnetic bearings, which eliminate friction, vibrations, and noise

The flywheel system is made of a fiber-epoxy composite rotor, supported on magnetic bearings, rotating in a vacuum, and mechanically coupled with an electrical machine that can work as a motor or a generator. Two counter-rotating wheels are placed side by side where gyroscopic effects must be eliminated, such as in a city transit bus, train, or an automobile.

13.10.1 ENERGY RELATIONS

The energy stored in a flywheel having the moment of inertia J and rotating at an angular speed ω is given by the following:

$$E = \frac{1}{2}J\omega^2 \qquad (13.5)$$

The centrifugal force in the rotor material of density ρ at radius r is given by $\rho(r > \omega)^2$, which is supported by the hoop stress in the rotor rim. Because the linear velocity $V = 2 = \pi r\omega$, the maximum centrifugal stress in the rotor is proportional to the square of the outer tip velocity. The allowable stress in the material places an upper limit on the rotor tip speed. Therefore, a smaller rotor can run at a high speed and *vice versa*. The thin-rim-type rotor has a high inertia-to-weight ratio and stores more energy per kilogram weight. For this reason, the rotor, in all practical flywheel system designs, is a thin-rim configuration. For such a rotor with inner radius R_1 and outer radius R_2, it can be shown that the maximum energy that can be stored for an allowable rotor tip velocity V is as follows:

$$E_{max} = K_1 V^2 \left[1 + \left(\frac{R_1}{R_2}\right)^2\right] \quad (13.6)$$

where K_1 is the proportionality constant. The thin-rim flywheel with R_1/R_2 ratio approaching unity results in a high specific energy for a given allowable stress limit. The higher the ultimate strength of the material, the higher the specific energy. The lower the material density, the lower the centrifugal stress produced, which leads to a higher allowable speed and specific energy. The maximum energy storage E_{max}, therefore, can be expressed as follows:

$$E_{max} = K_2 \frac{\sigma_{max}}{\rho} \quad (13.7)$$

where

K_2 = another proportionality constant
σ_{max} = maximum allowable hoop stress
ρ = density of the rotor material

A good flywheel design therefore has a high σ_{max}/ρ ratio for high specific energy. It also has a high E/ρ ratio for rigidity, where E is the Young's modulus of elasticity.

The metallic flywheel has low specific energy because of a low σ_{max}/ρ ratio, whereas high-strength polymer fibers such as graphite, silica, and boron, having much higher σ_{max}/ρ ratio, store an order of magnitude higher energy per unit weight. Table 13.7 compares the specific energy of various metallic and polymer fiber composite rotors. In addition to a high specific energy, the composite rotor has a safe mode of failure, as it disintegrates to fluff rather than fragmenting like the metal flywheel.

TABLE 13.7
Maximum Specific Energy Storable in a Thin-Rim Flywheel with Various Rim Materials

Rotor Wheel Material	Maximum Specific Energy Storable (Wh/kg)
Aluminum alloy	25
Maraging steel	50
E-glass composite	200
Carbon fiber composite	220
S-glass composite	250
Polymer fiber composite	350
Fused silica fiber composite	1000
Lead-acid battery	30–40
Lithium-ion battery	90–120

Figure 13.21 shows a rotor design recently developed at the Oakridge National Laboratory. The fiber-epoxy composite rim is made of two rings. The outer ring is made of high-strength graphite, and the inner ring of low-cost glass fiber. The hub is made of single-piece aluminum in the radial spoke form. Such a construction is cost-effective because it uses the costly material only where it is needed for strength, that is, in the outer ring where the centrifugal force is high, resulting in a high hoop stress.

Figure 13.22 shows a prototype 5-kWh flywheel weight and specific energy (watthour per pound) vs. σ_{max}/ρ ratio of the material. It is noteworthy that the weight decreases inversely and the specific energy increases linearly with the σ_{max}/ρ ratio.

13.10.2 FLYWHEEL SYSTEM COMPONENTS

The complete flywheel energy storage system requires the following components:

- High-speed rotor attached to the shaft via a strong hub
- Bearings with good lubrication system or with magnetic suspension in high-speed rotors

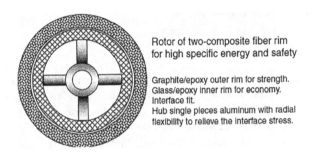

FIGURE 13.21 Flywheel rotor design using two composite rings. (Adapted from the DOE/Oakridge National Laboratory's prototype design.)

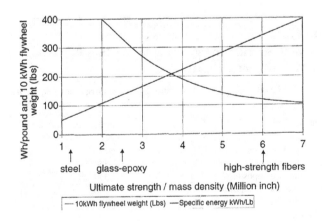

FIGURE 13.22 Specific energy vs. specific strength in flywheel design.

Energy Storage

- Electromechanical energy converter, usually a machine that can work as a motor during charging and as a generator while discharging the energy
- Power electronics to drive the motor and to condition the generator power
- Control electronics for controlling the magnetic bearings and other functions

Good bearings have low friction and vibration. Conventional bearings are used up to speeds in a few tens of thousands rpm. Speeds approaching 100,000 rpm are possible only by using magnetic bearings, which support the rotor by magnetic repulsion and attraction. The mechanical contact is eliminated, thus eliminating friction. Running the rotor in a vacuum eliminates windage.

The magnetic bearing comes in a variety of configurations using permanent magnets and dynamic current actuators to achieve the required restraints. A rigid body can have 6 degrees of freedom. The bearings retain the rotor in 5 degrees of freedom, leaving 1 degree free for rotation. The homopolar configuration is depicted in Figure 13.23. Permanent magnets are used to provide free levitation support for the shaft

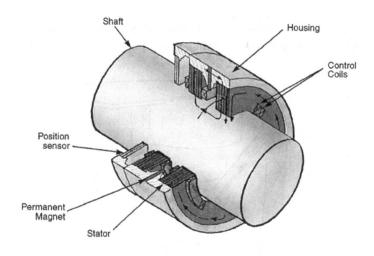

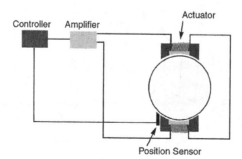

FIGURE 13.23 Avcon's patented homopolar permanent-magnet active bearing. (From Avcon Inc., Woodland Hills, CA. With permission.)

and to help stabilize the shaft under a rotor drop. The electromagnet coils are used for stabilization and control. The control coils operate at low-duty cycle, and only one servo-controller loop is needed for each axis. The servo-control coils provide active control to maintain shaft stability by providing restoring forces as needed to maintain the shaft in the centered position. Various position and velocity sensors are used in an active feedback loop. The electric current variation in the actuator coils compels the shaft to remain centered in position with desired clearances.

Small flux pulsation as the rotor rotates around the discrete actuator coils produces a small electromagnetic loss in the metallic parts. This loss, however, is negligible compared to the friction loss in conventional bearings.

In the flywheel system configuration, the rotor can be located radially outward, as shown in Figure 13.24. It forms a volume-efficient packaging. The magnetic bearing has permanent magnets inside. The magnetic flux travels through the pole shoes on the stator and a magnetic feedback ring on the rotor. The reluctance lock between the pole shoes and the magnetic feedback ring provides the vertical restraint. The horizontal restraint is provided by the two sets of dynamic actuator coils. The currents in the coils are controlled in response to a feedback loop controlling the rotor position.

The electromechanical energy conversion in both directions is achieved with one electrical machine, which works as a motor for spinning up the rotor for energy

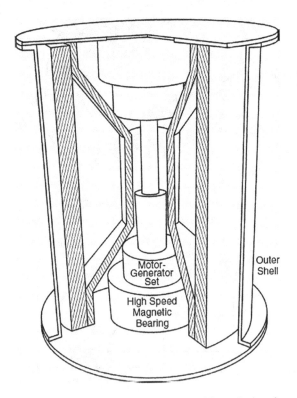

FIGURE 13.24 Flywheel configuration with rotor outside enclosing the motor generator and the bearing.

Energy Storage

charge, and as a generator while decelerating the rotor for a discharge. Two types of electrical machines can be used, the synchronous machine with variable-frequency converter or the permanent-magnet brushless DC machine.

The machine voltage varies over a wide range with speed. The power electronic converters provide an interface between the widely varying machine voltage and the fixed bus voltage. It is possible to design a discharge converter and a charge converter with input voltage varying over a range of 1 to 3. This allows the machine speed to vary over the same range. That is, the low rotor speed can be one-third of the full speed. Because the energy storage is proportional to the speed squared, the flywheel *SOC* at low speed can be as low as 0.10. This means 90% of the flywheel energy can be discharged with no hardship on the power electronics, or other components of the system.

As to the number of charge–discharge cycles the flywheel can withstand, the fatigue life of the composite rotor is the limiting factor. Experience indicates that the polymer fiber composites in general have a longer fatigue life than solid metals. A properly designed flywheel, therefore, can last much longer than a battery and can discharge to a much deeper level. Flywheels made of composite rotors have been fabricated and tested to demonstrate more than 10,000 cycles of full charge and discharge. This is an order of magnitude more than any battery can deliver at present.

13.10.3 BENEFITS OF FLYWHEEL OVER BATTERY

The main advantages of the flywheel energy storage over the battery are as follows:

- High energy storage capacity per unit of weight and volume
- High DoD
- Long cycle life, which is insensitive to the DoD
- High peak-power capability without concerns about overheating
- Easy power management, as the *SOC* is simply measured by the speed
- High round-trip energy efficiency
- Flexibility in design for a given voltage and current
- Improved quality of power as the electrical machine is stiffer than the battery

These benefits have the potential of making the flywheel the least-cost energy storage alternative per watthour delivered over its operating life.

13.11 SUPERCONDUCTING MAGNET

The superconductor technology for storing energy has started yielding highly promising results. In its working principle, the energy is stored in the magnetic field of a coil, and is given by the following expression:

Energy E stored in a coil carrying current I is given by the following:

$$E = \tfrac{1}{2} B^2 / \mu \left(J/m^3 \right) \text{ or } E = \tfrac{1}{2} I^2 L \left(J \right) \quad (13.8)$$

where

  B = magnetic field density produced by the coil (T)
  μ = magnetic permeability of air = $4\pi \cdot 10^{-7}$ (H/m)
  L = inductance of the coil (H)

The coil must carry current to produce the required magnetic field. The current requires a voltage to be applied to the coil terminals. The relation between the coil current I and the voltage V is as follows:

$$V = RI + L\frac{di}{dt} \tag{13.9}$$

where R and L are the resistance and inductance of the coil, respectively. For storing energy in a steady state, the second term in Equation 13.9 must be zero. Then the voltage required to circulate the needed current is simply $V = RI$.

The resistance of the coil is temperature dependent. For most conducting materials, it is higher at higher temperatures. If the temperature of the coil is reduced, the resistance drops as shown in Figure 13.25. In certain materials, the resistance abruptly drops to a precise zero at some critical temperature. In the figure, this point is shown as T_c. Below this temperature, no voltage is required to circulate current in the coil, and the coil terminals can be shorted. The current continues to flow in the shorted coil indefinitely, with the corresponding energy also stored indefinitely in the coil. The coil is said to have attained the superconducting state, one that has zero resistance. The energy in the coil then "freezes."

Although the superconducting phenomenon was discovered decades ago, the industry interest in developing practical applications started in the early 1970s. In the U.S., the pioneering work has been done in this field by the General Electric Company, Westinghouse Research Center, University of Wisconsin, and others. During the 1980s, a grid-connected 8-kWh superconducting energy storage system was built with funding from the Department of Energy, and was operated by the Bonneville

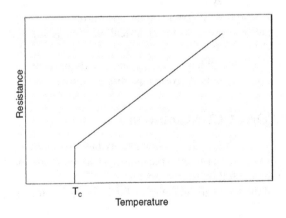

FIGURE 13.25 Resistance vs. temperature with abrupt loss of resistance at the critical superconducting temperature.

Energy Storage

Power Administration in Portland, OR. The system demonstrated over one million charge-discharge cycles, meeting its electrical, magnetic, and structural performance goals. Conceptual designs of large superconducting energy storage systems up to 5000 MWh energy for utility applications have been developed.

The main components in a typical superconducting energy storage system are shown in Figure 13.26. The superconducting magnet coil is charged by an AC-to-DC converter in the magnet power supply. Once fully charged, the converter continues providing the small voltage needed to overcome losses in the room temperature parts of the circuit components. This keeps a constant DC current flowing (frozen) in the superconducting coil. In the storage mode, the current is circulated through a normally closed switch.

The system controller has three main functions.

- It controls the solid-state isolation switch.
- It monitors the load voltage and current.
- It interfaces with the voltage regulator that controls the DC power flow to and from the coil.

If the system controller senses the line voltage dropping, it interprets that the system is incapable of meeting the load demand. The switch in the voltage regulator opens in less than 1 msec. The current from the coil now flows into a capacitor bank until the system voltage recovers the rated level. The capacitor power is inverted into 60- or 50-Hz AC and is fed to the load. The bus voltage drops as the capacitor energy is depleted. The switch opens again, and the process continues to supply energy to the load. The system is sized to store the required energy to power a specified load for a specified duration.

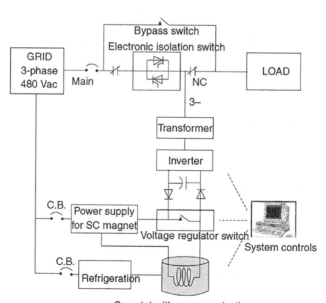

FIGURE 13.26 Superconducting energy storage system schematic.

The superconducting energy storage has several advantages over other technologies:

- The round-trip efficiency of the charge-discharge cycle is high at 95%. This is higher than that attainable by any other technology.
- It has a much longer life, up to about 30 years.
- The charge and discharge times can be extremely short, making it attractive for supplying large power for a short time if needed.
- It has no moving parts in the main system, except in the cryogenic refrigeration components.

In the superconducting energy storage system, a major cost is to keep the coil below the critical superconducting temperature. Until now, the niobium–titanium alloy has been extensively used, which has a critical temperature of about 9°K. This requires liquid helium as a coolant at around 4°K. The 1986 discovery of high-temperature superconductors has accelerated the industry interest in this technology. Three types of high-temperature superconducting materials are available now, all made from bismuth or yttrium–cuprate compounds. These superconductors have the critical temperature around 100°K. Therefore, they can be cooled by liquid nitrogen, which needs significantly less refrigeration power. As a result, numerous programs around the world have started to develop commercial applications. Toshiba of Japan, GEC-Alsthom along with Electricite de France, and many others are actively pursuing development in this field.[3,4]

13.12 COMPRESSED AIR

Compressed air stores energy in a pressure–volume relation. It can store excess energy of a power plant—thermal, nuclear, wind, or PV—and supply it when needed during lean periods or peak demands. The compressed air energy storage system consists of:

- Air compressor
- Expansion turbine
- Electric motor–generator
- Overhead storage tank or an underground cavern

If P and V represent the air pressure and volume, respectively, and if the air compression from pressure P_1 to P_2 follows the gas law PV^n = constant, then the work required during this compression is the energy stored in the compressed air. It is given by the following:

$$\text{Energy stored} = \frac{n(P_2 V_2 - P_1 V_1)}{n-1} \qquad (13.10)$$

And the temperature at the end of the compression is given by the following:

$$\frac{T_2}{T_1} = \left(\frac{P_2}{P_1}\right)^{\frac{n-1}{n}} \qquad (13.11)$$

Energy Storage

The energy stored is smaller with a smaller value of n. The isentropic value of n for air is 1.4. Under normal working conditions, n is about 1.3. When air at an elevated temperature after constant-volume pressurization cools down, a part of the pressure is lost with a corresponding decrease in the stored energy.

Electric power is generated by venting the compressed air through an expansion turbine that drives a generator. The compressed air system may work under a constant volume or constant pressure.

In constant-volume compression, the compressed air is stored in pressure tanks, mine caverns, depleted oil or gas fields, or abandoned mines. One million cubic feet of air stored at 600 psi provides an energy storage capacity of about 0.25 million kWh_e. This system, however, has a disadvantage. The air pressure reduces as compressed air is depleted from the storage, and the electric power output decreases with decreasing air pressure.

In constant-pressure compression, the air storage may be in an above-ground variable-volume tank or an underground aquifer. One million cubic feet of air stored at 600 psi provides an energy storage capacity enough to generate about 0.10 million kWh. A variable-volume tank maintains a constant pressure by a weight on the tank cover. If an aquifer is used, the pressure remains approximately constant whereas the storage volume increases because of water displacement in the surrounding rock formation. During electrical generation, water displacement of the compressed air causes a decrease of only a few percent in the storage pressure, keeping the electrical generation rate essentially constant.

The operating-energy cost would include cooling the compressed air to dissipate the heat of compression. Otherwise, air temperature may rise as high as 1000°C—in effect, shrinking the storage capacity and adversely affecting the rock wall of the mine. Energy is also lost to the cooling effect of expansion when the energy is released.

The energy storage efficiency of the compressed-air storage system is a function of a series of component efficiencies, such as the compressor efficiency, motor–generator efficiency, heat losses, and compressed air leakage. The overall round-trip energy efficiency of about 50% has been estimated.

The compressed air can be stored in the following:

- Salt caverns
- Mined hard rock
- Depleted gas fields
- Buried pipes

The current capital cost estimate of a compressed air power system varies between $1500 and 2000 per kW, depending on the air storage systems used.

Compressed air power plants of 300-MW capacities have also been built in Israel, Morocco, and other countries. Two 150-MW plants, one in Germany and one in Alabama, have been in operation for more than a decade. Both these installations operate reliably, although on a single generator. They occupy salt caverns created by dissolving salt and removing brine.

A 200-MW compressed air energy storage plant is being planned by a consortium of electrical and gas utilities near Fort Dodge, IA, to support the new rapidly developing wind power generation in the area. This $200-million project planned in 2007 to use energy from a 100-MW wind farm to store compressed air in an underground aquifer and then blend it with natural gas to fire turbines for power generation.

A Houston-based company, CAES, had in 2001 proposed the use of abandoned limestone mines in Norton, OH (Figure 13.27). The 10-million-m^3 mines can store enough compressed air to drive 2700-MW-capacity turbines. The Ohio power sitting board approved the proposal for the operation to start before 2005. A Sandia National Laboratory study found the rock structure dense enough to prevent air leakage and solid enough to handle the working pressures from 11 MPa down to 5.5 MPa. The air from the mine after the expansion cooling is heated with natural gas to drive the turbines at an optimal temperature. This operation would use less than one-third of the fuel of a gas-fired generator and would reduce the energy cost and emission levels. The Ohio system is designed with nine 300-MW generators and uses 18 compressors to pressurize the mines.

13.13 TECHNOLOGIES COMPARED

For a very large-scale energy storage, Enslin et al.[5] of KEMA investigated the feasibility of a 20-MWh energy storage system for a 100-MW Dutch wind farm. A minimum of 10-MWh capacity was found necessary for this wind farm, and a 10-MWh additional capacity was allowed to meet the losses during discharge with some margin. Water pump storage, compressed air pumped storage, and Pb-acid batteries were

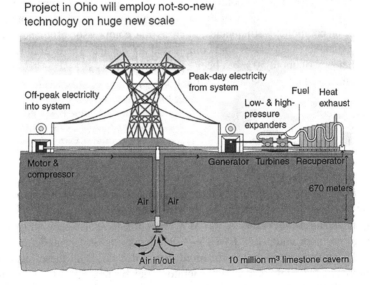

FIGURE 13.27 2700-MW power system to be installed for utility load leveling using compressed air energy storage at an Ohio mine. (From *IEEE Spectrum*, August 2001, p. 27. © IEEE. With permission.)

Energy Storage   223

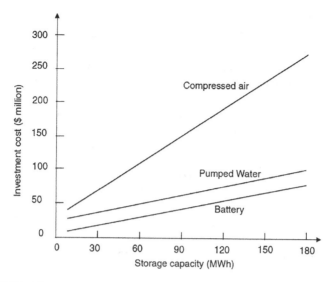

FIGURE 13.28 Energy storage capacity vs. investment cost of various alternative technologies. (Adapted from Enslin et al., *Renewable Energy World*, James & James, London, January–February 2004, p. 108.)

considered and compared for this project. They also conducted a parametric study with the storage capacity for potential applications in the 20 to 160 MWh range. The results of this study are shown in Figure 13.28. It shows that the Pb-acid battery is the least-cost option, followed by the pumped water, and then the compressed air systems. These trades, however, are extremely site-specific. The nature of air storage and its cost (e.g., free use of old mines) can alter the trades and the final installation decision.

13.14 MORE ON LITHIUM-ION BATTERY

Various energy storage technologies for power applications have been discussed earlier in the chapter. However, the technology which can meet large power and energy storage needs in a variety of applications, small and large, is most likely to be the Lithium-ion electrochemistry. This section therefore presents features of Lithium-ion batteries not covered earlier.

A variety of rechargeable lithium-ion (Li-ion) batteries are widely used in numerous industrial, commercial, and consumer applications. They are used in small power range in portable electronics, medium power range in electric vehicles, and large power range in utility energy storage for grid stability and in large solar and wind energy installations to maintain power availability to the users at all times. They are also increasingly used in military and aerospace applications.

Because the Li-ion battery has so significant contributions in everyday life of people around the world, from farmers in India to high tech industries in the silicon valleys of every country, the 2019 Nobel Prize in chemistry was awarded to its three inventors with citation: "Lithium-ion batteries have revolutionized our lives and are used in everything from mobile phones to laptops and electric vehicles. Through their work, this year's Chemistry Laureates have laid the foundation of a wireless, fossil fuel-free society".

The basic performance features of the Li-ion battery depend on the cell type and its component design, but their typical values are in the following range (Table 13.8 and Figure 13.29):

As we discussed earlier in the chapter, the actual cycle life of a battery depends primarily on the inverse of the depth of discharge (DoD), and to some extent on the operating temperature and the charge/discharge rates. The battery lasts longer (more charge/discharge cycles) at shallow DoD than at deep DoD. For example, if a given battery design lasts 3000 cycles at 100 % DoD (full capacity discharge) in very cycle at certain operating temperature and certain charge/discharge rates, then if the battery

TABLE 13.8
Basic Performance Features of the Li-ion Battery

Cell voltage when fully charged: 4.0 to 4.1 V
Cell voltage when fully discharged: 2.70 to 2.75 V
Cell voltage average during discharge: 3.5 V
Energy density: 300 to 600 W·h per liter
Specific energy: 200 to 300 W·h per kilogram
Specific power: 250 to 350 watts per kilogram
Charge/discharge efficiency: 80 to 90%
Self-discharge rate: 5–10 % per month at 25° C
Cost: $100–$150 per kWh (in 2020 and declining)
Charge/discharge cycle life:
1500 to 3500 cycles at 100 % depth of discharge in every cycle.

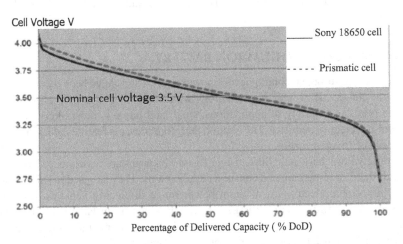

Charge at C/2 rate to 4.1 V CCCV, C/20 trickle charge at RT
Discharge at C/2 rate to 2.7 V at RT

FIGURE 13.29 Lithium-ion cell voltage vs. % DoD when fully charged at C/2 rate to 4.1 V at 25°C and fully discharged at C/2 rate to 1.7 V at 25°C.

Energy Storage

is discharged to only 25 % DoD in every cycle, it will last 100/25 = 4 times longer, i.e. about 3000 × 100 / 25 = 12,000 cycles. So, in the first approximation, we have the flowing relation for the battery life (true for all electrochemistries):

$$Battery\ cycle\ life\ at\ X\%DoD = \frac{Cycle\ Life\ at\ 100\%DoD\ from\ vendor's\ data \times 100}{X\%} \quad (13.12)$$

Like all batteries, the Li-ion battery is also made in a modular manner using the commercial off-the-shelf (COTS) cells readily available in the market at a mass-produced price. One such Li-ion cell No. 18650 made by Sony is widely used in the industry to assemble large batteries by connecting numerous such cells in series-parallel combination to obtain the required voltage and ampere-hour capacity of the assembled stack. Each No.18650 Sony cell is 18 mm diameter × 65 mm high and provides 1.5 Ah charge at 3.7 V average voltage. Large cubical (prismatic) cells up to 150 Ah are also available from other vendors, if desired for a compact design.

Large batteries of any desired capacity can be assembled using numerous small cells in modules. There is no real limit on the battery size as the design is highly modular. As shown in Figure 13.30, several small cells make a string, several strings make a block, several blocks make a battery module (pack), and several modules make a large battery.

This way, very large batteries up to 100 MWh capacity have been built, tested, and placed in operation on some grid power substations. For example, Tesla's 200 kWh *PowerPack-2* battery unit can power 50 kW load for 4 hours. Each such unit is 4.3' × 2.7' × 7.1' high in size, which weighs 3568 pounds. At Mira Loma utility power substation in the Los Angeles area, 400 such units are installed to store 80 MWh

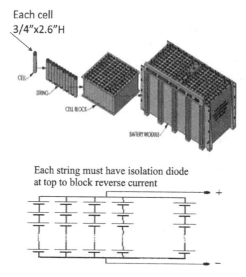

FIGURE 13.30 Modular battery assembly starting from cell to string to block to battery module (Source: ABSL Energy Systems Inc.).

TABLE 13.9
Large Li-ion Batteries Installed for Day and Night Load Leveling on Utility-Scale Power Plants

Substation	Energy Storage (MWh)	Power (MW)	Duration (hours)	Country
Minamisoma	40	10	4	Japan
Pomona	80	20	4	USA
Mira Loma	80	20	4	USA
Escondido	120	30	4	USA

energy for delivering 20 MW peak power for 4 hours. Each 200-kWh unit cost in 2017 was $75,000, including the inverter that interconnects the battery dc with the power grid ac lines. The battery is charged from the grid power at night to discharge back into the grid when the power demand exceeds the grid generation capacity, typically in the afternoon. Table 13.9 lists such large utility-scale batteries recently installed in the USA and Japan.

The Li-ion battery is now used in most new electrical vehicles made in the world. A safety issue of its exploding and causing fire due to an internal short has come to surface several times in the last few years, including in the Boeing Dreamliner planes in early 2019. The GM corporation and others have addressed this issue by developing ceramic-coated separators for lithium-ion cells. Such separators provide greater thermal stability than non-coated versions. The ceramic coating prevents the separators from shrinking at higher temperatures, thus minimizing the risk of a short in the cell.

There is no single energy storage technology that can meet all the desirable attributes for an application on hand, but a hybrid energy storage system can. For example, Chugach Electric Association in Anchorage, Alaska, has installed a hybrid battery and flywheel energy storage system for grid stabilization. The flywheel provides rapid injection of power to stabilize voltage and frequency and can deliver 18 MW-secs of energy within 1 millisecond. With the flywheel providing fast-response services, the 2 MW/0.5 MWh lithium-ion battery provides longer term storage services to enable greater integration of a 17 MW wind farm on Fire Island, Alaska[15].

REFERENCES

1. Riezenman, M.J., In Search of Better Batteries, *IEEE Spectrum*, pp. 51–56, April 1995.
2. Wicks, F. and Halls, S., *Evaluating Performance Enhancement of Lead-acid Batteries by Force Circulation of the Electrolytic*, Proceedings of the Intersociety Engineering Conference on Energy Conversion, Paper No. 180, 1995.
3. DeWinkel, C.C. and Lamopree, J.D., Storing Power for Critical Loads, *IEEE Spectrum*, pp. 38–42, June 1993.
4. Balachandran, U., Super Power, Progress in Developing the New Superconductors, *IEEE Spectrum*, pp. 18–25, July 1997.
5. Enslin, J., Jansen, C., and Bauer, P., In Store for the Future, Interconnection and Energy Storage for Offshore Wind Farms, *Renewable Energy World*, James & James Ltd., London, January–February 2004, pp. 104–113.

6. Omid Beik, An HVDC Off-shore Wind Generation Scheme with High Voltage Hybrid Generator, Ph.D. thesis, McMaster University, 2016.
7. Omid Beik, Ahmad S. Al-Adsani, *DC Wind Generation Systems, Design, Analysis, and Multiphase Turbine Technology*, Springer, 2020.
8. Omid Beik, Ahmad S. Al-Adsani, *Wind Energy Systems*, pp. 1–9, Springer, 2020.
9. Omid Beik, Ahmad S. Al-Adsani, *DC Wind Generation System*, pp. 33–69, Springer, 2020.
10. D. Schumacher, O. Beik and A. Emadi, "Standalone Integrated Power Electronics System: Applications for Off-Grid Rural Locations," *IEEE Electrification Magazine*, vol. 6, no. 4, pp. 73–82, December 2018.
11. O. Beik and N. Schofield, "High-Voltage Hybrid Generator and Conversion System for Wind Turbine Applications," *IEEE Transactions on Industrial Electronics*, vol. 65, no. 4, pp. 3220–3229, April 2018.
12. Beik, O., Schofield, N., 'High Voltage Generator for Wind Turbines', *8th IET International Conference on Power Electronics, Machines and Drives (PEMD 2016)*, 2016.
13. O. Beik and N. Schofield, "An Offshore Wind Generation Scheme with a High-Voltage Hybrid Generator, HVDC Interconnections, and Transmission," *IEEE Transactions on Power Delivery*, vol. 31, no. 2, pp. 867–877, April 2016.
14. O. Beik and N. Schofield, "*Hybrid Generator for Wind Generation Systems*," 2014 IEEE Energy Conversion Congress and Exposition (ECCE), Pittsburgh, PA, 2014, pp. 3886–3893.
15. U.S. Department of Energy "Potential Benefits of High-power High-capacity Batteries", DoE Report, January 2020

14 Power Electronics

Power electronic equipment in wind and photovoltaic (PV) power systems basically perform the following functions:

- Convert alternating current (AC) to direct current (DC)
- Convert DC to AC
- Control voltage
- Control frequency
- Convert DC to DC

These functions are performed by solid-state semiconductor devices periodically switched on and off at a desired frequency. Device costs have declined to less than a tenth of those three decades ago, fueling an exponential growth in applications throughout the power industry. No other technology has brought about a greater change in power engineering or holds a greater potential for bringing improvements in the future, than power electronic devices and circuits. In this chapter, we review the power electronic equipment used in modern wind and PV power systems.

14.1 BASIC SWITCHING DEVICES

Among a great variety of solid-state devices available in the market, some commonly used devices are as follows:

- Bipolar junction transistor (BJT)
- Metal-oxide semiconductor field-effect transistor (MOSFET)
- Insulated gate bipolar transistor (IGBT)
- Silicon controlled rectifier (SCR), also known as the thyristor
- Gate turn-off thyristor (GTO)

For a specific application, the choice of device depends on the power, voltage, current, and frequency requirement of the system. A common feature of these devices is that all are three-terminal devices, as shown in their (generally used) circuit symbols in Figure 14.1. The two power terminals 1 and 0 are connected in the main power circuit, and the third terminal G, known as the gate terminal, is connected to an auxiliary control circuit. In a normal conducting operation, terminal 1 is generally at a higher voltage than terminal 0.

Because these devices are primarily used for switching power on and off as required, they are functionally represented by a gate-controlled switch (shown in the last row in Figure 14.1). In the absence of a control signal at the gate, the device resistance between the power terminals is large—a functional equivalent of an open switch. When the control signal is applied at the gate, the device resistance approaches

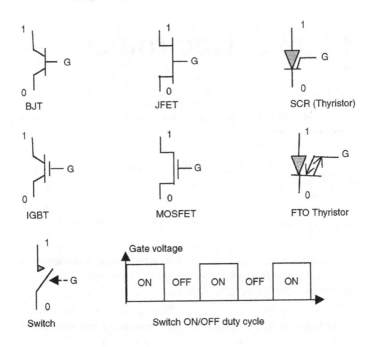

FIGURE 14.1 Basic semiconductor switching devices.

zero, making the device behave like a closed switch. The device in this state lets the current flow freely through it.

The characteristics of switching devices commonly available in the market are listed in Table 14.1. The maximum voltage and current ratings along with unique operating features of transistors and SCRs commonly used in high-power applications, such as in wind and PV power systems, are listed in Table 14.2. Thyristor technology has advanced dramatically into a variety of devices such as forced-commutated and line-commutated thyristors. GTO and static induction thyristors (SITHs) get turned on by a positive pulse to the thyristor gate and turned off by a negative pulse. They offer good forced-commutation techniques. GTOs are available in ratings of up to 4500 V/3000 A and SITHs up to 1200 V/300 A. Both have high flexibility and can be easily controlled. SITHs have a high-frequency switching capability greater than that of the GTOs, which are limited to about 10 kHz. The SCRs and IGBTs are limited to 100 Hz at present. IGBTs are less common but could give good control flexibility. In general, BJTs (1200 V/400 A) have lower power-handling capabilities than thyristors, but they have good control characteristics and high-frequency switching capabilities. MOSFETs are controlled by a gate voltage as opposed to other transistors controlled by a gate current and can be used at even higher switching frequencies but in low power ranges.

Power diode ratings can be as high as 5000 V/5000 A, with leakage currents in the off state reaching up to 100 A.

MOS controlled thyristor (MCT), reverse conducting thyristor (RCT), gate-assisted turn-off thyristor (GATT), and light-activated silicon controlled rectifier (LASCR) are other example of specialty devices, each having its niche application.

TABLE 14.1
Characteristics of Power Electronic Semiconductor Devices

Type	Function	Voltage	Current	Upper Frequency (kHz)	Switching Time (μsec)	On-State Resistance (mΩ)
Diode	General purpose	5000	5000	1	100	0.1–0.2
	High speed	3000	1000	10	2–5	1
	Schottky	<100	<100	20	0.25	10
Forced-turned-off thyristor	Reverse blocking	5000	5000	1	200	0.25
	High speed	1200	1500	10	20	0.50
	Reverse blocking	2500	400	5	40	2
	Reverse conducting	2500	1000	5	40	2
	GATT	1200	400	20	8	2
	Light triggered	6000	1500	1	200–400	0.5
TRIAC	—	1200	300	1	200–400	3–4
Self-turned-off thyristor	GTO	4500	3000	10	15	2–3
	SITH	1200	300	100	1	1–2
Power transistor	Single	400	250	20	10	5
		400	40	20	5	30
		600	50	25	2	15
	Darlington	1200	400	10	30	10
Power MOSFET	Single	500	10	100	1	1
		1000	5	100	1	2
		500	50	100	1	0.5
IGBT	Single	1200	400	100	2	60
MCT	Single	600	60	100	2	20

TABLE 14.2
Maximum Voltage and Current Ratings of Power Electronic Switching Devices

Device	Voltage Rating (volts)	Current Rating (amperes)	Operating Features
BJT	1500	400	Requires large current signal to turn on
IGBT	1200	400	Combines the advantages of BJT, MOSFET, and GTO
MOSFET	1000	100	Higher switching speed
SCR	6000	3000	Once turned on, requires heavy turn-off circuit

Switching transistors of any type in power electronic equipment are triggered periodically on and off by a train of gate signals of suitable frequency. The gate signal may be of rectangular or any other wave shape and is generated by a separate triggering circuit, which is often called the *firing circuit*. Although the firing circuit has a distinct identity and many different design features, it is generally incorporated in the main component assembly.

The output frequency and voltage of a variable-speed wind power generator vary with the wind speed. The output is converted into a fixed voltage at 60 or 50 Hz to match with the utility requirement. In modern plants, this is accomplished by the power electronics scheme shown in Figure 14.2. The variable frequency is first rectified into a DC, which is then inverted back into a fixed-frequency AC. The increase in energy production from the variable-speed wind turbine over the lifetime of the plant more than offsets the added cost of the power electronic equipment.

In a PV power system, the DC power output of the PV array is inverted into 60 or 50 Hz AC using the inverter. The inverter circuit in the PV system is essentially the same as that used in the variable-speed wind power system.

The main power electronic components of the wind and PV power systems are, therefore, the rectifier and the inverter. Their circuits and the AC and DC voltage and current relationships are presented in the following sections.

14.2 AC–DC RECTIFIER

The circuit diagram of the full-bridge, three-phase, AC–DC rectifier is shown in Figure 14.3. The power switch generally used in the rectifier is the SCR. The average DC output voltage is given by:

$$V_{dc} = \frac{3\sqrt{2}}{\pi} V_L \cos \alpha \qquad (14.1)$$

where
V_L = line-to-line voltage on the three-phase AC side of the rectifier and
α = angle of firing delay in the switching.

The delay angle is measured from the zero crossing in the positive half of the AC voltage wave. Equation 14.1 shows that the output DC voltage can be controlled by varying the delay angle α, which in turn controls the conduction (on-time) of the switch.

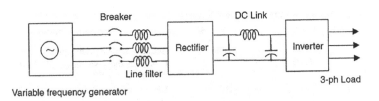

FIGURE 14.2 Variable-speed constant-frequency wind power system schematic.

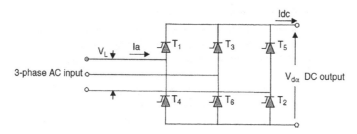

FIGURE 14.3 Three-phase, full-bridge, AC–DC controlled rectifier circuit.

Superimposed on the DC voltage at the rectifier output are high-frequency AC harmonics (ripples). A harmonic filter is, therefore, needed to reduce the AC component of the output voltage and increase the DC component. An L–C filter does this with an inductor connected in series and a capacitor in parallel with the rectified output voltage.

The load determines the DC-side current as:

$$I_{DC} = \frac{DC\,load\,power}{V_{DC}} \tag{14.2}$$

In steady-state operation, the balance of power must be maintained on both AC and DC sides. That is, the power on the AC side must be equal to the sum of the DC load power and the losses in the rectifier circuit. The AC-side power is therefore:

$$P_{AC} = \frac{DC\,load\,power}{Rectifier\,efficiency} \tag{14.3}$$

Moreover, the three-phase AC power is given by

$$P_{AC} = \sqrt{3} V_L L_L \cos\phi \tag{14.4}$$

where $\cos\Phi$ is the power factor on the AC side. With a well-designed power electronic converter, the power factor on the AC side is approximately equal to that of the load.

From Equation 14.3 with Equation 14.4, we obtain the AC-side line current I_L.

14.3 DC–AC INVERTER

The power electronic circuit used to convert DC into AC is known as the inverter. The term "converter" is often used to mean either the rectifier or the inverter. The DC input to the inverter can be from any of the following sources:

- Variable-speed wind power system (rectified DC output)
- PV power modules
- Battery used in the wind or PV power system

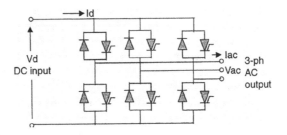

FIGURE 14.4 DC-to-three-phase AC inverter circuit.

Figure 14.4 shows the DC-to-three-phase AC inverter circuit diagram. The DC source current is switched successively in a 60-Hz three-phase time sequence so as to power the three-phase load. The AC current contains significant harmonics, as discussed in Section 14.7. The fundamental frequency (60 or 50 Hz) phase-to-neutral voltage is as follows:

$$V_{ph} = \frac{2\sqrt{2}}{\pi} \cos\left(\frac{\pi}{6}\right) V_{dc} \qquad (14.5)$$

The line-to-line AC voltage is given by $\sqrt{3} \cdot V_{ph}$.

Unlike in BJT, MOSFET, and IGBT, the thyristor current once switched on must be forcefully switched off (commutated) to terminate conduction. If the thyristor is used as a switching device, the circuit must incorporate an additional commutating circuit to perform this function. The commutating circuit is a significant part of the inverter circuit. There are two main types of inverters, the line-commutated and the forced-commutated.

The line-commutated inverter must be connected to the AC system they feed power to. The design method is well developed and has been extensively used in high-voltage DC (HVDC) transmission line inverters. This inverter is simple and inexpensive, and can be designed for any size. The disadvantage is that it acts as a sink for reactive power and generates high content of high-frequency harmonics. Therefore, its output needs a heavy-duty harmonic filter to improve the quality of power at the AC output. This is done by an inductor connected in series and a capacitor in parallel to the inverted output voltage, similar to that done in the rectification process.

The poor power factor and high harmonic content in the line-commutated inverter significantly degrade the quality of power at the grid interface. This problem has been recently addressed by a series of design changes. Among them are the 12-pulse inverter circuit and increased harmonic filtering. These new design features have resulted in today's inverter operating with near-unity power factor and less than 3 to 5% total harmonic distortion. The quality of power at the utility interface at many modern wind power plants now exceeds that of the grid they interface.

The forced-commutated inverter does not have to supply load and can be free-running as an independent voltage source. The design is relatively complex and expensive. The advantage is that it can be a source of reactive power and the harmonic content is low.

Among the inverters commonly used for high-power applications, the 12-pulse line-commutated bridge topology prevails. However, with the advent of GTOs and high-power IGBTs, the voltage source inverter with shunt capacitors in the DC link is emerging as a preferred topology.

There are three basic approaches to inverter design:

- One inverter inverts all DC power (central inverter).
- Each string or multistring unit has its own inverter (string inverter).
- Each module has a built-in inverter (module inverter).

Economies of scale would dictate the most economical design for a given system. Present inverter prices are about $1500/kW for ratings below 1 kW, $1000/kW for 1 to 10 kW, $600/kW for 10 to 100 kW, and $400/kW for ratings near 1000 kW. The DC–AC efficiency with the transformer at full load is typically 85 to 90% in small ratings and 92 to 95% in large ratings.

Most PV inverters incorporate active and/or passive islanding protection. Islanding, in which a section of the PV system is disconnected from the grid and still supplies the local loads, is undesirable for personnel safety and quality of power, and because of the possibility of equipment damage in the event of automatic or manual reclosure of the power island with the grid. With islanding, an electrical generating plant essentially operates without an external voltage and frequency reference. Operating in parallel is the opposite of islanding. Islanding prevention is therefore included in the inverter design specification. A grid computer could offer an inexpensive and efficient means for participants to cooperate in reliable operation.

The inverter is a key component of the grid-connected PV system. In addition to high efficiency in DC–AC conversion and peak-power tracking, it must have low harmonic distortion, low electromagnetic interference (EMI), and high-power factor. Inverter performance and testing standards are IEEE 929-2000 and UL 1741 in the U.S., EN 61727 in the EU, and IEC 60364-7-712 (the international standards). The total harmonic distortion (THD) generated by the inverter is regulated by international standard IEC-61000-3-2. It requires that the full-load current THD be less than 5% and the voltage THD be less than 2% for the harmonic spectra up to the 49th harmonic. At partial loads, the THD is usually much higher.

14.4 IGBT/MOSFET-BASED CONVERTERS

The power electronics converters based on IGBT/MOSFET switches provide a full control over electrical variables. The IGBT/MOSFET-based inverters, which are sometimes referred to as switch-mode inverters, convert DC voltage to a single-, three- or multi-phase square-wave or sinusoidal voltage with controllable magnitude, frequency, and phase angle. These converters are used in a variety of applications such as wind turbines, solar power systems, and electric vehicles. The output of the IGBT/MOSFET-based converters are pulsed voltages and currents. Therefore, filters such as LCL, LC, or active filters are required to eliminate the high order harmonics and extract the sinewave from the voltage and current before they are processed to the loads. A desired IGBT/MOSFET-based converter has bi-directional power flow, it has a high-quality output waveform and it has high fidelity.

There are two main types of these converters:

- Voltage source converter, and
- Current source converter.

In the voltage source converter (VSC) the input is a DC voltage, either from a large capacitor, or a DC source such as a battery system or a rectified voltage from an AC source. The voltage source converters are the most common power electronics converter found in a variety of applications. Figure 14.5(a) shows a schematic of a voltage source converter.

In the current source converters (CSC) the input is a DC current. These converters are less popular, with their main application being active power filters and the applications where long cables are used. Figure 14.5(b) shows a schematic of a current source converter with a large inductor to hold a constant current.

Unlike the voltage source converters where the free-wheeling anti-parallel have used the didoes in the current source converter in series with the active switches to

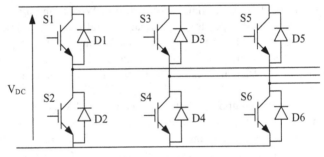

(a) Voltage source converter

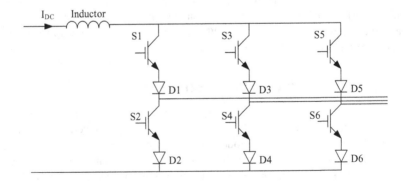

(b) Current source converter

FIGURE 14.5 Power electronics IGBT/MOSFET-based converters. (a) Voltage source converter. (b) Current source converter.

make up for the lack of reverse voltage withstand capability of switches. Note in both voltage and current source converters the output voltage and current are sinusoidal after filtering.

14.5 CONTROL SCHEMES

Different control schemes may be applied to power electronics converters depending on the application, cost, and other factors. A pulse width modulation (PWM) may be used to control the IGBT/MOSFET switches. In the PWM approach a DC voltage is applied to the converter input and the local average, i.e., average per switching period, of the output voltage (1-ph, 3-ph, or multi-ph) is controlled by modulation of width of the pulses applied to each IGBT/MOSFET. Different PWM techniques such as sinusoidal PWM (SPWM) and space vector PWM (SVM) have been studied extensively [4–17].

14.5.1 SPWM

In the SPWM, which is the most common PWM technique, the local average of the output voltage has a sinusoidal waveform (also called fundamental component) whose magnitude, phase, and frequency are controlled by pulse width modulation of the converter switches. The higher the switching frequency, the higher the quality of the resulting waveforms, and the smaller the filter capacitors and inductors. However, the higher the switching frequency, the higher the switching losses. High-switching frequency PWM is not appropriate for high-power applications. Figure 14.6 shows output voltage of a converter with PWM modulation. As seen the output voltage is pulsed before filtering.

In SPWM a sinusoidal control signal (modulating signal) is compared with a triangular carrier signal of constant amplitude and frequency to generate the switch control signals. The fundamental component of the output voltage is controlled and is proportional to the control signal. The information about the fundamental component of the output voltage is embedded (modulated) in the widths of the output voltage pulses. The demodulation takes place in the output low-pass filter, where the switching harmonics are separated from the fundamental component, or in the inductive load, where the pulsed voltage waveform is transformed to a sinusoidal current at the fundamental frequency.

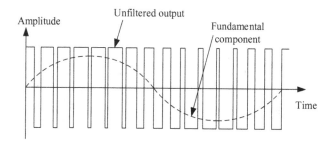

FIGURE 14.6 Converter output voltage with PWM modulation.

An amplitude modulation index (or ratio) may be calculated as follows:

$$m_a = \frac{\hat{v}_{control}}{\hat{v}_{tri}} \quad (14.6)$$

where, $\hat{v}_{control}$ is the amplitude of control signal, and $\hat{v}_{tri}$ is the amplitude of triangular signal. A frequency modulation index is calculated as follows:

$$m_f = \frac{f_s}{f_1} \quad (14.7)$$

where, f_s is the frequency of triangular signal or switching frequency, and f_1 is the frequency of control signal, and fundamental frequency of output voltage.

Based on the comparison of control voltage with the triangular voltage, the switch control signals and consequently the output voltage are generated. Generation of the output voltage is independent of the load current. Figure 14.7 compares the control and triangular voltages and shows the output voltage of converter. The comparison is performed such that where the control voltage is larger than the triangular signal the output pulse is at its saturated maximum, otherwise the output voltage is at its saturated minimum. This results in a train of pulses with different width as output voltage as shown in Figure 14.7. The output voltage waveform shows that local average of output voltage is a sinusoidal waveform proportional to the control voltage. Due to the linearity of PWM process, the ratio of the peak value of the fundamental component of the output voltage to the input DC voltage is equal to the modulation

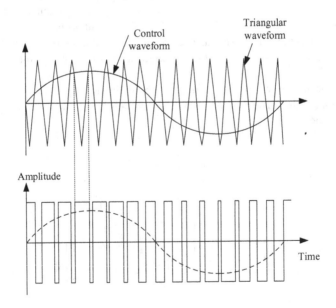

FIGURE 14.7 Generation of PWM and output voltage.

index. If the frequency modulation ratio is an odd number, the resulting output voltage waveform has half-wave symmetry and, as a result, does not contain even harmonics.

14.5.2 SQUARE WAVE

In the square-wave method, the output voltage has a square waveform whose the magnitude of the fundamental component is not controllable. To be able to control the magnitude of the fundamental component of the output voltage, the input DC voltage needs to be controlled. The inverter switching controls only the frequency and the phase angle of the output voltage. The switching frequency is minimal, i.e., each switch turns on and off once per period of the output voltage. Due to low switching frequency, and thus low switching losses, this scheme is appropriate for high-power applications. The square wave control is also achieved when the amplitude modulation index in the SPWM is sufficiency greater than 1, i.e. $m_a \gg 1$.

If the modulation index in the SPWM scheme is less than 1 ($m_a < 1$) the peak value of output voltage varies linearly with the modulation index. If the modulation index is greater than 1 ($m_a > 1$) the linearity is lost and the peak value of output voltage depends on the frequency modulation index, m_f. The PWM with a $m_a > 1$ is commonly referred to as over-modulation, where the output voltage waveform of the inverter degenerates from pulse-width modulated to square-wave, however, it has higher amplitude than the PWM case. Figure 14.8 shows the overmodulated, i.e., square-wave output voltage.

14.6 MULTILEVEL CONVERTERS

The solid-state switches, such as IGBTs and MOSFETs, that are building blocks of power electronics converters are rated at specific voltages, and currents. For high current applications, a number of switches are connected in parallel, while to use the switches in a converter beyond their ratings they are connected in series. This results in an AC voltage being synthesized from several levels of converter DC input voltage, which is referred to as a multilevel converter. In voltage source converter multilevel topologies capacitors on the DC link are used to create multilevel voltages. Figure 14.9 shows a schematic of a 3-level converter that uses two capacitors

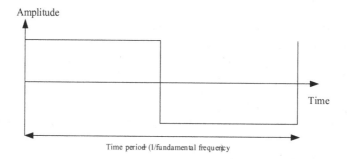

FIGURE 14.8 Output voltage in square wave method (over-modulation).

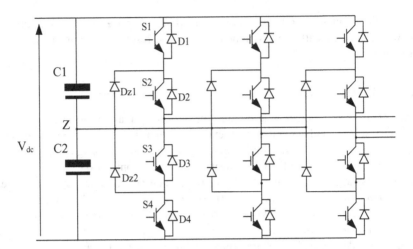

FIGURE 14.9 Three-level voltage source converter referred to as neutral point clamped (NPC).

connected in series to achieve required voltage levels. The midpoint of the capacitors provides a neutral point. In this scheme, commonly referred to as neutral point clamped (NPC), there are four switches connected in series with two diodes clamping the mid-point. The main challenge with these topologies is maintaining a balanced voltage between different levels. Voltage clamping and capacitor charge control are two common methods to balance the voltage between levels. Other multilevel topologies such as flying capacitors and cascade inverters with isolated DC sources have also been proposed [8–9].

For the three-level topology shown in Figure 14.9 each capacitor voltage under-balanced condition is half of the input DC voltage. If the upper switches S1 and S2 are turned ON the phase-A voltage with respect to the neutral point Z is ½ Vdc, while when the two bottom switches, S3 and S4, are turned ON the phase voltage is –½ Vdc. If the switches S2 and S3 are ON the phase voltage is clamped to zero, and one of the didoes is on depending on the direction of the load current. Note that switches S1 and S3 operate in the opposite manner, therefore if one is ON the other one must be OFF. Similarly, switches S2 and S4 operate the same way. Note, a deadtime is required between the complementary switches.

By connecting more switches in series higher level topologies can be achieved. Figure 14.10 depicts a diode clamped converter. In this topology, there are six switches and four diodes per converter leg. There are three capacitors connected in series, each having a 1/3 Vdc in balanced condition. When the three top switches, S1, S2, and S3 are turned ON the voltage across phase-A equals to Vdc, while when the three bottom switches are ON the voltage across the phase is zero. If the phase-A terminal is connected to either Z1 or Z2 via conduction of three middle switches and diodes, the phase voltage is either 2/3 Vdc or 1/3 Vdc. Note, the phase voltage in the four-level converter has four levels, i.e., Vdc, 2/3 Vdc, 1/3 Vdc, and 0. The switch pairs (S1, S1'), (S2, S2'), and (S3, S3') operate complementary, i.e., when one if ON the other one is OFF and vice versa.

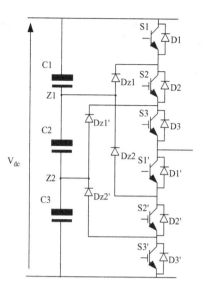

FIGURE 14.10 Four level voltage source converter referred to as neutral point clamped (NPC).

In the multilevel topologies given the same voltage ratings for the switches, the converter output voltage and hence power is proportional to the number of switches. However, as the voltage levels increase the number of clamping diodes is significantly increased. A three-level converter has six clamping diodes while a four-level converter has eighteen didoes. This together with the increased number of capacitors and voltage balancing challenges limits the application of higher voltage level converters.

14.7 HVDC CONVERTERS

The use of high-voltage DC (HVDC) systems for transmitting power has increased in the past decade with the offshore wind generation systems being a primary application for HVDC systems. In an HVDC offshore wind system the power from each turbine is sent to an offshore substation, where the aggregated power is sent to the shore using an HVDC transmission system. On each side of the HVDC transmission systems there exist a converter station that converts AC to DC or DC to AC in a controlled manner. Due to the high current and voltage these converters are constructed by series and parallel connection of several modules to achieve the required voltage and power level. A typical HVDC transmission system is in the range of 150 kV–800 kV DC, with a processing power in the order of several hundred Mega Watts or Giga Watts. The input to the converter station is a 3-phase system from the collector grid, however, a step-up transformer is usually used to further step-up the voltage to suitable levels for transmission.

In the early stages of high-voltage HVDC converters the thyristor-based line-commutated converters (LCC) were introduced. These LCC converters are to-date used in many of the HVDC applications as they are a cost-effective solution for very high voltages and powers with a relatively low loss. However, the LCC converters are not fully controllable and rely on the line current, and a constant feed of voltage for commutation. Additionally, they consume a variable amount of reactive power and require a reactive power control compensation. Over the years different approaches have been applied to improve the performance of the LCC converters, these include use of two six-pulse converters in parallel with a phase shift, 12-pulse and 18-pulse converters, and different control schemes. Figure 14.11 shows schematic of a typical 12-pulse HVDC LCC converter using a phase-shift transformer. There is a 30° phase shift between the two sets of 3-phase AC. This arrangement significantly reduces the harmonics content generated by the converter.

Voltage source converters (VSC) which are fully controllable have been used in modular and multilevel configuration for HVDC applications. Building blocks of these VSC converters are usually high-voltage IGBTs modules placed together to form a valve. As there are several IGBTs connected in series in these topologies they need to be switched simultaneously, which requires detailed considerations for gate drivers. Compared to LCC converters the VSCs are capable of independent active and reactive power flow control, hence no need for reactive power compensation. Another difference from LCC is the ability of VSC to control direction of power by reversing the current while in LCC the voltage polarity is required to be switched. The LCC converters are less expensive but bulkier due to the large passive filtering requirements, while the VSCs generate higher frequency harmonics which are easier to filter.

Multi-level modular converters (MMC) have been introduced for the HVDC systems in the past decade. An MMC consists of a number of identical submodules that are controlled independent from one another. The submodules can be, (i) two-level VSC half bridge, and (ii) three-level VSC full bridge. The half-bridge configuration

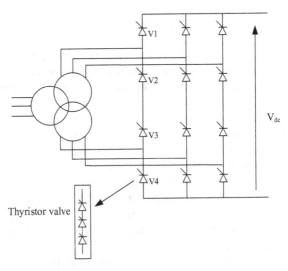

FIGURE 14.11 A 12-pulse HVDC LCC converter.

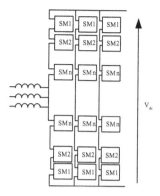

FIGURE 14.12 An MMC Converter.

can generate a voltage of +V and 0, while the full bridge generates ±V and 0. Therefore, the MMC is effectively a VSC with a desired voltage level depending on the number of submodules. The MMC converters have the capability to be scaled, an important factor in HVDC systems, and the issue of simultaneous switching does not exist as the submodules are switched independent. The effective switching frequency of the MMC converter is combination of submodules switching, hence can reach high frequencies giving a low harmonics content. Figure 14.12 shows structure of an MMC converter. As seen, there are no DC-link capacitors, however, each submodule has its own capacitor.

14.8 MATRIX CONVERTERS

A matrix converter performs a controlled direct AC-to-AC conversion using active power electronics active switches. To convert an AC to another controlled AC, traditionally an AC-to-DC conversion using rectifier, and then a DC-to-AC conversion using inverter may be applied. However, the matrix converter offers a direct conversion, although usually with higher number of active switches.

Schematic of a 3-phase AC-AC matrix converter is shown in Figure 14.13. The converter has nine switches, that allow each input phase to be connected to an output phase. The converter also requires a 3-phase input capacitor and an output inductive filter. Depending on the load the output inductive filter may not be required. The 3-phase input may be the grid while the output may be a 3-phase load such as an electric motor. Due to its topology a matrix converter is inherently bi-directional, hence allowing the flow of the current from input-to-output and vice versa. In a matric converter at each time instant only one switch per phase may be switched, hence there exist 27 different switching combination.

In a matrix converter, the output voltage is synthesized by a sequential piecewise sampling of the input voltage, where the sampling frequency is higher than the input voltage frequency. The duty cycle, i.e. sampling period, is controlled such that a desired output voltage is obtained. Similarly, the input currents in a matrix converter are synthesized from the output currents.

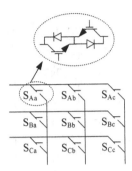

FIGURE 14.13 A 3-phase matric converter.

14.9 CYCLOCONVERTER

The cycloconverter, a direct frequency changer, converts AC power of one frequency to AC power of another frequency by AC–AC conversion. By adding control circuitry capable of measuring the voltage and frequency of the bus bar and by comparing them with the desired voltage and frequency, we can obtain a constant voltage and frequency output. The cycloconverter used to be more economical than the DC-link inverter but has been gradually replaced by the rectifier–inverter setup because of its high harmonic content and the decreasing price of power semiconductors. However, advances in fast switching devices and control microprocessors are increasing the efficiency and power quality of the cycloconverter.

Large three-phase cycloconverters use GTOs as semiconductors. The same thyristor commutating techniques are used in the cycloconverter as in the AC–DC rectifier. The cycloconverter uses one SCR for positive wave voltages and another for negative wave voltages. The firing angle of the thyristors controls the AC root mean square (rms) voltage, and the firing frequency controls the output frequency. However, the frequency control is complex.

14.10 GRID INTERFACE CONTROLS

At the utility interface, the power flow direction and magnitude depend on the voltage magnitude and the phase relation of the site voltage with respect to the grid voltage. The grid voltage being fixed, the site voltage must be controlled both in magnitude and phase in order to feed power to the grid when available and to draw from the grid when needed. If the inverter is already included in the system for frequency conversion, the magnitude and phase control of the site voltage is done with the same inverter at no additional hardware cost. The controls are accomplished as described in the following subsections.

14.10.1 Voltage Control

For interfacing with the utility grid lines, the renewable power system output voltage at the inverter terminals must be adjustable. The voltage is controlled by using one of the following two methods:

The first is controlling the alternating voltage output of the inverter using a tap-changing autotransformer at the inverter output. The tap changing is automatically obtained in a closed-loop control system. If the transformer has a phase-changing winding also, complete control of the magnitude and phase of the site voltage can be achieved. The advantages of this scheme are that the wave shape of the site output voltage does not vary over a wide range, and a high input power factor is achieved by using uncontrolled diode rectifiers for the DC-link voltage. The added cost of the transformer, however, can be avoided by using the following method.

Because the magnitude of the AC voltage output from the static inverter is proportional to the DC voltage input from the rectifier, voltage control can be achieved by operating the inverter with a variable DC-link voltage. Such a system also maintains the same output voltage, frequency, and wave shape over a wide range. However, in circuits deriving the load current from the commutating capacitor voltage of the DC link, the commutating capability decreases when the output voltage is reduced. This could lead to an operational difficulty when the DC-link voltage varies over a wide range, such as in a motor drive controlling the speed in a ratio exceeding 4:1. In renewable power applications, such a commutation difficulty is unlikely because the speed varies over a narrow range.

The variable DC-link voltage is obtained in two ways. One is to connect a variable-ratio transformer on the input side of the rectifier. The secondary tap changing is automatically obtained in a closed-loop control system. The other way is to use the phase-controlled rectifier in place of the uncontrolled rectifier in Figure 14.14. At reduced output voltages, this method gives a poor power factor and high harmonic content and requires filtering of the DC voltage before feeding to the inverter.

14.10.2 Frequency Control

The output frequency of the inverter solely depends on the rate at which the switching thyristors or transistors are triggered into conduction. The triggering rate is determined by the reference oscillator producing a continuous train of timing pulses, which are directed by logic circuits to the thyristor gating circuits. The timing-pulse train is also used to control the turn-off circuits. The frequency stability and accuracy

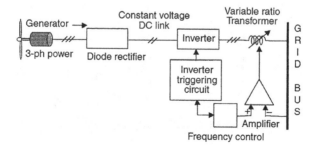

FIGURE 14.14 Voltage control by means of uncontrolled rectifier and variable-ratio tap-changing transformer.

requirements of the inverter dictate the selection of the reference oscillator. A simple temperature-compensated R–C relaxation oscillator gives frequency stability within 0.02%. When better stability is needed, a crystal-controlled oscillator and digital counters may be used, which can provide stability of .001% or better. Frequency control in a stand-alone power system is an open-loop system. Steady-state or transient load changes do not affect the frequency. This is a major advantage of the power electronic inverter over the old electromechanical means of frequency control.

14.11 BATTERY CHARGE/DISCHARGE CONVERTERS

The stand-alone PV power system uses the DC–DC converter for battery charging and discharging. DC–DC conversion techniques have undergone rapid development in recent decades. Over 500 different topologies currently exist, with more being developed each year. However, we will briefly cover a few basic topologies, leaving the details to advanced books dedicated to the subject.

14.11.1 Battery Charge Converter

Figure 14.15 is the most widely used DC–DC battery charge converter circuit, also called the *buck converter*. The switching device used in such a converter may be a BJT, MOSFET, or IGBT. The buck converter steps down the input bus voltage to the battery voltage during charging. The transistor switch is turned on and off at a high frequency (in tens of kHz). The duty ratio D of the switch is defined as:

$$D = \frac{\text{On-time}}{\text{Period}} = \frac{T_{on}}{T} = T_{on} \times \text{Switching frequency} \qquad (14.8)$$

The operation of a charge converter during one complete cycle of the triggering signal is shown in Figure 14.16. During the on-time, the switch is closed and the circuit operates as in Figure 14.16(a). The DC source charges the capacitor and supplies power to the load via an inductor. During the off-time, the switch is opened and the circuit operates as in Figure 14.16(b). The power drawn from the DC source is zero.

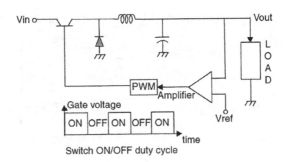

FIGURE 14.15 Battery charge converter for PV systems (DC–DC buck converter).

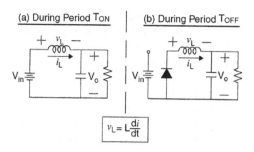

FIGURE 14.16 Charge converter operation during switch on-time and off-time.

However, full-load power is supplied by the energy stored in the inductor and the capacitor, with the diode providing the return circuit. Thus, the inductor and the capacitor provide the required energy storage to ride through the off-period of the switch. A simple analysis of this circuit is carried out in the following text. It illustrates the basic methodology of analyzing all power electronic circuits of this nature. It is based on the energy balance over one period of the switching signal as follows:

Energy supplied to the load over one complete period $(i.e., T_{on} + T_{off}) =$
  Energy drawn from the source during the on-time.
Energy supplied to the load during the off-time =
  Energy drawn from the inductor and capacitor during off-time.

Alternatively, the volt-second balance method is used. In essence, it gives the energy balance stated in the preceding text.

The voltage and current waveforms over one complete period are displayed in Figure 14.17. In the steady-state condition, the inductor volt-second balance during the on- and off-period must be maintained. Because the voltage across the inductor must equal $L\, dI_L/dt$,

$$\text{During on-time, } \Delta I_L L = (V_{in} - V_{out}) T_{on} \tag{14.9}$$

$$\text{and during off-time, } \Delta I_L L = V_{out} T_{off} \tag{14.10}$$

If the inductor is large enough, as is usually the case in a practical design, the change in the inductor current is small, and the peak value of the inductor current is given by the following:

$$I_{peak} = I_o + \tfrac{1}{2} \Delta I_L \tag{14.11}$$

where the load current $I_o(V_{out}/R_{load})$ is the average value of the inductor current.

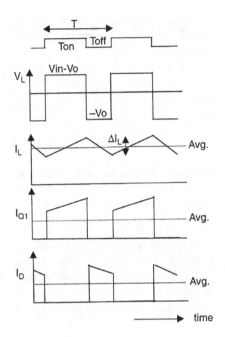

FIGURE 14.17 Current and voltage waveforms in the buck converter.

The algebraic manipulation of the preceding equations leads to:

$$V_{out} = V_{in} D \qquad (14.12)$$

It is seen from Equation 14.12 that the output voltage is controlled by varying the duty ratio D. This is done in a feedback control loop with the required battery charge current as the reference. The duty ratio is controlled by modulating the pulse width of T_{on}. Such a converter is, therefore, also known as the *pulse-width-modulated (PWM) converter*.

14.11.2 BATTERY DISCHARGE CONVERTER

The battery discharge converter circuit is shown in Figure 14.18. It steps up the sagging battery voltage during discharge to the required output voltage. When the transistor switch is on, the inductor is connected to the DC source. When the switch is off, the inductor current is forced to flow through the diode and the load. The output voltage of the boost converter is derived again from the volt-second balance in the inductor. With a duty ratio D of the switch, the output voltage is given by the following:

$$V_{out} = \frac{V_{in}}{1-D} \qquad (14.13)$$

For all values of D less than 1, the output voltage is always greater than the input voltage. Therefore, the boost converter can only step up the voltage. On the other

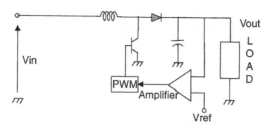

FIGURE 14.18 Battery discharge converter circuit for PV system (DC–DC boost converter).

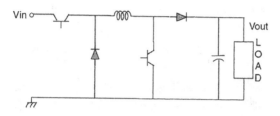

FIGURE 14.19 Buck–boost converter circuit (general DC–DC converter for PV systems).

hand, the buck converter presented in the preceding section can only step down the input voltage. Combining the two converters in a cascade, therefore, gives a buck–boost converter, which can step down or step up the input voltage. A modified buck–boost converter often used for this purpose is shown in Figure 14.19. The voltage relation is obtained by combining the buck–boost converter voltage relations:

$$V_{out} = \frac{V_{in}D}{1-D} \qquad (14.14)$$

Equation 14.12 for the buck–boost converter shows that the output voltage can be higher or lower than the input voltage, depending on the duty ratio D (Figure 14.20).

In addition to its use in battery charging and discharging, the buck–boost converter is capable of four-quadrant operation with a DC machine when used in the variable-speed wind power systems. The converter is used in the step-up mode during the generating operation and in the step-down mode during the motoring operation.

14.12 POWER SHUNTS

In a stand-alone PV system, the power generation in excess of the load and battery-charging requirements must be dissipated in a dump load in order to control the output bus voltage. The dump load may be resistance heaters. However, when heaters cannot be accommodated in the system operation due to thermal limitations, dissipation of excess power can pose a problem. In such situations, shorting (shunting) the

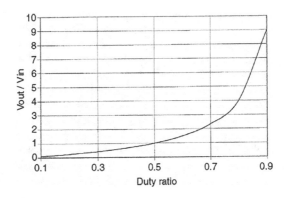

FIGURE 14.20 Buck–boost converter output-to-input ratio vs. duty ratio.

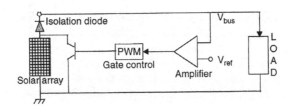

FIGURE 14.21 Power shunt circuit for shorting the module.

PV module to ground forces the module to operate under the short-circuit condition, delivering I_{sc} at zero voltage. No power is delivered to the load. The solar power remains on the PV array, raising the array temperature and ultimately dissipating the excess power in the surrounding air. The array area is essentially used here as a heat dissipater.

The circuit used for shunting the PV array is shown in Figure 14.21. A transistor is used as a switch. When the excess power is available, the bus voltage will rise above the rated value. This is taken as a signal to turn on the shunt circuits across the required number of PV modules. The shunt circuit is generally turned on or off by a switch controlled by the bus voltage reference. An electromechanical relay can perform this function, but the moving contacts have a much shorter life than a transistor switch. Therefore, such a relay is seldom used except in a small system.

Another application of the power shunt circuit is in a PV module dedicated to directly charging the battery without a charge regulator. When the battery is fully charged, the module is shunted to ground by shorting the switch. This way, the battery is always protected from overcharging.

For an array with several modules in parallel, the basic configuration shown in Figure 14.21 is used for each module separately, but the same gate signal is supplied to all modules simultaneously. For shunting large amounts of power, multiple shunt circuits can be switched on and off in sequence to minimize switching transients and EMI with neighboring equipment. For fine power control, one segment can be operated in the PWM mode.

Power electronics has a very wide scope and is a developing subject. The intent of this chapter is to present an overview of the basic circuits used in wind and PV power systems. Further details can be obtained from many excellent books on power electronics.[1-3]

REFERENCES

1. Bose, B., *Power Electronics and Variable Frequency Drives*, IEEE Press, New York, 1996.
2. Dewan, S.B. and Straughen, A., *Power Semiconductor Circuits*, John Wiley & Sons, New York, 1975.
3. Mohan, N. and Undeland, T., *Power Electronics Converters, Applications and Design*, IEEE and John Wiley & Sons, New York, 1995.
4. Omid Beik, An HVDC Off-shore Wind Generation Scheme with High Voltage Hybrid Generator, Ph.D. thesis, McMaster University, 2016.
5. Omid Beik, Ahmad S. Al-Adsani, *DC Wind Generation Systems, Design, Analysis, and Multiphase Turbine Technology*, Springer, 2020.
6. Omid Beik, Ahmad S. Al-Adsani, *Wind Energy Systems*, pp. 1–9, Springer, 2020.
7. Omid Beik, Ahmad S. Al-Adsani, *DC Wind Generation System*, pp. 33–69, Springer, 2020.
8. Beik, O., Dekka A., Narimani, M., 'A new modular neutral point clamped converter with space vector modulation control,' *Proc. IEEE Int. Conf. Ind. Tech.*, Lyon, France, 2018, pp. 591–595.
9. A. Dekka, O. Beik and M. Narimani, "Modulation and Voltage Balancing of a Five-Level Series-Connected Multilevel Inverter With Reduced Isolated Direct Current Sources," *IEEE Transactions on Industrial Electronics*, vol. 67, no. 10, pp. 8219–8230, October 2020.
10. O. Beik and A. S. Al-Adsani, "Parallel Nine-Phase Generator Control in a Medium-Voltage DC Wind System," *IEEE Transactions on Industrial Electronics*, vol. 67, no. 10, pp. 8112–8122, October 2020.
11. D. Schumacher, O. Beik and A. Emadi, "Standalone Integrated Power Electronics System: Applications for Off-Grid Rural Locations," *IEEE Electrification Magazine*, vol. 6, no. 4, pp. 73–82, December 2018.
12. O. Beik and N. Schofield, "High-Voltage Hybrid Generator and Conversion System for Wind Turbine Applications," *IEEE Transactions on Industrial Electronics*, vol. 65, no. 4, pp. 3220–3229, April 2018.
13. Beik, O., Schofield, N., 'High Voltage Generator for Wind Turbines', *8th IET International Conference on Power Electronics, Machines and Drives (PEMD 2016)*, 2016.
14. O. Beik and N. Schofield, "An Offshore Wind Generation Scheme With a High-Voltage Hybrid Generator, HVDC Interconnections, and Transmission," *IEEE Transactions on Power Delivery*, vol. 31, no. 2, pp. 867–877, April 2016.
15. O. Beik and N. Schofield, "*Hybrid Generator for Wind Generation Systems*," *2014 IEEE Energy Conversion Congress and Exposition (ECCE)*, Pittsburgh, PA, 2014, pp. 3886–3893.
16. M. R. Aghamohammadi, and A. Beik-Khormizi, "*Small Signal Stability Constrained Rescheduling using Sensitivities Analysis by Neural Network as a Preventive Tool*," in *Proc. 2010 IEEE PES Transmission and Distribution Conference and Exposition*, pp. 1–5.
17. A. B. Khormizi and A. S. Nia, "*Damping of Power System Oscillations in Multi-machine Power Systems Using Coordinate Design of PSS and TCSC*," in *2011 10th International Conference on Environment and Electrical Engineering (EEEIC)*, Rome. pp. 1–4

15 Stand-Alone Systems

The stand-alone power system is used primarily in remote areas where utility lines are uneconomical to install due to the terrain, right-of-way difficulties, or environmental concerns. Building new transmission lines is expensive even without these constraints. A 230-kV line costs more than $1 million per mile. A stand-alone wind system could be more economical for remote villages that are farther than a couple of miles from the nearest transmission line.

Solar and wind power outputs can fluctuate on an hourly or daily basis. The stand-alone system, therefore, must have some means of storing excess energy on a sunny day for use on a rainy day. Alternatively, wind, PV, or both can be used in a hybrid configuration with a diesel engine generator in remote areas or with a fuel cell in urban areas.

According to the World Bank, about 2 billion people still live in villages that are not yet connected to utility lines. These villages are the largest potential market for stand-alone hybrid systems using a diesel generator with wind or PV for meeting their energy needs. Additionally, wind and PV systems create more jobs per dollar invested than many other industries. This, on top of bringing much-needed electricity to rural areas, helps minimize migration to already strained cities in most countries.

Because power sources having significantly different performance characteristics must be used in parallel, the stand-alone hybrid system is technically more challenging and expensive to design than the grid-connected system that simply augments the existing utility network.

15.1 PV STAND-ALONE

The typical PV stand-alone system consists of a solar array and a battery connected as shown in Figure 15.1. The PV array supplies power to the load and charges the battery when there is sunlight. The battery powers the load otherwise. An inverter converts the DC power of the array and the battery into 60 or 50 Hz power. Inverters are available in a wide range of power ratings with efficiencies ranging from 85 to 95%. The array is segmented with isolation diodes for improving reliability. In such a design, if one string of the solar array fails, it does not load or short the remaining strings. Multiple inverters are preferred for reliability. For example, three inverters, each with a 35% rating, are preferred to one with a 105% rating. If one such inverter fails, the remaining two can continue supplying most loads until the failed one is repaired or replaced. The same design approach also extends to using multiple batteries.

Most stand-alone PV systems are installed in developing countries to provide basic necessities such as lighting and pumping water. Others go a step further (Figure 15.2).

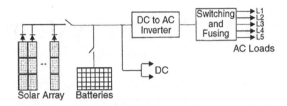

FIGURE 15.1 PV stand-alone power system with battery.

FIGURE 15.2 A traveling clinic uses photovoltaic electricity to keep vaccines refrigerated in the African desert area. (From Siemens Solar Industries. With permission.)

15.2 ELECTRIC VEHICLE

The solar electric car developed in the U.S. and in many other countries is an example of a stand-alone (or rather a "move-alone") PV power system. The first solar car was built in 1981 and driven across the Australian outback by Hans Tholstrup of Australia. A modern-day solar car has been developed and is commercially available, although it is more expensive than the conventional car at present. However, continuing development is closing the price gap every year.

A new sport at U.S. universities these days is the biennial solar car race. The Department of Energy (DOE) and several car manufacturers sponsor the race every 2 yr. It is open to all engineering and business students, who design, build, and run their cars across the heartland of America. The first U.S. solar car race was organized in 1990. Figure 15.3 shows one such car built at the University of Michigan in 1993. It finished first in the 1100-mi "Sunrayce" that started from Arlington, TX, and cruised through Oklahoma, Kansas, Missouri, and Iowa, and ended in Minnesota. It covered 1102 mi in six and a half days. In 1994, the same solar car finished the 1900-mi-long World Solar Challenge from Darwin to Adelaide across the Australian outback. Several dozens of teams participate in these races. Much more than just holding a race, the goal of this DOE program is to provide a hands-on, "minds-on"

Stand-Alone Systems 255

FIGURE 15.3 The University of Michigan solar car raced 1100 mi in the U.S. and 1900 mi in Australia. (From Patel, Ketan M., 1993 Sunrayce Team, University of Michigan.)

experience for young students in renewable power sources to display their creativity and use of the latest technology on wheels across the country. This biennial DOE-sponsored solar car race taps the bright young brains across the country.

The solar cars entering such university races generally have a wide range of design characteristics as listed in Table 15.1. The Michigan car was designed with silicon cells and a lead-acid battery, which was changed to a silver-zinc battery later on. With a permanent-magnet (PM) brushless DC motor, it reached a peak speed of 50 mph. The Gallium Arsenide PV cells are often used by many in the Australian race. A few race cars have been designed with a Stirling engine driven by helium heated by solar energy instead of using PV cells. The design considerations include trading off among hundreds of technical parameters covered in this book and meeting their constraints *vis-a-vis* a vehicle design. But certain key technical elements are essential for winning. They are as follows:

- PV cells with high conversion efficiency
- Peak-power-tracking design

TABLE 15.1
Design Characteristic Range of Solar Race Cars Built by U.S. University Students for the Biennial 1100-mi "Sunrayce"

Design Parameters	Parameter Range
Solar array power capability (silicon crystalline, gallium arsenide)	750–1500 W
Battery (lead-acid, silver-zinc)	3.5–7 kWh
Electric motor (DC PM brushless)	4–8 hp
Car weight	500–1000 lb
Car dimensions	≈20-ft long × 7-ft wide × 3.5-ft high

- Lightweight battery with high specific energy
- Energy-efficient battery charging and discharging
- Low aerodynamic drag
- High reliability without adding weight

The zinc-air battery discussed in Chapter 10 is an example of a lightweight battery. Chapter 9 covered the use of isolation diodes in a solar array for achieving reliability without adding weight, as well as the peak-power-tracking principle for extracting the maximum power out of the PV array for a given solar radiation.

After the car is designed and tested, developing a strategy to optimize solar energy capture and to use it efficiently, while maintaining the energy balance for the terrain and the weather on the day of the race, becomes the final test. The energy balance analysis methods for sizing the solar array and the battery are described later in this chapter.

15.3 WIND STAND-ALONE

A simple stand-alone wind system using a constant-speed generator is shown in Figure 15.4. It has many features that are similar to the PV stand-alone system. For a small wind system supplying local loads, a PM DC generator makes the system simple and easier to operate. The induction generator, on the other hand, gives AC power, which is used by most consumers these days. The generator is self-excited by shunt capacitors connected to the output terminals. The frequency is controlled by controlling the turbine speed. The battery is charged by an AC–DC rectifier and discharged through a DC–AC inverter.

The wind stand-alone power system is often used for powering farms (Figure 15.5). In Germany, nearly half the wind systems installed on farms under the "250 MW Wind" program were owned either by individual farmers or by an association. The performance of turbines was monitored and published by ISET, the Institute of Solar Energy and Technology at the University of Kassel.[1] The reports listed all installations, their performance, and any technical problems for analysis.

The steady-state performance of the electrical generator is determined by the theory and analyses presented in Chapter 5. In stand-alone wind systems, this includes determining the capacitor rating needed to self-excite the generator for the desired voltage and frequency. The power factor of load has a great effect on both the steady-state and the transient performance of the induction generator. The load power factor can be unity, lagging, or leading, depending on the load's being resistive, inductive, or capacitive, respectively. Most loads in the aggregate are inductive with a power factor of about 0.9 lagging. Unlike in the synchronous generator, the induction generator

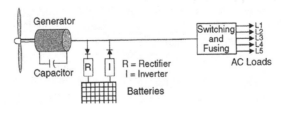

FIGURE 15.4 Stand-alone wind power system with battery.

FIGURE 15.5 Remote farms are a major market for the stand-alone power systems. (From World Power Technologies, Duluth, MN. With permission.)

output current and power factor for a given load are determined by the generator parameters. Therefore, when the induction generator delivers a certain load, it also supplies a certain in-phase current and a certain quadrature current. The quadrature current is supplied by the capacitor bank connected to the terminals. Therefore, the induction generator is not suitable for supplying a low-power-factor load.

The transient performance of the stand-alone, self-excited induction generator, on the other hand, is more involved. The generalized d-q axis model of the generator is required. Computer simulation using a d-q axis model shows the following general transient characteristics:[2]

- Under sudden loss of self-excitation due to tripping-off of the capacitor bank, the resistive and inductive loads cause the terminal voltage to quickly reach the steady-state zero. A capacitive load takes a longer time before the terminal voltage decays to zero.
- Under sudden loading of the generator, resistive and inductive loads result in a sudden voltage drop, whereas a capacitive load has little effect on the terminal voltage.
- Under sudden loss of resistive and inductive loads, the terminal voltage quickly rises to its steady-state value.
- At light load, the magnetizing reactance changes to its unsaturated value, which is large. This makes the machine performance unstable, resulting in terminal-voltage collapse. To remedy this instability problem, the standalone induction generator must always have a minimum load, a dummy if necessary, permanently connected to its terminals.

15.4 HYBRID SYSTEMS

15.4.1 Hybrid with Diesel

The certainty of meeting load demands at all times is greatly enhanced by hybrid systems, which use more than one power source. Most hybrids use a diesel generator

FIGURE 15.6 A 300-kW PV–diesel hybrid system in Superior Valley, California. (From ASE Americas, Inc., Billerica, MA. With permission.)

with PV or wind, because diesel provides more predictable power on demand. A battery is used in addition to the diesel generator in some hybrids. The battery meets the daily load fluctuation, and the diesel generator takes care of the long-term fluctuations. For example, the diesel generator is used in the worst-case weather condition, such as an extended period of overcast skies or when there is no wind for several weeks.

Figure 15.6 shows a large PV–diesel hybrid system installed in California. The project was part of the Environmental Protection Agency's PV-Diesel program. Figure 15.7 is a schematic layout of a wind–diesel–battery hybrid system. The power connection and control unit (PCCU) provides a central place to make organized connections of most system components. In addition, the PCCU houses the following components:

- Battery charges and discharge regulators
- Transfer switches and protection circuit breakers
- Power flow meters
- Mode controller

Figure 15.8 shows a commercially available PCCU for hybrid power systems. The transient analysis of the integrated wind–PV–diesel requires an extensive model that takes the necessary input data and event definitions for computer simulation.[3]

15.4.2 HYBRID WITH FUEL CELL

In stand-alone renewable power systems of hybrid design, the fuel cell has the potential to replace the diesel engine in urban areas, where the diesel engine is undesirable due to its high carbon emissions. The airborne emission of a fuel cell power plant is

Stand-Alone Systems

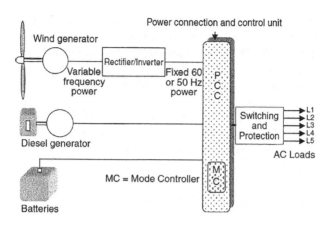

FIGURE 15.7 Wind–diesel–battery hybrid system.

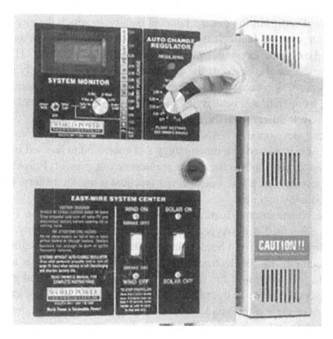

FIGURE 15.8 Integrated power connection and control unit for wind–PV–battery hybrid system. (From World Power Technologies, Duluth, MN. With permission.)

negligible, about 25 g per MWh delivered. Factories and hospitals have started replacing the diesel generator with the fuel cell in their uninterruptible power supply (UPS) systems. Electric utility companies are considering the use of the fuel cell for meeting peak demand and for load leveling between day and night, and during the week.

On a small scale, a fuel cell about the size of a dishwasher can power a 3000 to 5000 ft^2 house without a grid connection. The mass-production cost of such a unit could be $2000 to $3000. The first fuel cells for use in residential areas were installed in 80 locations in New York state around 2000. On a larger scale, New York Power Authority has installed 12 2.4-MW grid-connected fuel cell plants.

On a grander scale, Reykjavik, home to about 120,000 people in Iceland, plans to be the world's first hydrogen city. The first commercial hydrogen filling station opened in Iceland at an existing Shell retail station in Reykjavik. It dispenses hydrogen produced by electrolysis technology using renewable energy. The first to use this station were three Daimler-Chrysler hydrogen-powered buses placed in Reykjavik's mass transit in 2003. As proposed by the University of Iceland, the government hopes to transform the entire country into a hydrogen economy by 2050.

German car manufacturer Daimler-Benz and Ballard Power Systems of Canada have developed a solid polymer fuel cell for automobiles as an alternative to the battery. The first commercial fuel-cell-powered vehicle is already on the market.

The fuel cell is an electrochemical device that generates electricity by a chemical reaction that does not alter the electrodes and the electrolyte materials. This distinguishes the fuel cell from the electrochemical battery, in which the electrodes wear out. The concept of the fuel cell is the reverse of the electrolysis of water, in that hydrogen and oxygen are combined to produce electricity and water. Thus, the fuel cell is a static device that converts the chemical energy of fuel directly into electric energy. Because this process bypasses the thermal-to-mechanical conversion, and also because its operation is isothermal, the conversion efficiency is not Carnot limited. In this way, it differs from the diesel engine.

The fuel cell, developed by NASA as an intermediate-term power source for space applications, was first used in a moon buggy and continued to be used in the fleet of space shuttles. It also finds other niche applications at present. Providing electric power for a few days or a few weeks is not practical using the battery, but is effectively done with the fuel cell.

The basic constructional features of the fuel cell are shown in Figure 15.9. Hydrogen is combined with oxygen from the air to produce electricity. The hydrogen "fuel," however, does not burn as in the internal combustion engine; rather, it produces electricity by an electrochemical reaction. Water and heat are the only byproducts of this reaction if the fuel is pure hydrogen. With natural gas, ethanol, or methanol as the source of hydrogen, the byproducts include some carbon dioxide and traces of carbon monoxide, hydrocarbons, and nitrogen oxides. However, they are all less than 1% of that emitted by the diesel engine. The superior reliability, with no moving parts, is an additional benefit of the fuel cell as compared to the diesel generator. Multiple fuel cells stack up in a series–parallel combination for the required voltage and current, just as the electrochemical cells do in a battery.

The several types of fuel cells are as follows:

Phosphoric acid: This is the most established type of fuel cell for relatively small applications, such as in hospitals.

Alkaline: This was NASA's space shuttle fuel cell, too expensive for commercial use.

Stand-Alone Systems

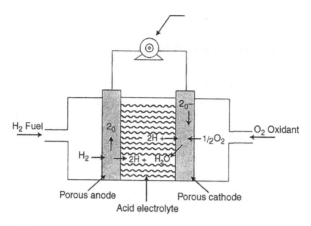

FIGURE 15.9 Fuel cell principle: hydrogen and oxygen in, electric power and water out.

Proton-exchange membrane: This is the most promising fuel cell for small-scale applications such as homes.

Molten carbonate: This fuel cell works at high temperature (so it can be used for heating as well) and has a high efficiency of around 80%. It uses zirconium oxide. Although expensive, it is the most likely fuel cell to be used in large-scale applications.

The operating characteristics of various fuel cell technologies under development at present are summarized in Table 15.2. Among the various alternative fuel cell types available, the optimum fuel cell for a specific application depends on the specific requirements, including the capital cost limitations.

Low-temperature (under 200°C) fuel cells are now commercially available from several sources. They use phosphoric acid as the electrolytic solution between the electrode plates. A typical low-temperature fuel cell with a peak-power rating of 200 kW costs under $1000 per kW at present, which is over twice the cost of the diesel engine. The fuel cell price, however, is falling with new developments being implemented every year.

High-temperature (over 1000°C) fuel cells yield a high electric energy per kilogram of fuel, but at a relatively high cost, limiting its use to special applications at present. Solid-oxide and molten carbonate fuel cells fall in this category. The industry interest in such cells is in large capacity for use in utility-scale power plants. The Fuel Cell Commercialization Group in the U.S. recently field-tested 2-MW molten carbonate direct fuel cells. The test results were a qualified success. Based on the results, multimegawatt commercial fuel cell power plants have been designed and installed in New York and other cities.

Solid-oxide fuel cells of several different designs, consisting of essentially similar materials for the electrolyte, the electrodes, and the interconnections, are being investigated worldwide. Most of the success to date has been achieved with the tubular geometry being developed by Westinghouse Electric in the U.S. and Mitsubishi Heavy Industries in Japan. The cell element in this geometry consists of two porous electrodes separated by a dense oxygen ion-conducting electrolyte as depicted in

TABLE 15.2
Characteristics of Various Fuel Cells

Fuel Cell Type	Working Temperature °C	Power Density W/cm²	Projected Life Hours
PAFC	150–200	0.2–0.25	40,000
AFC	60–100	0.2–0.3	10,000
MCFC	600–700	0.1–0.2	40,000
SOFC	900–1000	0.25–0.3	40,000
SPFC	50–100	0.35–0.6	40,000
DMFC	50–100	0.05–0.25	10,000

Note: PAFC = Phosphoric acid fuel cell; AFC = Alkaline fuel cell; MCFC = Molten carbonate fuel cell; SOFC = Solid-oxide fuel cell; SPFC = Solid-polymer fuel cell (also known as proton-exchange membrane fuel cell); and DMFC = Direct methanol fuel cell.
Source: From *IEEE Power and Energy*, November–December 2004, pp. 27–28.

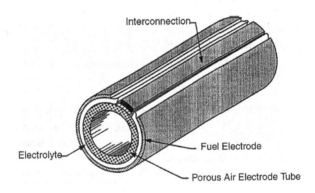

FIGURE 15.10 Air-electrode-supported-type tubular solid-oxide fuel cell design. (From Westinghouse Electric Co., a division of CBS Corp., Pittsburgh, PA. Reprinted with permission.)

Figure 15.10.[4] It uses a ceramic tube operating at 1000°C. The fuel cell is an assembly of such tubes. SureCELL™ (Trademark of Westinghouse Electric Corporation, Pittsburgh, PA) is a solid-oxide high-temperature tubular fuel cell and is shown in Figure 15.11. It has been developed for multimegawatt combined cycle gas turbine and fuel cell plants, and is targeted at distributed power generation and cogeneration plants of up to 60-MW capacity. It is well suited for utility-scale wind and PV power plants. Inside SureCELL, natural gas or other fuels are converted to hydrogen and carbon monoxide by internal reformation. No external heat or steam is needed. Oxygen ions produced from an air stream react with the hydrogen and carbon monoxide to generate electric power and high-temperature exhaust gas.

Because of the closed-end tubular configuration, no seals are required and the relative cell movement due to differential thermal expansions is not restricted. This enhances the thermal cycle capability. The tubular configuration solves many of the

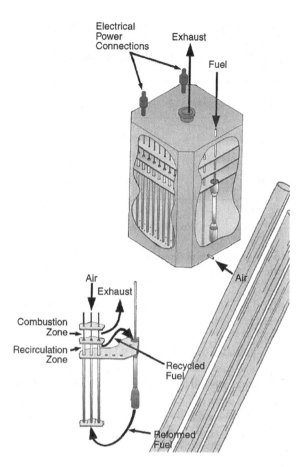

FIGURE 15.11 Seal-less solid-oxide fuel cell power generator. (From Westinghouse Electric Co., Pittsburgh, PA. Reprinted with permission.)

design problems facing other high-temperature fuel cells. The overall efficiency of the SureCELL design is about 75%, compared to the 60% maximum possible using only the gas turbine (Figure 15.12). Environmentally, the solid-oxide fuel cell produces much lower CO_2, NOx, and virtually zero SOx compared with other fuel cell technologies.

These cells have shown good voltage stability with operating hours in service. During the 8 yr of failure-free steady-state operation of the early prototypes, they maintained the output voltage within 0.5% per 1,000 h of operation. The second generation of the Westinghouse fuel cell shows even less voltage degradation, less than 0.1% per 1,000 h of operation. The SureCELL prototype has been tested for over 1,000 thermal cycles with zero performance degradation, and for 12,000 h of operation with less then 1% performance degradation. The estimated life is in tens of thousands of hours of operation.

The transient electrical performance model of the fuel cell includes electrochemical, thermal, and mass flow elements that affect the electrical output.[5] The electrical

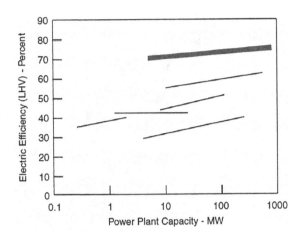

FIGURE 15.12 Natural gas power generation system efficiency comparison. (From Westinghouse Electric Co., Pittsburgh, PA. Reprinted with permission.)

response of the cell to a load change is of primary interest. To design for the worst case, the performance is calculated for both constant reactant flow and constant inlet temperature.

When hydrogen is passed across the anode and a catalyst, it is possible to separate hydrogen into protons and electrons. The electrons will pass through an external circuit with a current and a voltage while the protons will go through the electrolyte. The electrons come back from the electrical external circuit and they recombine with the protons and oxygen to produce water and heat. The reactions are described as follows:[6]

Anode reaction: $H_2 \longrightarrow 2H^+ + 2e^-$
Cathode reaction: $\frac{1}{2}O_2 + 2H^+ \longrightarrow H_2O$
Fuel cell efficiency is described by the thermodynamic efficiency as follows:

$$\eta_{th} = \Delta G / \Delta H \qquad (15.1)$$

where
$\Delta G = nFE$ = Gibb's free energy of reaction (kJ/mol),
ΔH = reaction enthalpy (kJ/mol),
n = number of electrons involved,
F = Faraday's constant = 96485 C/mol, and
E = theoretical cell potential.

The number of electrons involved with hydrogen–oxygen is 4, and the theoretical cell potential 1.23 V. After meeting the various losses in the reaction, we get less than 1 V at the output terminals under load. The current, however, is proportional to the electrode area, which can be made large to provide a high current capability. Thus, each cell generates a low voltage and a high current. High operating voltage and current are obtained by placing multiple fuel cells in series–parallel combination, known as the *fuel cell stack*.

Stand-Alone Systems

Hydrogen fuels can be stored in any one of the following ways, depending on the amount and duration of the storage:

- Underground, for large quantities of gas and also for long-term storage
- As liquid hydrogen for large quantities of gas, long-term storage, low electricity costs, or for applications requiring liquid hydrogen
- As compressed gas for small quantities of gas, high cycle times, or short storage times
- As metal hydrides for small quantities of gas

15.4.3 Mode Controller

The overall stand-alone power system must be designed for a wide performance range to accommodate the characteristics of the diesel generator (or fuel cell), the wind generator, and the battery. As and when needed, switching to the desired mode of generation is done by the mode controller. Thus, the mode controller is the central monitor and controller of the system. It houses the microcomputer and software for the source selection, the battery management, and load-shedding strategy. The mode controller performs the following functions:

- Monitors and controls the health and state of the system
- Monitors and controls the battery state of charge
- Brings up the diesel generator when needed and shuts it off when not needed
- Sheds low-priority loads in accordance with the set priorities

The battery comes online by an automatic transfer switch, which takes about 5 msec to connect to the load. The diesel, on the other hand, is generally brought online manually or automatically after going through a preplanned strategy algorithm stored in the system computer. Even with an automatic transfer switch, the diesel generator takes a relatively long time to come online, typically about 20 sec.

The mode controller is designed and programmed with deadbands to avoid frequent changeover between sources for correcting small variations on the bus voltage and frequency. The deadbands avoid chatter in the system. Figure 15.13 is an example of a 120-V hybrid system's voltage-control regions. The deadbands are along the horizontal segments of the control line.

As a part of the overall system controller, the mode controller may incorporate the maximum power extraction algorithm. The dynamic behavior of the closed-loop system, following common disturbances such as changes in sunlight due to clouds, wind fluctuation, sudden load changes, and short-circuit faults, is taken into account in a comprehensive design.[7]

15.4.4 Load Sharing

Because the wind, PV, battery, and diesel (or fuel cell) in various combinations are designed to operate in parallel, as is often the case, the load sharing between them is

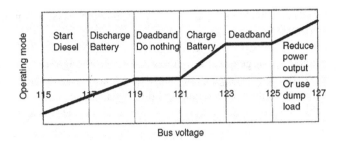

FIGURE 15.13 Mode controller deadbands eliminate system chatter.

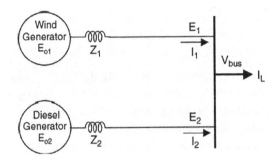

FIGURE 15.14 Thevenin's equivalent model of two sources in a hybrid power system.

one of the key design aspects of the hybrid system. For example, in the wind–diesel hybrid system (Figure 15.14), the electrical properties of the two systems must match so that they share the load in proportion to their rated capacities.

For determining the load sharing, the two systems are first reduced to their respective Thevenin equivalent circuit model, in which each system is represented by its internal voltage and the series impedance. This is shown in Figure 15.14. The terminal characteristics of the two generators are then given by the following:

$$E_1 = E_{01} - I_1 Z_1$$
$$E_2 = E_{02} - I_2 Z_2 \qquad (15.2)$$

where subscripts 1 and 2 represent systems 1 and 2, respectively,
E_0 = internally generated voltage,
Z = internal series impedance, and
E = terminal voltage of each system.

If the two sources are connected together, their terminal voltages E_1 and E_2 must be equal to the bus voltage V_{bus}. Additionally, the sum of the component loads I_1 and I_2 must be equal to the total load current I_L. Thus, the conditions imposed by the terminal connection are as follows:

$$E_1 = E_2 = V_{\text{bus}} \text{ and } I_1 + I_2 = I_L \qquad (15.3)$$

Stand-Alone Systems

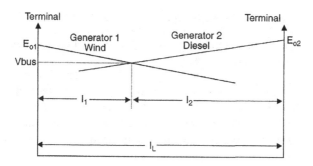

FIGURE 15.15 Graphical determination of load shared by two sources in a hybrid power system.

These imposed conditions, along with the machines' internal characteristics E_0 and Z, would determine the load sharing I_1 and I_2. The loading on an individual source is determined algebraically by solving the two simultaneous equations for the two unknowns, I_1 and I_2. Alternatively, the solution is found graphically as shown in Figure 15.15. In this method, the E vs. I characteristics of the two power sources are first individually plotted on the two sides of the current axis (horizontal). The distance between the two voltage axes (vertical) is kept equal to the total load current I_L. The power sources feeding power in parallel to a common bus share the load such that their terminal voltages are exactly equal. This condition, imposed by connecting them together at the bus, is met at the point of intersection of the two load lines. The point P in the figure, therefore, decides the bus voltage and the load sharing. The currents I_1 and I_2 in the two generators are then read from the graph.

Controlling the load sharing requires controlling the E vs. I characteristics of the sources. This may be easy in the case of the separately excited DC or the synchronous generator used with the diesel engine. It is, however, difficult in the case of the induction machine. Usually, the internal impedance Z is fixed once the machine is built. Care must be exercised in hybrid design to ensure that sufficient excitation control is built in for the desired load sharing between the sources.

The load-sharing strategy can vary depending on the priority of loads and the cost of electricity from alternative sources. In a wind–diesel system, for example, diesel electricity is generally more expensive than wind (~25 vs. 5 cents/kWh). Therefore, all priority-1 (essential) loads are met first by wind as far as possible and then by diesel. If the available wind power is more than priority-1 loads, wind supplies part of priority-2 loads and the diesel is not run. If wind power now fluctuates on the downside, the lower-priority loads are shed to avoid running the diesel. If wind power drops further to cut into the priority-1 load, the diesel is brought online again. Water pumping and heater loads are examples of priority-2 loads.

15.5 SYSTEM SIZING

For determining the required capacity of the stand-alone power system, estimating the peak-load demand is only one aspect of the design. Estimating the energy required over the duration selected for the design is discussed in the following subsections.

15.5.1 POWER AND ENERGY ESTIMATES

System sizing starts with compiling a list of all loads that are to be served. Not all loads are constant or are connected at all times. Therefore, all loads are expressed in terms of the peak watts they consume and the duty ratio. The peak power consumption is used in determining the wire size connecting to the source. The duty ratio is used in determining the contribution of an individual load in the total energy demand. If the load has clean on–off periods as shown in Figure 15.16(a), then the duty ratio D is defined as $D = T_o/T$, where T_o is the time the load is on and T is the period of repetition. For the irregularly varying loads shown in Figure 15.16(b), the duty ratio is defined as the actual energy consumed in one period over the peak power multiplied by the period, i.e.:

$$D = \frac{\text{energy in watthours consumed in one repetition period}}{\text{peak power in watts} \cdot \text{repetition period in hours}} \quad (15.4)$$

Once the peak power consumption and the duty ratio of all loads are compiled, the product of the two is the actual share of the energy requirement of that load on the system during one repetition period. If there are distinct intervals in the period, for instance, between the battery discharge and charge intervals, then the peak power and duty ratio of each load are computed over the two intervals separately. As a simple example of this in a solar power system, one interval may be from 8 a.m. to 6 p.m. and the other from 6 p.m. to 8 a.m. The power table is then prepared as shown in Table 15.3.

In a community of homes and businesses, not all connected loads draw power simultaneously. The statistical time staggering in their use results in the average power capacity requirement of the plant's being significantly lower than the sum of the individually connected loads. The National Electrical Code® provides factors for determining the average community load in normal residential and commercial areas (Table 15.4). The average plant capacity is then determined as follows:

$$\text{Required power system capacity} = \text{NEC}^® \text{ factor from Table 12.4} \times \text{Sum of connected loads} \quad (15.5)$$

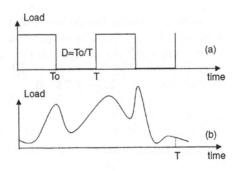

FIGURE 15.16 Duty ratio and peak power of intermittent loads.

Stand-Alone Systems

TABLE 15.3
Power and Energy Compilation Table for Energy Balance Analysis

Load	8 a.m. to 6 p.m. (Interval A) (Battery on Charge)			6 p.m. to 8 a.m. (Interval B) (Battery on Discharge)		
	Peak Watts	Duty Ratio	Energy per Period (Wh)	Peak Watts	Duty Ratio	Energy per Period (Wh)
Load$_1$	P_{1a}	D_{1a}	E_{1a}	P_{1b}	D_{1b}	E_{1b}
Load$_2$	P_{2a}	D_{2a}	E_{2a}	P_{2b}	D_{2b}	E_{2b}
...	...	...	...	...	...	...
Load$_n$	P_{na}	D_{na}	E_{na}	P_{nb}	D_{nb}	E_{nb}
Total	$\sum P_a$		$\sum E_a$	$\sum P_b$		$\sum E_b$

Note: Total battery discharge required = $\sum E_b$ wattshours.

TABLE 15.4
NEC® Demand Factors

Number of Dwellings	Demand Factor
3	0.45
10	0.43
15	0.40
20	0.38
25	0.35
30	0.33
40	0.28
50	0.26
>62	0.23

Source: Adapted from *National Electrical Code® Handbook*, 7th ed., 1996, Table 220–232.

15.5.2 BATTERY SIZING

The battery ampere-hour capacity to support the load energy requirement of E_{bat} as determined in Table 15.3 or equivalent is given by the following:

$$Ah = \frac{E_{bat}}{\eta_{disch}\left[N_{cell}V_{disch}\right]DoD_{allowed}N_{bat}} \qquad (15.6)$$

where

E_{bat} = energy required from the battery per discharge,
η_{disch} = efficiency of discharge path, including inverters, diodes, and wires, etc.,
N_{cell} = number of series cells in one battery,

V_{disch} = average cell voltage during discharge,
$DoD_{allowed}$ = maximum DoD allowed for the required cycle life, and
N_{bat} = number of batteries in parallel.

Example: The following example illustrates the use of this formula to size the battery. Suppose we want to design a battery for a stand-alone power system, which charges and discharges the battery from a 110-V DC solar array. For the DC–DC buck converter that charges the battery, the maximum available battery-side voltage is 70 V for it to work efficiently in the pulse-width-modulated (PWM) mode. For the DC–DC boost converter discharging the battery, the minimum required battery voltage is 45 V. Assuming that we are using NiMH electrochemistry, the cell voltage can vary from 1.55 V when fully charged to 1.1 V when fully drained to the maximum allowable DoD. Then, the number of cells needed in the battery is less than 45 (70/1.55) and more than 41 (45/1.1). Thus, the number of cells in the battery, estimated from voltage considerations, must be between 41 and 45. It is generally more economical to use fewer cells of a higher capacity than to use more lower-capacity cells. We, therefore, select 41 cells in the battery design.

Now, let us assume that the battery is required to discharge a total of 2 kW load for 14 h (28,000 Wh) every night for 5 yr before replacement. The lifetime requirement is, therefore, 5 × 365 = 1,825 cycles of deep discharge. For the NiMH battery, the cycle life at full depth of discharge is 2,000. Because this is greater than the 1,825 cycles required, we can fully discharge the battery every night for 5 yr. If the discharge efficiency is 80%, the average cell discharge voltage is 1.2 V, and we require three batteries in parallel for reliability, each battery ampere-hour capacity calculated from the above equation is as follows:

$$Ah = \frac{28000}{0.80 \times [41 \times 1.2] \times 1.0 \times 3} = 237 \quad (15.7)$$

Three batteries, each having 41 series cells of ampere-hour capacity 237, therefore, will meet the system requirement. We must also allow some margin to account for the uncertainty in estimating the loads.

15.5.3 PV Array Sizing

The basic tenet in sizing a stand-alone "power system" is to remember that it is really a stand-alone "energy system." It must, therefore, maintain the energy balance over the specified period. The energy drained during lean times must be made up by the positive balance during the remaining time of the period if there is no other source of energy in parallel. A simple case of a constant load on a PV system using solar arrays perfectly pointed toward the sun for 10 h of the day is shown in Figure 15.17 to illustrate the point. The solar array is sized such that the two shaded areas on the two sides of the load line must be equal. That is, the area *oagd* must be equal to the area *gefb*. The system losses in the round-trip energy transfers, e.g., from and to the battery, adjust the available load to a lower value.

In general, the stand-alone system must be sized so as to satisfy the following energy balance equation over one period of repetition:

Stand-Alone Systems

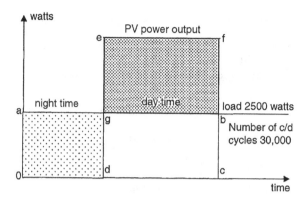

FIGURE 15.17 Energy balance analysis over one load cycle (not to scale).

$$\int_{8\,\text{a.m.}}^{6\,\text{p.m.}} \left(\text{solar radiation} \cdot \text{conversion efficiency}\right) dt =$$

$$\int_{8\,\text{a.m.}}^{6\,\text{p.m.}} \left(\text{loads} + \text{losses} + \text{charge power} + \text{shunt power}\right) dt \quad (15.8)$$

$$+ \int_{8\,\text{a.m.}}^{6\,\text{p.m.}} \left(\text{loads} + \text{losses}\right) dt$$

Or, in discrete time intervals of constant load and source power,

$$\sum_{8\,\text{a.m.}}^{6\,\text{p.m.}} \left(\text{solar radiation} \cdot \text{conversion efficiency}\right) \Delta t =$$

$$\sum_{8\,\text{a.m.}}^{6\,\text{p.m.}} \left(\text{loads} + \text{losses} + \text{charge power} + \text{shunt power}\right) \Delta t \quad (15.9)$$

$$+ \sum_{6\,\text{p.m.}}^{8\,\text{a.m.}} \left(\text{loads} + \text{losses}\right) \Delta t$$

15.6 WIND FARM SIZING

In a stand-alone wind farm, selecting the number of towers and the battery size depends on the load power availability requirement. A probabilistic model can determine the number of towers and the size of the battery storage required for meeting the load with a required certainty. Such a model can also be used to determine the energy to be purchased from, or injected into, the grid if the wind power plant were connected to the grid. In the probabilistic model, wind speed is taken as a random

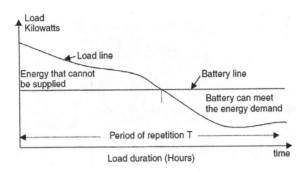

FIGURE 15.18 Effect of battery size on load availability for given load duration curve.

variable. The load is treated as an independent variable. The number of wind turbines and the number of batteries are also variables in the analysis. Each turbine in a wind farm may or may not have the same rated capacity and the same outage rate. In any case, the hardware failure rates in individual turbines are independent of each other. The resulting model has a joint distribution of the available wind power (wind speed variations) and the operating mode (each turbine working or not working). The events of these two distributions are independent. For a given load duration curve over a period of repetition, the expected energy not supplied to the load by the hybrid system clearly depends on the size of the battery, as shown in Figure 15.18. The larger the battery, the higher the horizontal line, thus decreasing the duration of the load not supplied by the system.

With such a probabilistic model, the time for which the load is not supplied by the system is termed as the *Expected Energy Not Supplied* (*EENS*). This is given by the shaded area on the left-hand side. The Energy Index of Reliability (*EIR*) is then given by the following:

$$EIR = 1 - \frac{EENS}{E_o} \qquad (15.10)$$

where E_o is the energy demand on the system over the period under consideration, which is the total area under the load duration curve. The results of such a probabilistic study[7,8] are shown in Figure 15.19, which indicates the following:

- The higher the number of wind turbines, the higher the *EIR*
- The larger the battery size, the higher the *EIR*
- The higher the *EIR* requirement, the higher the number of required towers and batteries, with a higher project capital cost

Setting an unnecessarily high *EIR* requirement can make the project uneconomical. For this reason, *EIR* must be set after a careful optimization of the cost and the consequences of not meeting the load requirement during some portion of the time period.

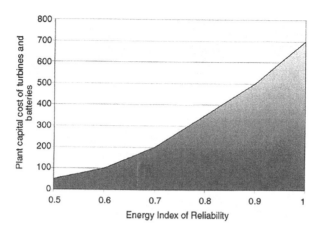

FIGURE 15.19 Relative capital cost vs. *EIR* with different numbers of wind turbines and battery sizes.

REFERENCES

1. *Institute of Solar Energy and Technology Annual Report,* Kassel, Germany, 1997.
2. Wang, L. and Su, J., Dynamic Performance of an Isolated Self-Excited Induction Generator under Various Load Conditions, IEEE Power Engineering Paper No. PE-230-EC-1-09, 1997.
3. Bonarino, F., Consoli, A., Baciti, A., Morgana, B., and Nocera, U., Transient Analysis of Integrated Diesel-Wind-Photovoltaic Generation Systems, IEEE Paper No. PE-425-EC-104, July 1998.
4. Singhal, S.C., *Status of Solid Oxide Fuel Cell Technology, Proceedings of the 17th Riso International Symposium on Material Science, High Temperature Electrochemistry, Ceramics and Metals, Roskilde, Denmark*, September 1996.
5. Hall, D.J. and Colclaser, R.G., Transient Modeling and Simulation of a Tubular Solid Oxide Fuel Cell, IEEE Paper No. PE-100-EC-004, July 1998.
6. Appleby, A.J., *Fuel Cells: Trends in Research and Applications*, Hemisphere Publishing, New York, 1985, p. 146.
7. Abdin, E.S., Asheiba, A.M., and Khatee, M.M., Modeling and Optimum Controllers Design for a Stand-Alone Photovoltaic-Diesel Generating Unit, IEEE Paper No. PE-1150-0-2, 1998.
8. Baring-Gould, E.I., Hybrid2, The Hybrid System Simulation Model User Manual, NREL Report No. TP-440-21272, June 1996.
9. D. Schumacher, O. Beik and A. Emadi, "Standalone Integrated Power Electronics System: Applications for Off-Grid Rural Locations," *IEEE Electrification Magazine*, vol. 6, no. 4, pp. 73–82, December 2018.

16 Grid-Connected Systems

Wind and photovoltaic (PV) power systems have made a successful transition from small stand-alone sites to large grid-connected systems. The utility interconnection brings a new dimension to the renewable power economy by pooling the temporal excess or the shortfall in the renewable power with the connecting grid that generates base-load power using conventional fuels. This improves the overall economy and load availability of the renewable plant site—the two important factors of any power system. The grid supplies power to the site loads when needed or absorbs the excess power from the site when available. A kWh energy meter is used to measure the power delivered to the grid, and another is used to measure the power drawn from the grid. The two meters are generally priced differently on a daily basis or on a yearly basis that allows energy swapping and billing the net annual difference.

Figure 16.1 is a typical circuit diagram of the grid-connected PV power system. It interfaces with the local utility lines at the output side of the inverter as shown. A battery is often added to meet short-term load peaks. In the U.S., the Environmental Protection Agency sponsors grid-connected PV programs in urban areas where wind towers would be impractical. In recent years, large building-integrated and residential roof-top PV installations have made significant advances by adding grid connections to the system design. For example, Figure 16.2 shows a grid-connected 18-kW building-integrated PV system on the roof of the Northeastern University Student Center in Boston. The project was part of the EPA PV DSP program. It collects sufficient research data using numerous instruments and computer data loggers. The vital data are sampled every 10 sec, and are averaged and stored every 10 min. The incoming data includes information about air temperature and wind speed. The performance parameters include direct current (DC) voltage and current generated by the PV roof and the alternating current (AC) power at the inverter output side.

In the U.K., a 390-m^2 building-integrated PV system has been in operation since 1995 at the University of Northumbria, Newcastle (Figure 16.3). The system produces 33,000 kWh electricity per year and is connected to the grid. The PV panels are made of monocrystalline cells with a photoconversion efficiency of 14.5%.

On the wind side, most grid-connected systems are large utility-scale power plants. A typical equipment layout in such a plant is shown in Figure 16.4. The wind generator output is at 690-V AC, which is raised to an intermediate level of 35 kV by a pad-mounted transformer. An overhead transmission line provides a link to the site substation, where the voltage is raised again to the grid level. The site computer, sometimes using multiplexers and remote radio links, controls the wind turbines in response to the wind conditions and load demand.

Large wind systems being installed now generally operate at variable speeds to maximize the annual energy yield. The power schematic of such a system is shown in Figure 16.5. The variable-frequency generator output is first rectified into DC and then inverted into a fixed-frequency AC. Before the inversion, harmonics in rectified DC are

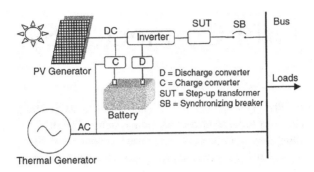

FIGURE 16.1 Electrical schematic of the grid-connected PV system.

FIGURE 16.2 18 kW grid-connected PV system on the Northeastern University Student Center in Boston. (ASE Americas, Billerica, MA. With permission.)

filtered out by an inductor and a capacitor. The frequency reference for inverter firing and the voltage reference for rectifier phase-angle control are taken from the grid lines. The rotor tip speed ratio (TSR) computed with the measured wind speed is continuously compared with the optimum reference value stored in the computer. The turbine speed is accordingly changed to assure maximum power generation at all times.

Most utility-scale wind turbines require a minimum wind speed of about 5 m/sec (12 mph). Therefore, 30 to 40% of the time, they run feeding the grid. Other times, the grid acts as a backup power source for local loads.

16.1 INTERFACE REQUIREMENTS

Both the wind and PV systems interface the grid at the output terminals of a synchronizing breaker after the inverter. The power flows in either direction depending on the site voltage at the breaker terminals. The fundamental requirements on the site voltage for interfacing with the grid are as follows:

Grid-Connected Systems

project	University of Northumbria
location	Newcastle
consulting engineer	Ove Arup and Partners
date completed	October 1995
area of solar facade	390m²
electricity generated	33,000 kWh/year
cell material	monocrystalline
efficiency	14.5%
number of panels	465
orientation	16° east of south
angle of orientation	25°
grid connected	yes

FIGURE 16.3 Grid-connected PV system at the University of Northumbria, Newcastle, U.K. The 390-m² monocrystalline modules produce 33,000 kWh per year. (From *Professional Engineer*, publication of the Institution of Mechanical Engineers, London. With permission.)

- The voltage magnitude and phase must equal that required for the desired magnitude and direction of the power flow. The voltage is controlled by the transformer turn ratio or the power electronic converter firing angle in a closed-loop control system.
- The frequency must be exactly equal to that of the grid or else the system will not work. To meet the exacting frequency requirement, the only effective means is to use the utility frequency as the inverter switching frequency reference.
- In the wind system, the base-load synchronous generators in the grid provide the magnetizing current for the induction generator.

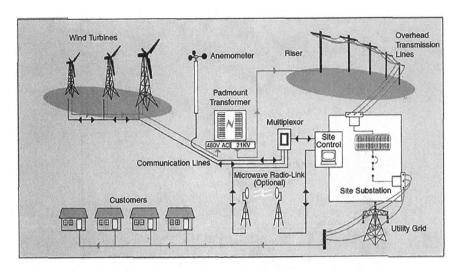

FIGURE 16.4 Electrical component layout of the grid-connected wind power system. (From AWEA/IEA/CADDET Technical Brochure).

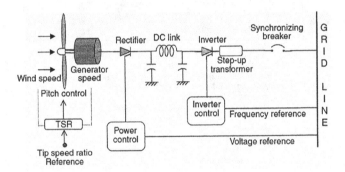

FIGURE 16.5 Electrical schematic of the grid-connected variable-speed wind power system.

The interface and control issues are similar in many ways in the PV and wind systems. The wind system, however, is more involved. The electrical generator and turbine, due to their large inertia, introduce certain dynamic issues not applicable in the static PV system. Moreover, wind plants generally have much greater power capacity than the PV plants. For example, many wind plants that have been already installed around the world have capacities of hundreds of megawatts each. More wind plants exceeding 300 MW capacity have been installed in the last decade, and even larger plants exceeding 1000 MW are being installed now.

16.2 SYNCHRONIZING WITH THE GRID

The synchronizing breakers in Figures 16.1 and 16.5 have internal voltage and phase-angle sensors to monitor the site and grid voltages and signal the correct instant for

Grid-Connected Systems

closing the breaker. As a part of the automatic protection circuit, any attempt to close the breaker at an incorrect instant is rejected by the breaker. Four conditions that must be satisfied before the synchronizing switch permits the closure are as follows:

1. The terminal-voltage frequency must be as close as possible to the grid frequency, preferably about one-third of a hertz higher.
2. The terminal-voltage magnitude must match with that of the grid, preferably a few percent higher.
3. The phase sequences of both three-phase voltages must be the same.
4. The terminal-voltage phase angle must be about same as the grid, preferably a few degrees leading.

Taking the wind power system as an example, the synchronizing process runs as follows:

1. With the synchronizing breaker open, the wind power generator is brought up to speed by using the electrical machine in the motoring mode.
2. The machine is changed to the generating mode, and the controls are adjusted so that the site and grid voltages match to meet the requirements as closely as possible.
3. The match is monitored by a synchroscope or three synchronizing lamps, one in each phase (Figure 16.6). The voltage across the lamp in each phase is the difference between the renewable site voltage and the grid voltage at any instant. All three lamps are dark when the site and grid voltages are exactly equal in all three phases. However, it is not enough for the lamps to be dark at any one instant. They must remain dark for a long time. This condition is met only if the generator and grid voltages have nearly the same frequency. If not, one set of the two three-phase voltages will rotate faster relative to the other, and the phase difference between the two voltages will light the lamps.
4. The synchronizing breaker is closed if the lamps remain dark for ¼ to ½ sec.

Following the closure, any small mismatch between the site voltage and grid voltage circulates an inrush current from one to the other until the two voltages equalize and come to a perfect synchronous operation.

16.2.1 Inrush Current

A small unavoidable difference between the site and grid voltages results in an inrush current flowing between the site and the grid. The inrush current eventually decays to zero at an exponential rate that depends on the internal resistance and inductance. The initial magnitude of this current at the instant of the circuit breaker closing depends on the degree of mismatch between the two voltages. This is not completely disadvantageous as it produces the synchronizing power, which brings the two systems into a synchronous lock. However, it produces a mechanical torque step, setting up electromechanical oscillations before the two sides come into synchronism and

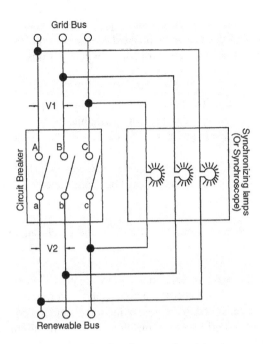

FIGURE 16.6 Synchronizing circuit using three synchronizing lamps or the synchroscope.

get locked with each other at one voltage. The magnitude of the inrush current is calculated as follows:

Let ΔV be the difference between the site voltage and the grid voltage at the instant of closing. Because this voltage is suddenly applied to the system, the resulting inrush current is determined by the subtransient reactance X_d'' of the generators:

$$I_{\text{inrush}} = \frac{\Delta V}{X_d''} \qquad (16.1)$$

The inrush current is primarily reactive as it is solely determined by X_d''. Its magnitude is kept within the allowable limit or else thermal or mechanical damage may result.

The synchronizing power produced by the inrush current brings the wind system and the grid in synchronism after the oscillations decay out. Once synchronized, the wind generator has a natural tendency to remain in synchronism with the grid, although it can fall out of synchronous operation if excessively loaded or a large load step is applied or during a system fault. Small load steps induce swings in the load angle decay out over time, restoring the synchronous condition. The magnitude of the restoring power, also known as the synchronizing power, is high if the wind generator is running at a light load and is low if it is running near the steady-state stability limit.

16.2.2 SYNCHRONOUS OPERATION

Once synchronized, the voltage and frequency of the wind system need to be controlled. The grid serves as the frequency reference for the generator output frequency when the induction generator is directly connected to the grid. The grid also acts as the excitation source supplying the reactive power. Because the torque vs. speed characteristic of the induction generator has a steep slope near zero slip (Figure 16.7), the speed of the wind turbine remains approximately constant within a few percentages. A higher load torque is met by a proportionately increased slip, up to a certain point beyond which the generator becomes unstable. If the load torque is immediately reduced, the generator returns to the stable operation. From the operating point of view, the induction generator is softer, as opposed to a relatively stiff operation of the synchronous generator, which works at an exact constant speed or falls out of stability.

If the synchronous generator is used, as in wind farms installed in California in the 1980s, the voltage is controlled by controlling the rotor field excitation current. The frequency control, however, is not required on a continuous basis. Once synchronized and connected with the lines, the synchronous generator has an inherent tendency to remain in synchronous lock with the grid. The synchronism can be lost only during severe transients and system faults. The generator must be resynchronized after such an event.

In the variable-speed induction generator system using the inverter at the interface, the inverter gate signal is derived from the grid voltage to assure synchronism. The inverter stability depends a great deal on the design. For example, there is no stability limit with a line-commutated inverter. The power limit in this case is the steady-state load limit of the inverter with any short-term overload limit.

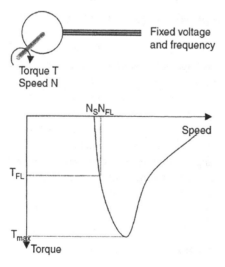

FIGURE 16.7 Resilience in the torque vs. speed characteristic of induction generator.

16.2.3 LOAD TRANSIENT

The grid will pick up the area load during steady-state operation if the renewable power system output is fully or partially lost. The effect of this is felt in two ways:

- The grid generators slow down slightly to increase their power angle under the increased load. This results in a momentary drop in frequency until the governors allow more fuel.
- The grid conductors now carry more load, resulting in a small voltage drop throughout the system.

The same effects are felt if a large load is suddenly switched on. Because starting the wind turbine as the induction motor draws a large current, it also results in the aforementioned effect. Such load transients are minimized by soft-starting large machines. In wind farms consisting of many generators, individual generators are started in sequence, one after another.

16.2.4 SAFETY

Safety is a concern when renewable power is connected to the utility grid lines. The interconnection may endanger the utility repair crew working on the lines by continuing to feed power into the grid even when the grid itself is down. This issue has been addressed by including an internal circuit that takes the inverter off-line immediately if the system detects a grid outage. Because this circuit is critical for human safety, it has a built-in redundancy.

The grid interface breaker can get suddenly disconnected either accidentally or to meet an emergency situation. The high-wind-speed cutout is a usual condition when the power is cut off to protect the generator from overloading. In a system with large capacitors connected at the wind site for power factor improvement, the site generator would still be in the self-excitation mode, drawing excitation power from the capacitors and generating terminal voltage. In the absence of such capacitors, one would assume that the voltage at the generator terminals would come down to zero. The line capacitance, however, can keep the generator self-excited. The protection circuit is designed to avoid both of these situations, which are potential safety hazards to unsuspecting site crew.

When the grid is disconnected for any reason, the generator will experience a loss of frequency regulation as the frequency-synchronizing signal derived from the grid lines is now lost. When a change in frequency is detected beyond a certain limit, the automatic control can shut down the system, cutting off all possible sources of excitation.

In new systems, however, the wind generator is designed to remain safely connected to the grid and ride through all transients. This is discussed later in this chapter.

16.3 OPERATING LIMIT

The link connecting a renewable power site with the area grid introduces an operating limit in two ways, the voltage regulation and the stability limit. In most cases, the

Grid-Connected Systems

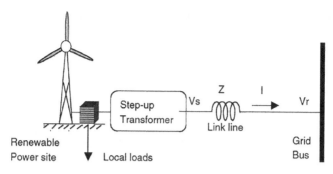

FIGURE 16.8 Equivalent circuit of renewable power plant connected to grid *via* transmission line link.

link can be considered as an electrically short transmission line. The ground capacitance and the ground leakage resistance are generally negligible. The equivalent circuit of such a line, therefore, reduces to a series leakage impedance Z (Figure 16.8). Such an approximation is valid in lines up to 50 mi long. The line carries the renewable power to the grid or from the grid to the renewable site during local peak demand. There are two major effects of the transmission line impedance, one on the voltage regulation and the other on the maximum power transfer capability of the link, as discussed in the following text.

16.3.1 Voltage Regulation

The phasor diagram of the voltage and current at the sending and receiving ends is shown in Figure 16.9. Because the shunt impedance is negligible, the sending-end current I_s is the same as the receiving-end current I_r, i.e., $I_s = I_r = I$. The voltage at the sending end is the vector sum of the receiving-end voltage and the voltage drop $I \cdot Z$ in the line:

$$V_s = V_r + I(R + jX) \qquad (16.2)$$

where Z is the line impedance, equal to $R + jX$ (per phase).

The voltage regulation is defined as the rise in the receiving-end voltage, expressed in percent of the full-load voltage, when full load at a specified power factor is removed, holding the sending-end voltage constant. That is as follows:

$$\text{percent voltage regulation} = \frac{V_{nl} - V_{fl}}{V_{fl}} \times 100 \qquad (16.3)$$

where
V_{nl} = magnitude of receiving-end voltage at no load = V_s
V_{fl} = magnitude of receiving-end voltage at full load = V_r

With reference to the phasor diagram of Figure 16.9, $V_{nl} = V_s$ and $V_{fl} = V_r$.

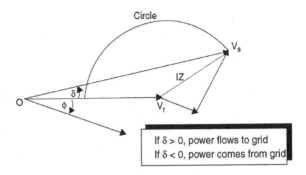

FIGURE 16.9 Phasor diagram of the link line carrying rated current.

The voltage regulation is a strongly dependent function of the load power factor. For the same load current at a different power factor, the voltage drop in the line is the same but is added to the sending-end voltage at a different phase angle to derive the receiving-end voltage. For this reason, the voltage regulation is greater for a more lagging power factor and is the least or even negative with a leading power factor.

In Figure 16.9, if the magnitude of V_r and I are held constant and the power factor of the load is varied from zero lagging to zero leading, the vector V_s varies such that its end point lies on a semicircle because the magnitude $I(R + jX)$ is constant. Such a circle diagram is useful for plotting the sending-end voltage vs. load power factor for a given load voltage and kVA.

If the voltages at both ends of the lines are held constant in magnitude, the receiving-end real power and reactive power points plotted for several loads would lie on a circle known as the power circle diagram. The reader is referred to Stevenson[1] for further information on the transmission line circle diagrams.

16.3.2 STABILITY LIMIT

The direction of power flow depends on the sending- and receiving-end voltages and the electrical phase angle between the two. However, the maximum power the line can transfer while maintaining a stable operation has a certain limit. We derive in the following text the stability limit, assuming that the power flows from the renewable site to the grid, although the same limit applies in the reverse direction as well. The series resistance in most lines is negligible and hence is ignored here.

The power transferred to the grid via the link line is as follows:

$$P = V_r I \cos\phi \tag{16.4}$$

Using the phasor diagram of Figure 16.9, the current I can be expressed as follows:

$$I = \frac{V_s - V_r}{jX} = \frac{V_s < \delta - V_r < 0}{jX} = \frac{V_s(\cos\delta + j\sin\delta) - V_r}{jX} \tag{16.5}$$

The real part of this current is as follows:

Grid-Connected Systems

$$I_{real} = \frac{V_s \sin \delta}{X} \tag{16.6}$$

This, when multiplied with the receiving-end voltage V_r, gives the following power:

$$P = \frac{V_s V_r}{X} \sin \delta \tag{16.7}$$

Thus, the magnitude of the real power transferred by the line depends on the power angle δ. If $\delta > 0$, the power flows from the site to the grid. On the other hand, if $\delta < 0$, the site draws power from the grid.

The reactive power depends on $(V_s - V_r)$. If $V_s > V_r$, the reactive power flows from the site to the grid. If $V_s < V_r$, the reactive power flows from the grid to the site.

Obviously, the power flow in either direction is maximum when δ is 90° (Figure 16.10). Beyond P_{max}, the link line becomes unstable and falls out of synchronous operation. That is, it loses its ability to synchronously transfer power from the renewable power plant to the utility grid. This is referred to as the *steady-state stability limit*. In practice, the line loading must be kept well below this limit to allow for transients such as sudden load steps and system faults. The maximum power the line can transfer without losing the stability even during system transients is referred to as the *dynamic stability limit* (P'_{max} in the figure). In a typical system, the power angle must be kept below 15 or 20° to maintain dynamic stability at all times.

Because the generator and the link line are in series, the internal impedance of the generator is added in the line impedance for determining the link's maximum power transfer capability, dynamic stability, and steady-state performance.

If an area has many large wind farms, each connecting to the grid via short links, the utility company needs a major transmission line to carry the bulk of the wind power to a major load center. The maximum power transfer and the dynamic stability limits we discussed for a short link also apply to the transmission line. New lines are

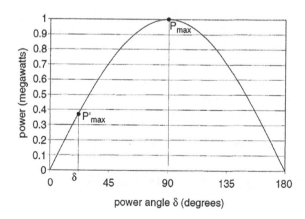

FIGURE 16.10 Power vs. power angle showing static and dynamic stability limits of the link line.

needed if the existing lines are near their limits. For example, the California Independent Systems Operators, which manage the flow of electric power through the grid, built a new transmission line in 2006 to carry wind power from the Tehachapi and Antelope Valleys to other parts of the state. This 40-km Antelope–Pardee line cost was about $100 million when built.

16.4 ENERGY STORAGE AND LOAD SCHEDULING

A wind farm sized to meet an average community load with average wind speed cannot meet 100% of the load demand at all times. The options for making the power availability nearly 100% are as follows:

- Interconnecting with the area grid to draw power from the grid or to feed power back to the grid, depending on the wind speed
- A diesel generator or a fuel cell of needed size, which would burn fuel only when the wind speed is below average. At high wind, the excess power is wasted by deflecting the blades away from the optimum position
- A rechargeable energy storage, such as a battery, pumped water, or compressed air

Even with a grid connection, large wind and PV plants may find it economical to store some energy locally in a battery. The short-term peak demand is met by the battery without drawing from the grid and paying the demand charge. For formulating an operating strategy for scheduling and optimizing the use of grid power, major system constraints are first identified. The usual constraints are the battery size, the minimum on/off times and ramp rates for thermal power plants, the battery charge and discharge rates, and the renewable power capacity limits. For arriving at the best short-term and long-term scheduling, the optimization problem is formulated to minimize the cost of all thermal and renewable plants combined subject to the constraints. Such an optimization process determines the hours for which the baseload thermal units of the utility company should be taken either off-line or online. The traditional thermal scheduling algorithms, augmented by Lagrangian relaxation, branch and bound, successive dynamic programming, or heuristic method (genetic algorithms and neural networks), can be used for minimizing the cost of operating the thermal units with a given renewable-battery system. Marwali et al.[2] have recently utilized the successive dynamic programming to find the minimum cost trajectory for battery and the augmented Langrangian to find thermal unit commitment. In a case study of a 300-MW thermal–PV–battery power plant, the authors have arrived at the total production costs shown in Table 16.1 that a battery hybrid system saves $54,000 per day compared to the thermal power plants alone.

16.5 UTILITY RESOURCE PLANNING TOOLS

Wind and PV power—in spite of their environmental, financial, and fuel diversity benefits—were initially slow to enter utility resource planning because of the planners' lack of familiarity with them, and analytical tools for nondispatchable sources

TABLE 16.1
Production Cost of 300-MW Thermal–PV–Battery System

System Configuration	Battery Depletion MWh/day	Production Cost $/day	Savings $/day
Thermal only	—	750,000	—
Thermal + PV	—	710,000	40,000
Thermal + PV + battery	344	696,000	54,000

of such power are not available on demand. The energy laboratory at the Massachusetts Institute of Technology (MIT) developed an analytical tool to analyze the impact of nondispatchable renewables on New England's power systems operation. Cardell and Connors[3] applied this tool for analyzing two hypothetical wind farms totaling 1500-MW capacity for two sites, one in Maine and the other in Massachusetts. The average capacity factor at these two sites is estimated to be 0.25. This is good, although many sites have achieved the capacity factor of 0.35 or higher. The MIT study showed that the wind energy resource in New England is comparable to that in California. The second stage of their analysis developed the product-cost model, demonstrating the emission and fuel cost risk mitigation benefits of the utility resource portfolios incorporating wind power.

16.6 WIND FARM–GRID INTEGRATION

With restructuring and technological changes in the utility sector, electric utilities have begun to include wind farms and PV parks in their resource mix. The issues the power industry must deal with in integration of these new power sources are the following:[4, 5]

- Branch power flows and node voltages
- Protection scheme and its ratings
- Harmonic distortion and flicker
- Power system dynamics and dynamic stability
- Reactive power control and voltage control
- Frequency control and load dispatch from conventional generators

The first three have primarily local impacts, whereas the last three have broad grid-level impacts. In addressing these issues, however, there is an increased need for independent analysis of the technical and economic aspects. Projects funded by the National Renewable Laboratory's (NREL) National Wind Technology Center (NWTC) and its partners in the utility and wind industries developed new information on integration and valuation issues and the reliability of new wind turbine products. The program output has become a catalyst in a national outreach effort (with investor-owned utilities, electric cooperatives, public power organizations, energy regulators, and consumers) encouraging the use of wind power in generation portfolios and the purchase of wind-generated power using market-based activities. Numerous reports[6–8] are available on these issues that can be downloaded from the Internet.

As for modeling the system performance, different wind farms are connected to different kinds of utility grids. The NWTC studies the behavior of power systems under different conditions to identify grid stability and power quality factors that enter into the development of wind farms throughout the U.S. Again, numerous reports[9–13] are available on these issues that can be downloaded from the Internet.

As for the planning models and operations, researchers are studying how multiple wind farms or multiple wind generators in one large farm can smooth out each other's output in a variable wind environment. Power output fluctuations are also being studied in the context of wind farm integration into utility grids. Hand and Madsen reports[14,15] are just two examples of such studies.

Certification and standards are of equal importance when the country as a whole must deal with a new technology. The NREL/NWTC conducts a certification process and provides guidelines to help users prepare for certification. Underwriters Laboratory (UL) is NREL's partner in this process. NREL has developed checklists to help designers understand what the certification body is likely to be looking for in their documentation. These are the same checklists that NREL and UL would use when evaluating their design documentation. Sign-offs on these checklists are used as a report of compliance or resolution on each design issue. Also offered is a checklist to help users comply with the International Electrotechnical Commission (IEC)'s requirements. NWTC has documented the general quality management, design evaluation, and testing procedures related to the certifications.

16.7 GRID STABILITY ISSUES

With today's wind turbine technology, wind power could supply 20% of the electricity needs of the U.S. A study by GE Energy concluded that the New York State power grid can handle 3300 MW of power from future utility-scale wind farms with proper procedures installed. Although the U.S. has some way to go in that direction, Europe has wind capacity already up to 20% in certain regions and more than 60% in some extremely large wind farm areas. As a result, grid stability is becoming a major concern in the European wind industry. Until recently, wind turbines were disconnected during faults on the grids, but this led to dynamic instability and blackouts. A new requirement is that the wind turbines remain connected with the grid, ride through the problems, and have a built-in capacity for active grid support. Functionally, this means that wind turbines must behave as conventional thermal and nuclear power plants.

16.7.1 LOW-VOLTAGE RIDE-THROUGH

Wind turbines in the past were designed to trip off-line in the event of major system disturbances such as lightning strikes, equipment failures, or downed power lines. However, this loss of generation impacts system stability and can lead to cascaded tripping and loss of revenue. Today, many utilities now require that wind farms ride

Grid-Connected Systems

through grid disturbances while remaining online to continue supporting the system. The new wind systems are designed to deliver ride-through capability at or below 15% of the grid voltage for up to 500 msec. This requires upgrading the wind turbine's main control cabinet, low-voltage distribution panel, pitch system, uninterruptible power supply, and power converter to ensure compliance with the low-voltage ride-through requirements.

With recent advances in power electronics, the low-voltage ride-through (LVRT) capability that enables wind turbines to stay connected to the grid during system disturbances is among the technologies introduced in the market by many manufacturers. For example, the LVRT capability is now built into all of GE Wind's new turbines, ranging from 1.5-MW units to the 12-MW units designed for offshore applications to meet more stringent transmission standards. On land, the 200-MW Taiban Mesa Wind Farm in New Mexico was the first project to install the ride-through capability.

With this new feature, wind turbines remain online and feed reactive power to the electric grid right through major system disturbances. This meets the transmission reliability standards similar to those demanded of thermal power plants. The LVRT adds significant new resiliency to wind farm operations at the time when more utilities require it.

Another power electronics technology that has been recently introduced is the dynamic power conversion system with optional reactive power control to provide support and control to local grid voltage, improve transmission efficiencies, provide the utility grid with reactive power (VARs), and increase grid stability. For example, GE's WindVAR® technology automatically maintains defined grid voltage levels and power quality within fractions of a second. This feature is particularly beneficial with weaker grids or larger turbine installations.

16.7.2 Energy Storage for Stability

Recent experience in the Scandinavian countries indicates that large-scale wind farms can cause grid stability problems if the wind capacity exceeds 20% of the instantaneous load. Innovative solutions are needed to avoid such possibilities, such as Flexible AC transmission systems (FACTS), high-voltage direct current (HVDC), and energy storage. The peak loading in the Netherlands at present approaches the threshold of the stability problem area. The fundamental issues to be addressed here are the dynamic stability, short-circuit power, and the grid upgrades. The power balancing between wind farms and other independent power producers, export and import, and natural loading on high-voltage grids are also of concern. Regulations are being prepared to ensure that these large fluctuating power sources can be handled without sacrificing power stability and quality. Large wind farms' output can vary more than ±50% around the average value over a 1-week period due to fluctuating wind speed. Such variation can be smoothed out to a constant power output by energy storage using one of the following options: electrochemical batteries (least expensive), pumped water storage (medium cost), and compressed air (most expensive). KEMA investigated the storage requirement for a large-scale offshore wind farm. Their study, reported by Enslin, Jansen, and Bauer,[16] estimated that a 100-MW

TABLE 16.2
10-MWh Energy Storage Alternatives for Grid Support

Technology	Round Trip Efficiency Percent	Storage Element Life Years	Power Equipment Cost (Euros/kWh)	Energy Storage Cost (Euros/KWh)
Water-pumped storage	75	40	1800	300
Compressed-air-pumped storage	60	30	1400	700
Lead-acid batteries	70	5	100	350

Source: From Enslin, J., Jansen, C., and Bauer, P., In Store for the Future, Interconnection and Energy Storage for Offshore Wind Farms, *Renewable Energy World*, James & James Ltd., London, January–February 2004, pp 104–113.

wind farm may require a 20-MW peak power and a 10-MWh energy storage capacity. The results of the plant economy are summarized in Table 16.2. The energy storage option, however, may become more economical if implemented for more than one function.

Rodriguez et al.[17] have addressed the impact of high wind power penetration in grid planning and operation, with a particular reference to the 15,000-MW wind capacity planned in the Spanish grid system over the next 5 yr. The key problems studied were those related to the stability of the power systems. Their analysis was based on results of the dynamic simulation. Dynamic models of the wind farms using squirrel cage and doubly fed induction generators have been developed and used in simulations for arriving at recommendations for integrating the new wind capacity of such a large magnitude.

16.8 DISTRIBUTED POWER GENERATION

In a classical power system, the power flows from the central generating station to the end users via transmission and distribution lines using many step-down transformers. The voltage drops gradually from the highest value at the generating end to the lowest value at the user end, with some boosts at the substation transformers by adjusting the turns ratio. However, with distributed power generation using PV parks and wind farms, power is injected all along the lines in a distributed pattern (Figure 16.11). In such a system, the voltage does not drop gradually; rather, it rises at the point of injection and then drops again (Figure 16.12). The European Standard EN-50160 requires that the supply voltage at the delivery point of the low-voltage network be within ±10% of the nominal value. However, generally acceptable voltage is typically between 90 and 106%. The voltage variation between the noload and full-load conditions is generally kept below ±2% by regulating the voltage by automatic tap-changing transformers or other means. Povlsen[18] has recently studied these problems in some detail.

Grid-Connected Systems

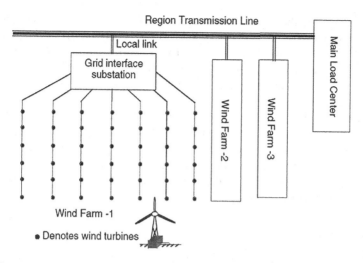

FIGURE 16.11 Distributed power generation with wind farms.

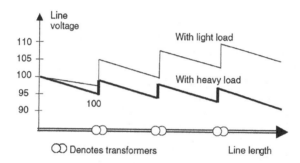

FIGURE 16.12 Voltage drops and rises along power lines with distributed generators and transformers for voltage boosts.

The system voltage can rise above the limits if the power is injected in between no- and full-load conditions by PV or wind farms. The two distributed power sources, however, influence the grid differently. Compared with wind farms, distributed PV parks generally hold the system voltage well, as its high output coincides with high demand on a hot, bright summer day. PV power injection is less suitable for grids in cold regions, where use of air conditioners is limited. In such regions, high-power output from PV farms generally coincides with low demands from both residential and industrial sectors, as during holiday seasons. Similar situations exist even in warm countries during summer holidays when the load is light.

It is possible that PV injection has to be limited to keep the voltage from rising above the specified limit when the system is operating at the minimum load. The minimum load is typically set at 25% of the maximum load in residential areas in many countries. No PV injection can be tolerated at the minimum load when the voltage is already at the maximum value, and the PV power injection would raise it even higher. On the other hand, at the maximum load, a significant amount of PV power (up to 70%) could be acceptable if well distributed, or even higher (up to 140% of load on that line) if injected into one low-voltage (LV) line. If the system is lightly loaded to only 25%, these limits may be 0 and 20%, respectively. The PV injection limits rise linearly from the lightly loaded to fully loaded systems.

It is uneconomical not to produce PV because the maximum allowable PV injected is limited by the systems load. The maximum PV injection can be increased by using tap changers on the distribution transforms. This can be done by off-load tap changing by hand on the LV side, or by automatic on-load on the high-voltage (HV) side. The automatic tap changers are expensive, whereas manual tap changing would require power interruption twice a year.

REFERENCES

1. Stevenson, W.D., *Elements of Power System Analysis*, McGraw-Hill Book Company, New York, 1962.
2. Marwali, M.K.C., Halil, M., Shahidehpour, S.M., and Abdul-Rahman, K.H., Short-Term Generation Scheduling in Photovoltaic-Utility Grid with Battery Storage, IEEE paper No. PE-184-PWRS-16-09, 1997.
3. Cardell, J.B. and Connors, S.R., Wind Power in New England, Modeling and Analyses of Nondispatchable Renewable Energy Technologies, IEEE Paper No. PE-888-PWRS-2-06, 1997.
4. Riso National Laboratory, Power Quality and Grid Connection of Wind Turbines Summary Report R-853, Denmark, October 1996.
5. Gardner, P., *Wind Farms and Weak Networks, Wind Directions, Magazine of the European Wind Energy Association* London, July 1997.
6. Smith, J.W., DOE/NREL Wind Farm Monitoring: Annual Report, NICH Report No. SR-500-31188, 2002.
7. Wan, Y.H., Wind Power Plant Monitoring Project Annual Report, NICH Report No. TP-500-30032, 2001.
8. Cadogan, J., Milligan, M., Wan, Y., and Kirby, B., Short-Term Output Variations in Wind Farms—Implications for Ancillary Services in the United States, NICH Report No. CP-500-29155, 2000.
9. Muljadi, E., Wan, Y., Butterfield, C.P., and Parsons, B., Study of a Wind Farm Power System, NICH Report No. CP-500-30814, 2002.
10. Muljadi, E. and McKenna, H.E., Power Quality Issues in a Hybrid Power System, NICH Report No. CP-500-30412, 2001.
11. Milligan, M.R., Modeling Utility-Scale Wind Power Plants, Part 1: Economics, NICH Report No. TP-500-27514, 2000.
12. Milligan, M.R., Modeling Utility-Scale Wind Power Plants, Part 2: Capacity Credit, NICH Reports No. TP-500-29701, 2002.
13. Schwartz, M. and Elliott, D., Remapping of the Wind Energy Resource in the Midwestern United States, NICH Report No. AB-500-31083, 2001.
14. Hand, M. M. and Balas, M. J., Systematic Controller Design Methodology for Variable-Speed Wind Turbines, NICH Report No. TP-500-29415, 2002.
15. Madsen, P. H., Pierce, K., and Buhl, M., Predicting Ultimate Loads for Wind Turbine Design, NICH Report No. CP-500-25787, 1998.

16. Enslin, J., Jansen, C., and Bauer, P., In Store for the Future, Interconnection and Energy Storage for Offshore Wind Farms, *Renewable Energy World*, James & James Ltd., London, January–February 2004, pp. 104–113.
17. Rodriguez, J.M. et al., Incidence on Power System Dynamics of High Penetration of Fixed-Speed and Doubly-Fed Wind Energy Systems: Study of the Spanish Case, IEEE Power & Energy, Paper No. PE-165PRS, September 2002.
18. Povlsen, A.F., Distributed Power Using PV, Challenges for the Grid, *Renewable Energy World*, James & James Ltd., London, April 2003, pp. 63–73.
19. Omid Beik, An HVDC Off-shore Wind Generation Scheme with High Voltage Hybrid Generator, Ph.D. thesis, McMaster University, 2016.
20. Omid Beik, Ahmad S. Al-Adsani, *DC Wind Generation Systems, Design, Analysis, and Multiphase Turbine Technology*, Springer, 2020.
21. Omid Beik, Ahmad S. Al-Adsani, *Wind Energy Systems*, pp. 1–9, Springer, 2020.
22. Omid Beik, Ahmad S. Al-Adsani, *DC Wind Generation System*, pp. 33–69, Springer, 2020.
23. O. Beik and A. S. Al-Adsani, "Parallel Nine-Phase Generator Control in a Medium-Voltage DC Wind System," *IEEE Transactions on Industrial Electronics*, vol. 67, no. 10, pp. 8112–8122, October 2020.
24. O. Beik and N. Schofield, "High-Voltage Hybrid Generator and Conversion System for Wind Turbine Applications," *IEEE Transactions on Industrial Electronics*, vol. 65, no. 4, pp. 3220–3229, April 2018.
25. Beik, O., Schofield, N., '*High Voltage Generator for Wind Turbines*', *8th IET International Conference on Power Electronics, Machines and Drives (PEMD 2016)*, 2016.
26. O. Beik and N. Schofield, "An Offshore Wind Generation Scheme With a High-Voltage Hybrid Generator, HVDC Interconnections, and Transmission," *IEEE Transactions on Power Delivery*, vol. 31, no. 2, pp. 867–877, April 2016.
27. O. Beik and N. Schofield, "*Hybrid Generator for Wind Generation Systems*," *2014 IEEE Energy Conversion Congress and Exposition (ECCE)*, Pittsburgh, PA, 2014, pp. 3886–3893.

17 Electrical Performance

17.1 VOLTAGE CURRENT AND POWER RELATIONS

Power systems worldwide are 3-phase, 60 Hz or 50 Hz AC. The three phases (coils) of the generator are connected in Y or Δ as shown in Figure 17.1. In balanced three-phase operation, the line-to-line voltage, the line current, and the three-phase power $P_{3\text{-ph}}$ in terms of the phase voltage V_{ph}, phase current I_{ph}, and load power factor pf are given by the following expression, with notations marked in Figure 17.1.

In a Y-connected system:

$$V_L = \sqrt{3}\, V_{ph}$$
$$I_L = I_{ph} \quad (17.1)$$
$$P_{3\ ph} = \sqrt{3}\, V_L I_L pf$$

In a Δ-connected system:

$$V_L = V_{ph}$$
$$I_L = \sqrt{3} I_{ph} \quad (17.2)$$
$$P_{3\ ph} = \sqrt{3}\, V_L I_L pf$$

For steady-state or dynamic performance studies, the system components are modeled to represent the entire system. The components are modeled to represent the conditions under which the performance is to be determined. Electrical generators, rectifiers, inverters, and batteries were discussed in earlier chapters. This chapter concerns system-level performance.

A one-line diagram often represents the three phases of the system. Figure 17.2 is an example of a one-line diagram of a 3-phase grid-connected system. On the left-hand side are two Y-connected synchronous generators, one grounded through a reactor and the other through a resistor, supplying power to load A. On the right-hand side is a wind power site with one Δ-connected induction generator supplying power to load B. The remaining power is fed to the grid via a step-up transformer, circuit breakers, and the transmission line.

The balanced three-phase system is analyzed on a single-phase basis. The neutral wire in the Y-connection does not enter the analysis in any way, because it is at zero voltage and carries zero current.

An unbalanced system, with balanced three-phase voltage on unbalanced three-phase load or with unbalanced faults, requires advanced methods of analyses, such as the method of symmetrical components, which is beyond the scope of this book.

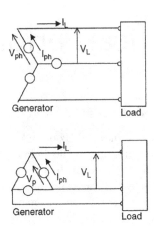

FIGURE 17.1 Three-phase AC systems connected in Y (top) and Δ (bottom).

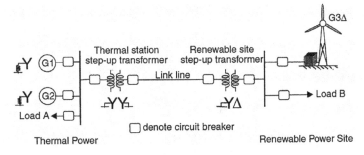

FIGURE 17.2 One-line schematic diagram of 3-phase grid-connected wind farm.

17.2 COMPONENT DESIGN FOR MAXIMUM EFFICIENCY

An important performance criterion of any system is the efficiency, which is the power output expressed as a percentage of the power input. Because the system is only as efficient as its components, designing an efficient system means designing each component to operate at its maximum efficiency.

The electrical and electronic components, while transferring power from the input side to the output side, lose some power in the form of heat. In practical designs, a maximum efficiency of 90 to 98% is typical in large power equipment of hundreds of kW ratings, and 80 to 90% in small equipment of tens of kW ratings. The component efficiency, however, varies with load as shown in Figure 17.3. The efficiency increases with load up to a certain point, and then it decreases. A good design maximizes the efficiency at the load that the equipment supplies most of the time. For example, if the equipment is loaded at 70% of its rated capacity most of the time, it is beneficial to have the maximum efficiency at 70% load. The method of achieving the maximum efficiency at a desired load level is presented in the following text.

The total loss in any power equipment generally has two components. The one that represents the quiescent no-load power consumption remains fixed. It

Electrical Performance

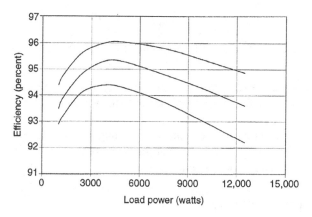

FIGURE 17.3 Power equipment efficiency varies with load with a single maximum.

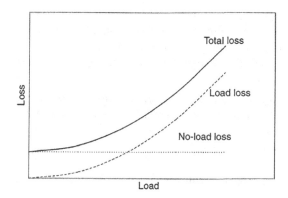

FIGURE 17.4 Loss components varying with load in typical power equipment.

primarily includes eddy and hysteresis losses in the magnetic parts. The other component, representing ohmic loss in the conductors, varies with the square of the current. For a constant voltage system, the conductor loss varies with the square of the load power. The total loss (Figure 17.4) is therefore expressed as the following:

$$\text{loss} = L_o + kP^2 \qquad (17.3)$$

where P is the output power delivered to the load, L_o the fixed loss, and k the proportionality constant. The efficiency is given by the following:

$$\eta = \frac{\text{output}}{\text{input}} = \frac{\text{output}}{\text{output} + \text{loss}} = \frac{P}{P + L_o + kP^2} \qquad (17.4)$$

For the efficiency to be maximum at a given load, its derivative with respect to the load power must be zero at that load:

$$\frac{d\eta}{dP} = \frac{P(1+2kP)-(P+L_o+kP^2)}{(P+L_o+kP^2)^2} = 0 \tag{17.5}$$

This equation reduces to $L_o = kP^2$. Therefore, the component efficiency is maximum at the load under which the fixed loss is equal to the variable loss. This is an important design rule, which can result in significant savings in electric energy in large power systems. Power electronic components using diodes have an additional loss that is linearly proportional to the current. Inclusion of such a loss in the preceding analysis does not change the maximum efficiency rule just derived.

17.3 ELECTRICAL SYSTEM MODEL

An electrical network of any complexity can be reduced to a simple Thevenin equivalent circuit consisting of a single internal source voltage V_s and impedance Z_s in series (Figure 17.5). The two parameters are determined as follows:

With the system operating at no load and with all other parameters at rated values, the internal voltage drop is zero and the voltage at the open-circuit terminals equals the internal source voltage; that is, V_s = open-circuit voltage of the system.

For determining the source impedance, the terminals are shorted together and the terminal current is measured. Because the internal voltage is now totally consumed in driving the current through the source impedance only, we have the following:

$$Z_s = \text{open-circuit voltage/short-circuit current.} \tag{17.6}$$

If Z_s is to be determined by an actual test, the general practice is to short the terminals with a small open-circuit voltage that is only a small percentage of the rated voltage. The low-level short-circuit current is measured and the full short-circuit current is calculated by scaling to the full-rated voltage. Any nonlinearity, if present, is accounted for.

The equivalent circuit is developed on a per-phase basis and in percent or per-unit basis. The source voltage, current, and impedance are expressed in units of their respective base values at the terminal. The base values are defined as follows:

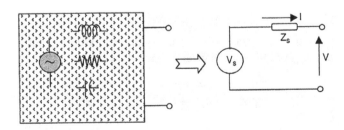

FIGURE 17.5 Thevenin equivalent circuit model of a complex power system.

V_{base} = rated output voltage,
I_{base} = rated output current, and
Z_{base} = rated output voltage/rated output current.

17.4 STATIC BUS IMPEDANCE AND VOLTAGE REGULATION

If the equivalent circuit model of Figure 17.5 is derived under steady-state static conditions, then Z_s is called the static bus impedance. The steady-state voltage rise on removal of full-load-rated current is then $\Delta V = I_{base} Z_s$.

Under a partial or full-load step transient, as in the case of a loss of load due to accidental opening of the load-side breaker, the bus voltage oscillates until the transient settles to a new steady-state value. If the load current rises in step as shown in Figure 17.6, the voltage oscillates before settling down to a lower steady-state value. The steady-state change in the bus voltage is then given by $\Delta V = \Delta I \, Z_s$. Then, the percent voltage regulation is given by the following:

$$\text{Voltage Regulation}(\%) = \frac{\Delta V}{V_{base}} \times 100 \qquad (17.7)$$

The feedback voltage-control system responds to bring the bus voltage deviation back to the rated value. However, in order not to flutter the system more than necessary, the control system is designed with suitable deadbands. For example, Figure 17.7 shows a 120-V photovoltaic (PV)-battery system with two deadbands in its control system.

Often we need to calculate the voltage drop across an impedance Z, such as that of a generator, transformer, and/or a transmission cable, to determine the required generator voltage. It is given by the following:

$$V_{drop} = IZ \qquad (17.8)$$

where V and I are voltage and current phasors, and $Z = Z_{gen} + Z_{transformer} + Z_{cable}$.

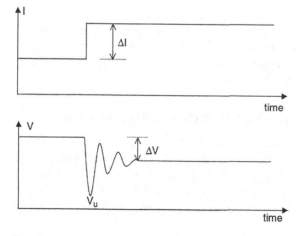

FIGURE 17.6 Transient response of the system voltage under sudden load step.

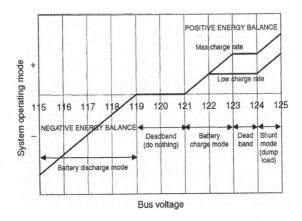

FIGURE 17.7 Deadbands in the feedback voltage-control system avoid system flutter.

Because the voltage drop in the cable is proportional to the length and inversely proportional to the cross-sectional area of the conductor, the cable size is selected to have a relatively low voltage drop, generally below 3%. Temperature limitation is another factor that may determine the cable size in some cases. Standard tables for cable gauge include the impedance, and the temperature and current limitations.

If the load is to be powered at voltage V_L and power factor $\cos\theta$, the generator must generate voltage given by the following phasor sum:

$$V_{gen} = V_L + V_{drop} \qquad (17.9)$$

If Z is expressed as resistance and reactance, i.e., $Z = R + jX$, the generator voltage is given by the following relationship:

$$V_{gen} = \sqrt{\left(V_L \cos\theta + IR\right)^2 + \left(V_L \sin\theta + IX\right)^2} \qquad (17.10)$$

All calculations are done in a per-phase analysis. Note that the generated voltage depends on the load power factor also. A power factor of 0.85 lagging is common in many systems. Capacitors can improve the power factor to close to 1.0, with a subsequent reduction in the voltage drop.

17.5 DYNAMIC BUS IMPEDANCE AND RIPPLES

If the circuit model of Figure 17.5 is derived for dynamic conditions, i.e., for an incremental load, the source impedance is called the dynamic bus impedance and is denoted by Z_d. It can be either calculated or measured as follows: With the bus in operational mode supplying rated load, a small high-frequency AC current I_h is injected into the bus using an independent grounded current source (Figure 17.8). The high-frequency voltage perturbation in the bus voltage is measured and denoted by V_h. The dynamic bus impedance at that frequency is given by:

Electrical Performance

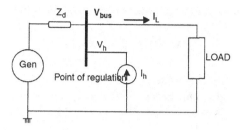

FIGURE 17.8 Harmonic and dynamic source impedance test measurement setup.

$$Z_d = \frac{V_h}{I_h} \quad (17.11)$$

Ripples is the term used to describe periodic high-frequency glitches in the current or the voltage. Ripples are commonly found in systems with power electronic components, such as rectifiers, inverters, and battery chargers. Transistors switching on and off cause the ripples, the frequencies of which are integral multiples of the switching frequency. The ripples are periodic but not sinusoidal, and are superimposed on the fundamental wave.

The ripple voltage induced on the bus due to ripple current is given by the following:

$$V_{ripple} = I_{ripple} Z_d \quad (17.12)$$

The ripple content is minimized by capacitors connected to the bus or preferably at the load terminals of the component causing ripples. The ripple current is then supplied or absorbed by capacitors, rather than by the bus, thus improving the quality of power.

17.6 HARMONICS

Harmonics is the term used to describe higher-frequency sine wave currents or voltages superimposed on the fundamental sine wave. Phase-controlled power switching is one source of harmonics. Harmonics are also generated by magnetic saturation in power equipment. The generator and transformer behave linearly, but not with saturation present in the magnetic circuit. The saturated magnetic circuit requires non-sine-wave magnetizing current.

The usual method of analyzing a system with harmonics is to determine the system performance for each harmonic separately and then to superimpose the results. For such an analysis, the system is represented by the equivalent circuit for each harmonic separately.

The fundamental equivalent circuit of the electrical generator is represented by the d-axis and q-axis.[1,2] The inductance L_n in the nth harmonic equivalent circuit, being for high frequency, is the average of the subtransient inductance in the d and q axes:

$$L_n = \frac{\left(L_d'' + L_q''\right)}{2} \quad (17.13)$$

The reactance for the harmonic of order n is given by the following:

$$X_n = 2\pi f_n L_n = 2\pi nf L_n \qquad (17.14)$$

where f_n is the nth harmonic frequency and f the fundamental frequency.

In AC current having symmetrical positive and negative portions of the cycle, an odd number of harmonics is always absent. That is, $I_n = 0$ for $n = 2, 4, 6, 8$, and so on. In three-phase load circuits fed by transformers having the primary windings connected in delta, all triple harmonics are also absent in the line currents, that is, $I_n = 0$ for $n = 3, 9, 15$, and so on. The m-pulse full-bridge inverter circuit contains harmonics of the order $n = mk \pm 1$, where $k = 1, 2, 3, 4$, and so on. For example, the harmonics present in a 6-pulse inverter are 5, 7, 11, 13, 17, and 19. On the other hand, the harmonics present in a 12-pulse inverter are 11, 13, 23, and 25. The approximate magnitude and phase of the harmonic currents are found to be inversely proportional to the harmonic order n:

$$I_n = \frac{I_1}{n} \qquad (17.15)$$

where I_1 is the fundamental current. This formula gives the approximate harmonic content in 6- and 12-pulse inverters, which are given in the first two columns of Table 17.1 and clearly show the benefits of using a 12-pulse converter. The actually measured harmonic currents are lower than those given approximately by Equation 17.15. IEEE Standard 519 gives the current harmonic spectrum in a typical 6-pulse converter as listed in the last column of Table 17.1.

The harmonic currents induce a harmonic voltage on the bus. A harmonic voltage of order n is given by $V_n = I_n Z_n$, where Z_n is the nth harmonic impedance. Harmonic impedance can be derived in a manner similar to dynamic impedance, in which a harmonic current I_n is injected into or drawn from the bus and the resulting harmonic voltage V_n is measured. A rectifier circuit drawing harmonic current I_n provides a simple circuit, which works as the harmonic current load. If all harmonic currents are measured, then the harmonic impedance of order n is given by the following:

$$Z_n = V_n / I_n, \text{ where } n = mk \pm 1, k = 1, 2, 3, \cdots \qquad (17.16)$$

TABLE 17.1
Harmonic Content of 6-Pulse and 12-Pulse Converters

Harmonic Order (n)	6-Pulse Converter (Equation 17.15)	12-Pulse Converter (Equation 17.15)	3-Pulse and 6-pulse Converters (IEEE Standard 519)
5	20	—	17.5
7	14.5	—	11.1
11	9.1	9.1	4.5
13	7.7	7.7	2.9
17	5.9	5.9	1.5
19	5.3	5.3	1.0

17.7 QUALITY OF POWER

The quality of power at the grid interface is a part of the power purchase contracts between the utility and the renewable power plant. The rectifier and inverter are the main components influencing power quality. The grid-connected power system, therefore, must use converters, which are designed to produce high-quality, low-distortion AC power. Power quality concerns become more pronounced when the renewable power system is connected to small-capacity grids using long low-voltage links.

There is no generally acceptable definition of the quality of power. However, the International Electrotechnical Commission and the North American Reliability Council have developed working definitions, measurements, and design standards. Broadly, power quality has three major components for measurement:

- Total harmonic distortion, primarily generated by power electronic rectifier and inverter
- Transient voltage sags caused by system disturbances and faults
- Periodic voltage flickers

17.7.1 Harmonic Distortion Factor

Any nonsinusoidal alternating voltage $V(t)$ can be decomposed by the following Fourier series:

$$V(t) = V_1 \sin \omega t + \sum_{n=2}^{\infty} V_n \sin(n\omega t + \alpha_n) \qquad (17.17)$$

The first component on the right-hand side of the preceding equation is the fundamental component, whereas all other higher-frequency terms ($n = 2, 3, \ldots \infty$) are harmonics.

The total harmonic distortion (*THD*) factor is defined as follows:

$$THD = \frac{\left[V_2^2 + V_3^2 + \cdots V_n^2\right]^{\frac{1}{2}}}{V_1} \qquad (17.18)$$

The *THD* is useful in comparing the quality of AC power at various locations of the same power system, or of two or more power systems. In a pure sine wave AC source, $THD = 0$. The greater the value of *THD*, the more distorted the sine wave, resulting in more I^2R loss for the same useful power delivered; this way, the quality of power and the efficiency are related.

As seen earlier, the harmonic distortion on the bus voltage caused by harmonic current I_n drawn by any nonlinear load is given by $V_n = I_n Z_n$. It is this distortion in the bus voltage that causes the harmonic current to flow even in a pure linear resistive load, called the *victim load*. If the renewable power plant is relatively small, a nonlinear electronic load may cause significant distortion on the bus voltage, which then supplies distorted current to the linear loads. The harmonics must be filtered out before feeding power to the grid. For a grid interface, having a *THD* less than 3% is generally acceptable. IEEE Standard 519 limits the *THD* for utility-grade power to less than 5%.

Harmonics do not contribute to the delivery of useful power but produce I^2R heating. Such heating in generators, motors, and transformers is more difficult to dissipate due to their confined designs, as compared to open conductors. The National Electrical Code® requires all distribution transformers to state their k ratings on the permanent nameplate. This is useful in sizing the transformer for use in a system having a large *THD*. The k-rated transformer does not eliminate line harmonics. The k rating merely represents the transformer's ability to tolerate harmonics. A k rating of unity means that the transformer can handle the rated load drawing pure sine wave current. A transformer powering only electronic loads may require a high k rating from 15 to 20.

A recent study funded by the Electrical Power Research Institute reports the impact of two PV parks on the power quality of the grid-connected distribution system.[3] The harmonic current and voltage waveforms were monitored under connection/disconnection tests over a 9-month period in 1996. The current injected by the PV park had a total distortion below the 12% limit set by IEEE standard 519. However, the individual even harmonics of orders between 18 and 48, except the 34th, exceeded IEEE standard 519. The total voltage distortion, however, was minimal.

A rough measure of quality of power is the ratio of the peak voltage to rms voltage measured by the true rms voltmeter. In a pure sine wave, this ratio is $\sqrt{2}$, i.e., 1.414. Most acceptable bus voltages will have this ratio in the 1.3 to 1.5 range, which can be used as a quick approximate check of the quality of power at any location in the system.

17.7.2 Voltage Transients and Sags

The bus voltage can deviate from the nominally rated value due to many reasons. The deviation that can be tolerated depends on its magnitude and the time duration. Small deviations can be tolerated for a longer time than large deviations. The tolerance band is generally defined by voltage vs. time (*V–t*) limits. Computers and business equipment using microelectronic circuits are more susceptible to voltage transients than rugged power equipment such as motors and transformers. The power industry has developed an array of protective equipment. Even then, some standard of power quality must be maintained at the system level. For example, the system voltage must be maintained within the *V–t* envelope shown in Figure 17.9, where the solid line is that specified by the American National Standard Institute (ANSI), and the dotted line is that specified by the Computer and Business Equipment Manufacturers Associations (CBEMA). The right-hand side of the band comes primarily from the steady-state performance limitations of motors and transformer-like loads, the middle portion comes from visible lighting flicker annoyance considerations, and the left-hand side of the band comes from the electronic load susceptibility considerations. The CBEMA curve allows larger deviations in the microsecond range based on the volt-second capability of the power supply magnetics. ANSI requires the steady-state voltage of the utility source to be within 5%, and short-time frequency deviations to be less than 0.1 Hz.

Capacitors are often used to economically create reactive power VARs. However, in wind farm applications, each capacitor-switching event creates a voltage transient that causes excessive torque on the wind turbine's gear mechanism, leading to excessive

Electrical Performance

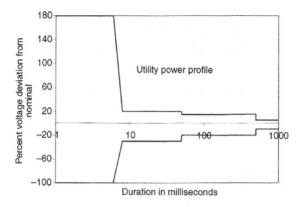

FIGURE 17.9 Allowable voltage deviation in utility-grade power vs. time duration of the deviation. (Adapted from the American National Standards Institute.)

wear and tear on the gearbox. Moreover, the high volume of switching necessary for system voltage maintenance leads to early failures of the capacitors, resulting in higher maintenance costs. A new product in the industry, such as GE's D-VAR™ system, when used in conjunction with capacitors, drastically reduces the number of required capacitor switchings. One such system has been used for the SeaWest project in Wyoming.

17.7.3 Voltage Flickers

Turbine speed variation under fluctuating wind conditions causes slow voltage flickers and current variations that are large enough to be detected as flickers in fluorescent lights. The relation between the fluctuation of mechanical power, rotor speed, voltage, and current is analyzed by using the dynamic d- and q-axis model of the induction generator or the Thevenin equivalent circuit model shown in Figure 17.10. If we assume:

R_1, R_2 = resistance of stator and rotor conductors, respectively,
x_1, x_2 = leakage reactance of stator and rotor windings, respectively,
x_m = magnetizing reactance,
x = open-circuit reactance,
x = transient reactance,
τ_o = rotor open-circuit time constant = $(x_2 + x_m)/(2\Pi f R_2)$,
s = rotor slip,
f = frequency,
E_T = machine voltage behind the transient reactance,
V = terminal voltage,
I_s = stator current,

then the value of E_T is obtained by integrating the following differential equation:

$$\frac{dE_T}{dt} = -j2\pi f s E_T - \frac{1}{\tau_o}\left[E_T - j(x - x')I_s\right] \qquad (17.19)$$

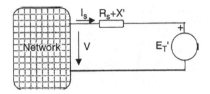

FIGURE 17.10 Thevenin equivalent circuit model of induction generator for voltage-flicker study.

And the stator current from:

$$I_s = \frac{V - E_T}{R_1 + jX'} \quad (17.20)$$

The mechanical equation, taking into account rotor inertia, is as follows:

$$2H\frac{ds}{dt} = P_e - \frac{P_m}{1-s} \quad (17.21)$$

where
P_e = electric power delivered by the generator,
P_m = mechanical power of the wind turbine,
H = rotor inertia constant in seconds = $\frac{1}{2}J\omega^2/P_{rated}$,
J = moment of inertia of the rotor, and
$\omega = 2\Pi f$.

The electric power is the real part of the product of E_T and I_s*:

$$P_e = \text{real part of}\left[E_T \cdot I_s^*\right] \quad (17.22)$$

where I_s* is the complex conjugate of the stator current.

The mechanical power fluctuations can be expressed by a sine wave superimposed on the steady value of P_{mo} as follows:

$$P_m = P_{mo} + \Delta P_m \sin \omega_1 t \quad (17.23)$$

where ω_1 = rotor speed fluctuation corresponding to the wind power fluctuation.

Solving these equations by an iterative process on the computer, Feijoo and Cidras[4] showed that fluctuations of a few hertz could cause noticeable voltage and current fluctuations. Fluctuations on the order of several hertz are too small to be detected at the machine terminals. The high-frequency fluctuations, in effect, are filtered out by the wind turbine inertia, which is usually large. That leaves only a band of fluctuations that can be detected at the generator terminals.

The flickers caused by wind fluctuations may be of concern in low-voltage transmission lines connecting to the grid. The voltage drop related to power swing is small in high-voltage lines because of the small current fluctuation for a given wind fluctuation.

The power quality regulations in Denmark require all grid-connected power plants—onshore or offshore—to be capable of reducing the power output by 20% within 2 sec of a fault to keep the network stable during a fault. German and Swedish grid authorities are also considering imposing such regulations on all grid interconnections.

The flickers are also caused by frequent switching of the wind generator on and off at the wind speed around the cut-in speed. This problem is aggravated when the area grid has many wind generators distributed along the lines. The grid company usually allows such switching no more often than three to four times per hour. A desirable solution is to keep the rotor switched off until the wind reaches a stable speed beyond the cut-in value, with some deadband hysteresis built into the control process.

The wind generator's low-frequency mode of oscillation may be excited by random wind fluctuation, wind shear, and tower shadow, producing large ripples in the driving torque as well as in generated electric power, which can be noticed as an approximately 0.5% voltage flicker at a frequency between 2 and 4 Hz. Several control strategies of static converters have been designed to provide damping of such oscillations and voltage flickers.

Flicker limits are specified in IEEE standards 141-1993 and 519-1992, which have served the industry well for many years. Cooperative efforts between IEC, UIE, EPRI, and IEEE have resulted in updated standards as documented in IEC standards 61000-3.

17.7.4 HARMONICS ELIMINATION WITH PASSIVE FILTERS

Harmonics mitigation is a necessary action enforced by the grid standards to prevent unwanted harmonics feeding to the grid. Rapid increase in the use of variable frequency drives (VFDs) as nonlinear loads has increased the needs for harmonics filtering.

Majority of the existing VFDs are fed from the 3-phase power grid, use a passive rectifier to converter the AC to DC that is supplied to an inverter. Figure 17.11 shows

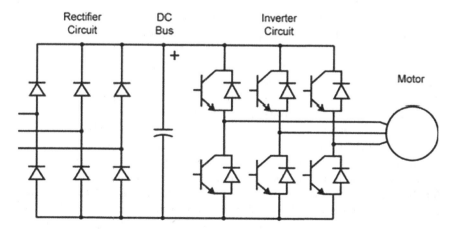

FIGURE 17.11 Schematic of a widely used DC-link VFD.

schematic of a widely used VFD. The passive rectifier in a VFD is a nonlinear device as it draws a non-sinusoidal current from the power grid. The harmonics in a 3-phase rectifier is expressed as:

$$h = 6n \pm 1, \quad n = 1, 2, 3, \ldots \tag{17.24}$$

Without any harmonic elimination method, the harmonics 5th, 7th, 11th, and 13th are generated in the line current on the grid side. The 5th and 7th harmonics are the dominant ones, and there are higher order harmonics present in the waveform. The modern power distribution systems supply many nonlinear loads and VFDs, which introduce heavy and excessive harmonics to the power grid. These harmonics could cause:

- Failure of power factor capacitors due to overloading
- Excessive losses and hence overheating of cables, transformers and other equipment, which leads to reduced life span and failures
- Unwanted tripping of circuit breakers and other protection devices
- Failures and performance degradation in electronics loads such as computers, etc.

Traditionally, to eliminate the harmonics, an AC reactor may be installed at the input terminals of a VFD. This is a relatively low-cost approach, however, the higher the amplitude of current harmonics, the larger the AC reactor required to filter the harmonics. The AC reactor also helps the VFD from overvoltage caused by inverter switching or capacitors. However, the addition of an AC reactor also causes additional voltage drop at the VFD input. As the size of AC reactor increases in high current loads, AC reactor is not a very effective approach. A reactor on the DC link of the VFD is another approach in harmonics reduction. The DC reactor does not cause a voltage drop on the AC side and have proven to be more effective. However, the DC reactors do not offer overvoltage protection for the drive and unlike the AC reactor, the DC reactor is a part of the VFD which make the VFD bulky.

Traditional LC filters with a tuned frequency are another device in harmonics treatment. These filters are usually tuned to a specific harmonic, often the 5th, and their effectiveness is limited. Low pass filters are widely used in harmonics elimination. Compared to single tuned filters and reactors, the low pass filters treat a wider range of harmonics, and hence are more effective. The series inductor in the low pass filter causes a voltage drop, but this voltage drop can be compensated by a parallel capacitor bank in the filter. A large capacitor bank leads to a leading power factor at the VFD input which could adversely affect the power grid and/or the performance of the other equipment in the system.

Another approach to harmonics elimination is to use multi-pulse (greater than 6-pulse) rectifiers in the VFD. Figure 17.12 shows a 12-pulse rectifier in a VFD. In this scheme a transformer with two secondary windings, shifted from one another, is used in the input. The two sets of the 3-phase windings are connected to two parallel 6-pulse rectifiers sharing the same DC-link. The inverter is fed from the same DC-link. An 18-pulse rectifier may also be used in a similar approach with an input transformer with three secondary windings. The 12-, 18- or higher pulse rectifiers significantly reduce the harmonics generated by the VFD. However, they are bulky

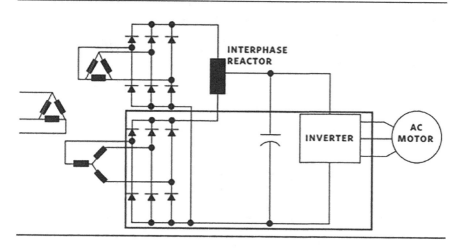

FIGURE 17.12 12-pulse rectifier in a VFD to reduce harmonics.

and more expensive than their 6-pulse counterparts. The transformer used in such scheme is commonly referred to as phase-shifting transformer. The phase-shifting transformer may also be used in a scheme where it feeds multiple 6-pulse VFD in a system with multiple loads.

17.7.5 Harmonics Elimination with Active Filters

Active power filters (APFs) are a relatively modern approach in harmonics elimination. An APF in principle is a power electronic converter that operates as a switch-mode current source amplifier by use of analog and digital logic to control fast switching devices (MOSFETs or IGBTs). The APF provides a compensating current to limit harmonic distortion and improve true power factor for the electrical systems.

Figure 17.13 shows schematic of an APF and a load. The APF is connected in parallel to the AC power lines, where it actively monitors the non-linear current demanded by the load in real-time by means of active filtering and digital signal processing. It extracts the harmonic current from the measured load current, calculates power factor, and electronically generates the proper compensating current waveform, which matches the shape and the phase-shift required to cancel out the non-fundamental frequency current and voltage components (such as harmonics). Such non-fundamental components are commonly produced by electronic power converters, VFDs, and other nonlinear loads. The APF also supplies a substantial amount of reactive current at the fundamental frequency, which provides the displacement power factor correction for the elements and loads connected to the AC power system. By providing this compensating harmonic and reactive current into the distribution bus, the APF almost completely cancels harmonic currents and improves the power factor at the point of connection. The current generated by the APF, combined with the non-linear current of the load, results in perfectly sinusoidal

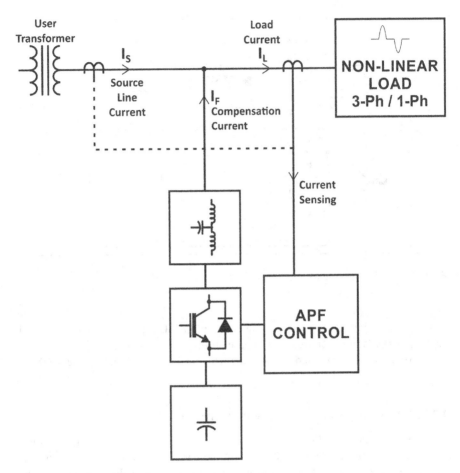

FIGURE 17.13 Active power filter (APF).

shape of the line current waveform. The APF in this mode of operation provides true harmonic current elimination.

The APFs can also provide beneficial non-fundamental frequency currents for systems with 4-wire connected loads with neutral returns. Although the neutral current cannot be affected by a 3-wire connection, the APF can be adapted so that positive- and negative-sequence harmonic currents are removed and only zero-sequence harmonic currents remain (harmonic currents that return thru the neutral line). The result is that each phase has a harmonic line current equal to one third of the harmonic current in the neutral line. The APFs may be paralleled for increased capacity. Each paralleled unit provides an equal proportion of the compensation current. The APFs can also work in redundant mode, in which the loss of one paralleled unit does not cause any loss of output compensating capacity.

For operation, the APF requires the use of external current transformers (CTs) to provide load current sensing signals proportional to the loads on the 3-phase lines. In the 3-wire load configuration, two CTs are sufficient for the APF operation. However, the three-wire APF adapted for 4-wire loading requires three CTs. Typically the APF operates with load side measurements (external CTs are installed downstream of the active filter, between the APF connection and the load), but alternatively it can be controlled with source side measurements (CTs installed upstream of the APF, between the source of power and the APF connection). In this SOURCE sensing configuration, the APF uses the internal monitoring of its own output currents to determine the necessary load current (source current + APF output current = load current).

Figure 17.14 shows a non-sinusoidal load current that is from a VFD. The APF injects a compensating current that contains the required harmonics with opposite angle, and when added to the load current it results in a perfectly shaped sinusoidal input current, i.e. the source current shown in Figure 17.14.

The APFs control system provides the ability to select the preferred control method between current waveform shape control—very fast analog method, often called full spectrum cancellation control, and discrete selective har-monic elimination—based on discrete Fast Fourier Transform (FFT) algorithm, which allows for only specific harmonics to be cancelled out. The control system also allows for reactive power compensation, which can be either enabled to control targeted displacement power factor, or disabled by the user. The preferred harmonic correction control method and power factor compensation is user programmable through an HMI interface.

The APFs usually draw a small amount of power at the fundamental frequency to make up for power losses in generating the compensation currents. The total power required for operation, which includes power for controls and fans/blowers, is typically about 1% of the APF rated power.

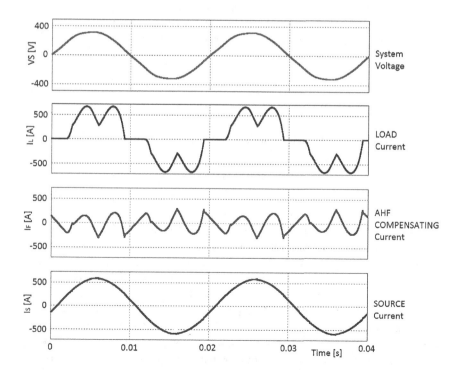

FIGURE 17.14 Operation of active power filter.

17.8 RENEWABLE CAPACITY LIMIT

A recent survey made by Gardner[5] in the European renewable power industry indicates that the grid interface issue is one of the economic factors limiting full exploitation of available wind resources. The regions with high wind-power potentials have weak existing electrical grids. In many developing countries such as in India, China, and Mexico, engineers are faced with problems in locating sites that are also suitable for interfacing with the existing grid from the point of view of power quality. The basic consideration in such decisions is the source impedance before and after making the connection. Another way of looking at this issue is the available short-circuit megavoltampere (*MVA*) at the point of proposed interconnection. The short-circuit *MVA* capacity is also known as the *system stiffness* or the *fault level*.

17.8.1 SYSTEM STIFFNESS

One way of evaluating the system stiffness after making the interconnection is by using the Thevenin equivalent circuit of the grid and the renewable plant separately, as shown in Figure 17.15.

V = network voltage at the point of the proposed interconnection,
Z_i = source impedance of the initial grid before the interconnection,
Z_w = source impedance of the wind farm,
Z_l = impedance of the interconnecting line from the wind farm to the grid, and
Z_T = total combined source impedance of the two systems,

Then, after connecting the system with the proposed wind capacity to the grid, the combined equivalent Thevenin network would be as shown in (b), where:

$$\frac{1}{Z_T} = \frac{1}{Z_i} + \frac{1}{Z_W + Z_l} \qquad (17.25)$$

The combined short-circuit *MVA* at the point of the interconnection is then:

$$MVA_{sc} = \frac{\text{Rated MVA Capacity}}{Z_{T \text{per unit}}} \qquad (17.26)$$

The higher the short-circuit *MVA*, the stiffer the network. A certain minimum grid stiffness in relation to the renewable power capacity is required to maintain the power quality of the resulting network. This consideration limits the total wind capacity that can be added at a site. A wind capacity exceeding that limit may be a difficult proposal to sell to the company in charge of the grid.

Not only is the magnitude of Z_T important, the resistance R and the reactance X, which are components of Z_T, have their individual importance. Fundamental circuit theory dictates that the real and reactive power in any electrical network must be separately maintained in balance. Therefore, the real and reactive components of the wind generator impedance would impact the network, more so in a weaker grid. The R/X

Electrical Performance

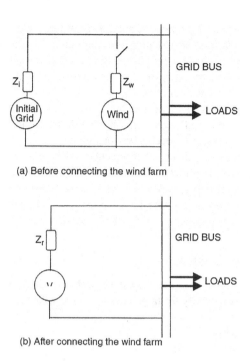

FIGURE 17.15 Thevenin equivalent circuit model of the grid and wind farm for evaluation of the grid interface performance for evaluating system stiffness at the interface.

ratio is often found to be 0.5, which is generally satisfactory for fixed-speed wind generators. Lower R/X ratios may pose another limitation when designing the system.

The fault current decays exponentially as $e^{-R/X}$. A low R/X ratio causes the fault current to persist, making fault protection relaying more difficult. Voltage regulation (variation from zero to full load on the wind farm) is yet another design consideration that is impacted by the R/X ratio. The estimated voltage regulation of the wind farm must be compared with the contractual limit of the utility. In doing so, the actual continuous maximum load the wind farm can deliver must be taken into account, and not the nominal rating, which can be much higher. The acceptable voltage regulation is typically 5% in industrialized countries and 7% in developing countries.

Starting the wind generator as induction motor causes an inrush of current from the grid, resulting in a sudden voltage drop for a few seconds. This can be outside the tolerance limits of the grid line. Most countries limit this transient voltage dip to 2 to 5%, the higher value prevailing in developing countries. Fortunately, for wind farms with many machines, the machines can be started in sequence to minimize this effect. For a very weak system, however, this issue can limit the number of machines that can be connected to the grid. The voltage-flicker severity increases as the square root of the number of machines. The flicker caused by starting one machine varies inversely with the fault level at the point of grid connection, and hence can be an issue on weaker grids.

Harmonics are generated by the power electronics employed for soft-start of large wind turbines and for speed control during energy-producing operations. The former can be generally ignored due to its short duration and because of the use of sequential starts.

The operating harmonics, however, need to be filtered out. The total harmonic content of the wind farm is empirically found to depend on the square root of the number of machines. Because the high-voltage grid side has a lower current, using pulse-width-modulated converters on the grid side can be advantageous. Utility grids around the world are not consistent in the way they limit harmonics. Some grids limit harmonics in terms of the absolute current in amperes, whereas others set limits proportional to the grid's short-circuit MVA at the point of interconnection.

Utilities find it convenient to meet power quality requirements by limiting the total renewable power rating to less than a small percentage of the short-circuit MVA of the grid at the proposed interface. The limit is generally 2 to 3% in developed countries and 5% in developing countries. This becomes restrictive because a strong wind area is usually available in a nonurbanized area, where the grid is usually weak. It is more restrictive than the overall power quality requirement imposed in accordance with national standards.

17.8.2 Interfacing Standards

Wind farms connected to the grid may pose problems in regard to the following system-level issues, especially for large-capacity wind farms on a weak grid:

- Electrical stiffness
- Short-circuit MVA capacity
- Quality of power
- National standards

The maximum capacity the renewable power plant can install is primarily determined from system stability and power quality considerations (presented in this chapter). A rough rule of thumb to control power quality has been to keep the renewable power plant capacity in MW less than the grid-line voltage in kV if the grid is stiff (large). On a weak grid, however, only 10 or 20% of this capacity may be allowed. The regulations generally imposed on renewable power farms these days are in terms of the short-circuit capacity at the proposed interface site.

The major power quality issues discussed in this chapter are as follows:

- Acceptable voltage variation range on the distribution system
- Step change in voltage due to step loading
- Steady-state voltage regulation
- Voltage flickers caused by wind speed fluctuation
- *THD* factor of harmonics

The generally acceptable voltage variations in the grid voltage at the distribution point in four countries are given in Table 17.2. The limit on the step change a customer can trigger on loading or unloading is listed in Table 17.3. It is difficult to determine the maximum renewable capacity that can be allowed at a given site that will meet all these complex requirements. It may be even more difficult to demonstrate such compliance. In this situation, some countries impose limits as percentages of the grid short-circuit capacity. For example, such limits in Germany and Spain are listed in Table 17.4. These limits, however, are continuously evolving in most countries at present.

TABLE 17.2
Acceptable Voltage Variation at Distribution Points

Country	Acceptable Range (%)
U.S.	±5
France	±5
U.K.	±6
Spain	±7

Note: Low-voltage consumers may see wider variations.

TABLE 17.3
Allowable Step Change in Voltages a Customer Can Trigger by Step Loading or Unloading

Country	Allowable Range (%)
France	±5
U.K.	±3
Germany	±2
Spain	±2 for wind generators ±5 for embedded generators

TABLE 17.4
Renewable Power Generation Limits as Percentages of the Grid Short-Circuit Capacity at the Point of Interface

Country	Allowable Limit (%)
Germany	2
Spain	5
Other countries	Evolving

The International Electrotechnical Commission (IEC) has drafted the power quality standards applicable specifically to grid-connected wind farms. It is expected that these standards will be more realistic than the rigid criteria based on the ratio of the wind turbine capacity to the short-circuit *MVA* of the grid. It may ultimately provide a consistent and predictable understanding of the power quality requirements for designing wind farms that comply with power quality requirements.

Voltage dips, interruptions, and variations in consumer power networks have been measured and quantified by several organizations, including Eurelectric. Table 17.5 is a typical assessment of power quality within European power distribution networks.

TABLE 17.5
Voltage Dips, Interruptions, and Variations on European Power Distribution Networks

Voltage Drop (%)	Duration of Disturbance and Number of Events per Year			
	10 msec to 100 msec	100 msec to ½ sec	1.2 sec to 1 sec	1 sec to 3 sec
10–30	61	66	12	6
30–60	8	36	4	1
60–100	2	17	3	2
100	0	12	24	5

Source: From Lutz, M. and Nicholas, W., *Conformity Magazine*, November 2004, p. 12.

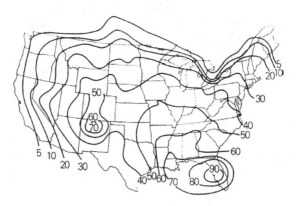

FIGURE 17.16 Thunderstorm frequency in the U.S. and southern Canada. The contour lines indicate the average number of days per year on which thunderstorms occur.

17.9 LIGHTNING PROTECTION

The risk of mechanical and thermal damage to the wind turbine blades and the electrical systems due to lightning is minimized by a coordinated protection scheme using lightning arresters and spark gaps[6] in accordance with international standards. IEC Standard 1024-1 covers the requirements for protection of structures against lightning. The electrical generator and transformer are designed with a certain minimum Basic Insulation Level (BIL) consistent with the lightning risk in the area. The risk is proportional to the number of thunderstorms per year in the area, which is depicted in Figure 17.12 for the U.S. and southern Canada.

Tall towers in open spaces are more vulnerable to lightning risk. A new development in the wind power industry is that offshore wind power towers experience a higher-than-average incidence of lightning strikes. System manufacturers have addressed this issue with design changes. Figure 17.13 is one such solution developed by Vestas Wind Systems (Denmark). The lightning current is conducted to the ground through a series of spark gaps and equipotential bonding at joints. The transformer is protected by placing it inside the tower.

Electrical Performance

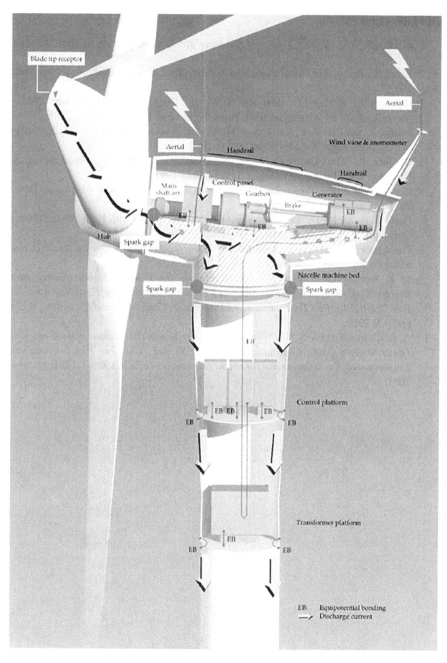

FIGURE 17.17 Lightning protection of the wind tower. (From Vestas Wind Systems, Denmark. With permission.)

REFERENCES

1. Adkins, B., *The General Theory of Electrical Machines*, Chapman and Hall, London, 1964.
2. Kron, G., *Equivalent Circuits of Electrical Machines*, Dover Publications, New York, 1967.
3. Oliva, A.R., Balda, J.C., McNabb, D.W., and Richardson, R.D., Power Quality Monitoring of a PV Generator, IEEE Paper No. PE-507-EC0-07, 1997.
4. Feijoo, A. and Cidras, J., Analyses of Mechanical Power Fluctuations in Asynchronous WECs, IEEE Paper No. PE-030-EC-0-1 0, 1997.
5. Gardner, P., Wind Farms and Weak Networks, *Wind Directions*, Magazine of the European Wind Energy Association, London, July 1997.
6. Lewis, W.W., *The Protection of Transmission Systems Against Lightning*, Dover Publications, New York, 1965.
7. Omid Beik, An HVDC Off-shore Wind Generation Scheme with High Voltage Hybrid Generator, Ph.D. thesis, McMaster University, 2016.
8. Omid Beik, Ahmad S. Al-Adsani, *DC Wind Generation Systems, Design, Analysis, and Multiphase Turbine Technology*, Springer, 2020.
9. M. R. Aghamohammadi, and A. Beik-Khormizi, "*Small Signal Stability Constrained Rescheduling using Sensitivities Analysis by Neural Network as a Preventive Tool*," in *Proc. 2010 IEEE PES Transmission and Distribution Conference and Exposition*, IEEE. pp. 1–5, 2010.
10. A. B. Khormizi and A. S. Nia, "*Damping of Power System Oscillations in Multi-machine Power Systems Using Coordinate Design of PSS and TCSC*," in *10th International Conference on Environment and Electrical Engineering (EEEIC), Rome*. IEEE. pp. 1–4, 2011.

18 Plant Economy

The economic viability of a proposed plant is influenced by several factors that contribute to its profitability. The plant proposal is generally initiated with some expected values of the contributing factors. As the expected profitability would vary with deviations in these factors, the sensitivity of profitability to such variances around their expected values is analyzed. Such a sensitivity analysis determines the range of profitability and raises the confidence level of potential investors. This is important for both wind and photovoltaic (PV) systems but more so for wind systems, in which profitability is extremely sensitive to wind speed variations. The primary factors that contribute to the economic viability of a wind farm or PV park are discussed in this chapter.

18.1 ENERGY DELIVERY FACTOR

The key economic performance measure of a power plant is the electric energy it delivers over the year. Not all power produced is delivered to the paying customers. A fraction of it is used internally to power the control equipment, in meeting power equipment losses, and in housekeeping functions such as lighting. In a typical wind farm or PV park, about 90% of the power produced is delivered to the paying customers, and the remaining is self-consumed in plant operations. The quantity of energy delivered depends on the peak power capacity of the site and how fully that capacity is utilized. A normalized measure of the power plant performance is the energy delivery factor (*EDF*). It is defined as the ratio of the electric energy delivered to customers to the energy that could have been delivered if the plant were operated at full installed capacity during all 8760 h of the year. It is expressed as follows:

$$\text{Average annual } EDF = \frac{\text{kWh delivered over the year}}{\text{Installed kW capacity} \times \text{number of hours in the year}} \quad (18.1)$$

EDF is usually determined by recording and summing the energy delivered over a continuous series of small discrete time intervals Δt as follows where P_{avg} = average power delivered over a small time interval Δt:

$$\text{Average annual } EDF = \frac{\sum^{\text{year}} P_{avg}(\Delta t)}{P_{max} \times 8760} \quad (18.2)$$

Thus, *EDF* is a figure of merit that measures how well the plant is utilized to deliver the maximum possible energy. Not only does it include the energy conversion efficiencies of various components, it also accounts for the overall reliability, maintainability, and availability of the plant over the entire year. Therefore, *EDF* is useful in comparing the economic utilization of one site over another, or to assess the annual performance of a given site. Wind plants operate at an annual average *EDF* of around 30% to 40, with some plants reporting an *EDF* as high as 50%. This compares with 40 to 80% for conventional power plants, the peak-power plants near 40%, and the base-load plants near 80%.

It must be taken into account that the wind farm *EDF* varies with the season. As an example, the quarterly average *EDF* of early wind plants in England and Wales ranged from 15 to 45%. The *EDF* is high in the first quarter and low in the last quarter of every year. Even with the same kWh produced per year, the average price per kWh that the plant may fetch could be lower if the seasonal variations are wide.

18.2 INITIAL CAPITAL COST

The capital cost depends on the size, site, and technology of the plant. Typical ranges for various types of large power plants are given in Table 18.1. The costs of technologically matured coal and gas turbine plants are rising with inflation, whereas wind and PV plant costs are falling with new developments in the field. Therefore, renewable-power-plant costs must be estimated along with their component costs current at the time of procurement, which may be lower than at the time of planning.

The percentage breakdown of the component costs in the total initial capital cost of a typical wind farm is given in Table 18.2. As expected, the single-largest cost item is the rotor assembly (blades and hub).

18.3 AVAILABILITY AND MAINTENANCE

The rates and effects of failure determine the maintenance cost of the plant and its availability to produce power. The data from past operating experiences are used for learning lessons and making improvements. In the wind power industry, failures, their causes, and their effects are recorded and periodically published by ISET, the solar energy research unit of the University of Kassel in Germany. Figure 18.1 is an example from ISET's early database.[1] It shows that the plants were nonoperational for 67% of the time during the reported period. Among the repairs needed to bring the plants back to operation, 20% was in the electric power equipment and 19% in the electronic controls. The major causes of failure were related to the control

TABLE 18.1
Capital Cost for Various Power Technologies

Plant Technology	Capital Cost (US$/kW)
Wind turbine	500–700
Solar PV	1500–2000
Solar thermal (Solar-II type)	3000–5000
	(land ~15 acres/MW)
Coal thermal (steam turbine)	400–600
	(land ~10 acres/MW)
Combined cycle (gas turbine)	800–1200

TABLE 18.2
Wind Power System Component Cost Contribution in Total Capital Cost

Cost Item	Contribution (%)
Rotor assembly	25
Nacelle structure and auxiliary equipment	15
Electric power equipment	15
Tower and foundation	10
Site preparation and roads	10
Ground equipment stations	8
Maintenance equipment and initial spares	5
Electrical interconnections	4
Other nonrecurring costs	3
Financing and legal	5
Total	100

Note: Land cost is not included.

systems (28%) and component defects (24%). The failure rates have declined from those levels with the design improvements made since then. The wind power industry has developed a database that records specific failures, their causes, and their effects. In modern systems, however, with many components incorporating computers of some sort, almost 80% of the problems are computer related.

The overall availability of the plant is defined as the ratio of hours in a year that the plant is available to deliver power to the total hours in the year. It is impacted by downtimes due to repairs and routine maintenance, which are reflected in the EDF. Significant improvements in reliability and maintainability have pushed the availability of modern renewable power plants up to 95% in recent years. However, data show that, for about 50% of the time during an average year, the plant undergoes an outage or is not operational because of low wind speeds.

The PV systems, being static, require a much lower level of maintenance relative to the wind systems, which have moving parts. With proper maintenance, most wind turbines are designed to operate for 20 to 30 yr, which translate to about 200,000 h of operation. However, the actual life and optimum maintenance schedule depend on the climatic conditions and the quality of design and construction. Climatic conditions include wind turbulence and the stress cycles on the turbine. Offshore wind turbines experience low wind turbulence compared to their counterparts on land.

As for maintaining wind turbines, many factors go into setting an optimum maintenance strategy and schedule. Too little maintenance can cost in downtime, whereas too much would in itself be expensive. The industry-average normal maintenance period has been once every 6 months for onshore turbines and once every year for offshore turbines. With regular, periodic oil change, the life of the gearbox is 5 to 7 yr, and the overall life of the turbine is 20 to 30 yr.

Maintenance cost is low when the machine is new and increases with time. For large wind turbines in megawatt capacities, the annual maintenance costs range from

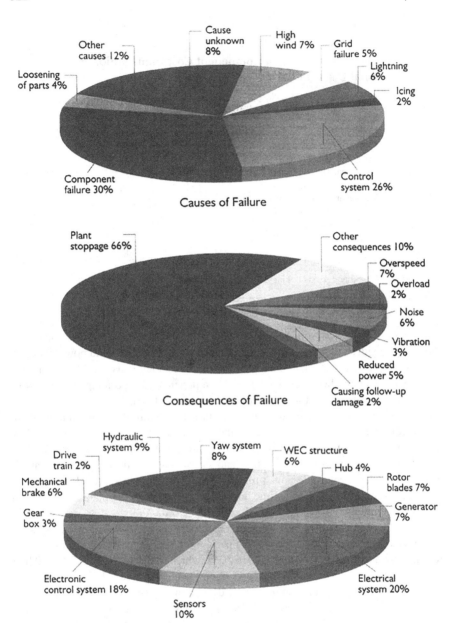

FIGURE 18.1 Wind power plant failure, cause, effect, and repair statistics. (From Institute of Solar Energy and Technology, University of Kassel, Germany. With permission.)

1.5 to 2% of the original turbine cost. In plants with EDF of 30 to 40%, this translates as ½ to 1 cent/kWh in energy costs.

When a wind turbine nears the end of its life, the overall structure, including the foundation and tower, may be in good shape, but the rotor and gearbox may need to be replaced. The price of a new set of rotor blades and gearbox is usually 15 to 20% of the turbine price.

18.4 ENERGY COST ESTIMATES

A key parameter of performance of an electric power plant is the unit cost of energy (*UCE*) per kWh delivered to the paying consumers. It takes into account all economic factors discussed in the preceding sections and is given by the following:

$$UCE = \frac{ICC(AMR + TIR) + OMC}{EDF \cdot KW \cdot 8760} \qquad (18.3)$$

where
  ICC = initial capital cost, including the land cost and startup cost up to the time the first unit of energy is sold
  AMR = amortization rate per year as a fraction of the ICC
  TIR = tax and insurance rate per year as a fraction of the ICC
  OMC = operating and maintenance costs per year
  EDF = energy delivery factor over 1 yr
  KW = kilowatt electric power capacity installed

The amortization rate reflects the cost of capital. It is generally taken as the mortgage interest rate applicable to the project. The *EDF* accounts for variations in the impinging energy (sun intensity or wind speed) at the proposed site and all downtimes—full or partial. It is also a strong function of the reliability and maintainability of the plant through the year.

The insurance on a wind farm project covers delays or damages during construction or operation whether because of technical reasons or the forces of nature. The costs of insurance necessary for financing wind projects have significantly increased in recent years. Offshore wind farms, being relatively new, are exposed to even more unknown factors and therefore charged higher premiums.

18.5 SENSITIVITY ANALYSIS

It is not sufficient to accurately estimate the cost of electricity produced. Project planners must also carry out a sensitivity analysis, in which the energy cost is estimated with a series of input parameters deviating on both sides of the expected values. The sensitivity of the energy cost to variations in two primary economic factors—wind speed in wind farms and solar radiation in PV parks—is discussed in the following subsections.

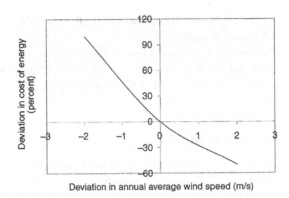

FIGURE 18.2 Sensitivity analysis of wind electricity cost with variation in wind speed around the expected annual mean of 10 m/sec.

18.5.1 EFFECT OF WIND SPEED

Because the energy output of a wind plant varies with the cube of wind speed, a change in speed by several percentage points can have a significant impact on plant economy. For example, if the annual average wind speed is 10 m/sec, the unit cost of energy is calculated for wind speeds ranging from 8 to 12 m/sec. The plant site is considered economically viable only if the plant can be profitable at the low end of the range. Figure 18.2 illustrates the result of a sensitivity study. It shows that the cost of electricity can be as low as half or as high as twice the expected value if the wind speed turns out to be higher or lower by 2 m/sec than the expected value of 10 m/sec.

18.5.2 EFFECT OF TOWER HEIGHT

The tower height varies from approximately four times the rotor diameter for small turbines (kW ratings) to a little over one diameter for large turbines (MW ratings). For small turbines, it is determined by the wind speed at the hub. The higher the hub, the greater the wind speed. For large turbines, it is primarily determined by the tower structure and the foundation design.

As seen in Table 18.2, the turbine cost constitutes 25% of the total wind farm cost. Because wind speed increases with tower height, a given turbine can produce more energy and reduce the cost of energy if installed on a taller tower. To study the influence of tower height on energy cost, we define E_1 as the contribution of the turbine cost in the total energy cost per square meter of the rotor-swept area. E_1 is, therefore, measured in \$/kWh·m². The power is proportional to the square of the rotor diameter and the cube of the wind speed as follows:

$$P \propto D^2 V^3 \qquad (18.4)$$

The energy captured over the life of *the* turbine is also proportional to the same parameters, i.e., D^2 and V^3. In a matured, competitive market, the rotor cost would be proportional to the swept area, hence D^2, although it is far from that stage in the rapidly evolving market at present. However, assuming a competitive market, E_1 is as follows:

Plant Economy

$$E_1 \propto D^2 / D^2 V^3 \propto 1/V^3 \qquad (18.5)$$

As seen earlier in Equation 3.34, the wind speed varies with tower height in the exponential relation:

$$V \propto H^\alpha \qquad (18.6)$$

where H = tower height at the hub and α = *terrain* friction coefficient.

Combining Equations 18.5 and 18.6, we obtain:

$$E_1 \propto \frac{1}{H^{3\alpha}} \qquad (18.7)$$

The parameter α varies with the terrain. Table 3.3 in Chapter 3 gives these parameter values—0.10 over the ocean, 0.40 over urban areas with tall buildings, and the often-used average of 0.15 (~1/7) in areas with foot-high grass on flat ground. Even a higher value of 0.43 was estimated for a New England site at Stratton, Mt. Vermont.[2] With such wide variations in α, the wind speed can vary with tower height over a wide range from $H^{0.3}$ to $H^{1.3}$.

Equations 18.7 clearly indicates that the turbine contribution in cost of energy per unit swept area decreases at least with the square root of the tower height. With the average value of $\alpha = 0.15$, the turbine cost contribution is approximately given by the following:

$$\text{turbine cost per kWh per m}^2 \quad E_1 = \frac{\text{constant}}{\sqrt{\text{tower height}}} \qquad (18.8)$$

TABLE 18.3
Example of Profitability of 1-MW Wind Plant for a School District

Total capital cost C	$1,125,000
Total yearly cost	$135,235
20-yr mortgage on C @ 4.5%	$86,485
Operating + Maintenance + Insurance costs @ 3 % of C	$33,750
Maintenance contract @ $15,000/MW	$15,000
Plant installed at a wind site in class 5–6 range	
Average wind 8 m/sec at 50-m height (to be verified by data)	
Yearly average power (with k = 2, 80-m hub)	346.6 kWe
Energy production per year with 90% availability	2,732 MWh
Swapping up to 3,150 MWh/yr consumption is allowed	
So, feedback to a grid with 1-MW plant is not needed	
Yearly savings in energy bill at 9 cents/kWh	$245,860
Yearly cost with 100% debt financing	$135,235
Net savings per year after debt servicing	$110,625
Yearly reward for risking capital C	9.8%
Yearly cost with 100% equity financing	$48,750
Net savings per year without debt servicing	$197,110
Payback period at 4.5% discount rate	6.7 yr

In rough terrains, the tower height can be *extremely* beneficial as shown by the following equation with α = 0.40:

$$\text{turbine cost per kWh per m}^2 \quad E_1 = \frac{\text{constant}}{H^{1.2}} \qquad (18.9)$$

Equation 18.8 and Equation 18.9 indicate that there is not much benefit in increasing the tower height in offshore installations that have low α values. However, on rough terrains *that* have high α, increasing the tower height from 30 to 60 m would decrease the contribution of the turbine cost per kWh/m² by 56%, a significant reduction. Therefore, determining the α parameter accurately for a specific site is extremely important in determining the tower height and plant economy. An uneconomical site using short towers or a conservative estimate of an average value of α can turn out to be profitable with tall towers and/or the site-specific value of α obtained by actual measurements. In the early data-collection period, therefore, anemometers are installed on multiple towers at two or more heights on the same tower to determine the α at the site.

18.6 PROFITABILITY INDEX

As with conventional projects, profitability is measured by the profitability index (*PI*), defined as follows:

$$PI = \frac{\text{present worth of future revenues} - \text{initial project cost}}{\text{initial project cost}} \qquad (18.10)$$

By definition, a *PI* of zero gives the break-*even* point.

Profitability obviously depends on the price at which a plant can sell the energy it produces. In turn, it depends on the prevailing market price that the utilities are charging the area customers. Installations in regions with high energy cost could be more profitable if the capital cost is not high in the same proportion. The average electricity prices in urban regions of the U.S. (NY, NJ, PA, NH, MA, ME) is around 20 cents/kWh, whereas that in the U.K. is about 10 pence/kWh.

Inputs to the renewable power plant profitability analysis include the following:

- Anticipated energy impinging the site, that is, the wind speed at hub height for a wind farm or the solar radiation rate for a PV park
- Expected initial capital cost of the installation
- Cost of capital (usually the interest rate on the loans)
- Expected economic life of the plant
- Operating and maintenance costs
- Average selling price of the energy generated by the plant

A detailed multivariable profitability analysis with these parameters is always required before making financial investments. Potential investors can make initial profitability assessments using screening charts.

18.7 PROJECT FINANCE

The project financing and economic viability of a proposed wind or PV power plant depends a great deal on the electricity prices prevailing in the region. The average electricity prices to household customers in various countries are 6 cents/kWh in India, 10 cents/kWh in Mexico, 17 cents/kWh in USA, 25 Cents/kWh in Germany, and 30 cents/kWh in Japan. The high prices in Germany and Japan partly explain the high rate of growth in wind and solar power in those countries.

The capital cost is also a factor. For large wind power plants in MW capacity, the total capital cost is about $1/W, including the cost of the last switch to turn on the power. For small wind plants of around 100-kW capacity, it is about $2/W.

As mentioned earlier, with the new net-metering law in the U.S., school districts are getting interested in installing wind power plants. An example of a 1-MW project for a school district of typical size is shown in Table 18.3. It involves a total capital cost of $1.125 million, produces 2732 MWh electricity per year, and has a payback period of 6.7 yr at a 4.5% project financing cost (discount rate).

As for the project funding, there have been strong financial incentives for renewable power in many countries around the world. However, the incentives have been declining because the renewables are becoming economically competitive on their own merits. More projects are being funded strictly on a commercial basis. Enviro-Tech Investment Funds in the U.S. is a venture capital fund supported by Edison Electric Institute (an association of investor-owned electric utility companies providing 75% of the nation's electricity). The World Bank now includes wind and PV projects in its lending portfolios. Several units in the World Bank group are jointly developing stand-alone projects in many developing countries.

Because wind power and PV power are intermittent, nondispatchable, and unpredictable, they have less economic value than power from fuel sources that can deliver steady, predictable power on demand. Utilities obligated to provide service with almost 100% availability must either back up the intermittent power at a premium, estimated to be between ½ and 1 cent/kWh, or penalize the provider of interruptible power by ½ to 1 cent/kWh price differential. Also, because of the uncertainty of output, lenders impose higher financing costs compared to those for more predictable power generation.

REFERENCES

1. Institute of Solar Energy and Technology, Annual Report, University of Kassel, Germany, 1997.
2. Cardell, J.B. and Connors, S.R., Wind Power in New England, Modeling and Analyses of Non-dispatchable Renewable Energy Technologies, IEEE Paper No. Power Engineering-888-PWRS-2-06, 1997.
3. Chabot, B., L'analyse economique de l'energie ecolienne, *Revue de l'Energie*, ADME, France, No. 485, February 1997.
4. Chabot, B., From Costs to Prices: Economic Analyses of Photovoltaic Energy and Services, *Progress in Photovoltaic Research and Applications*, Vol. 6, pp. 55–68, 1998.
5. Baring-Gould, E.I. *Hybrid-2, The Hybrid System Simulation Model User Manual*, NREL Report No. TP-440-21272, June 1996.

6. Omid Beik, An HVDC Off-shore Wind Generation Scheme with High Voltage Hybrid Generator, Ph.D. thesis, McMaster University, 2016.
7. Omid Beik, Ahmad S. Al-Adsani, *DC Wind Generation Systems, Design, Analysis, and Multiphase Turbine Technology*, Springer, 2020.
8. D. Schumacher, O. Beik and A. Emadi, "Standalone Integrated Power Electronics System: Applications for Off-Grid Rural Locations," *IEEE Electrification Magazine*, vol. 6, no. 4, pp. 73–82, December 2018.
9. O. Beik and N. Schofield, "High-Voltage Hybrid Generator and Conversion System for Wind Turbine Applications," *IEEE Transactions on Industrial Electronics*, vol. 65, no. 4, pp. 3220–3229, April 2018.
10. M. R. Aghamohammadi, and A. Beik-Khormizi, "*Small Signal Stability Constrained Rescheduling using Sensitivities Analysis by Neural Network as a Preventive Tool*," in *Proc. 2010 IEEE PES Transmission and Distribution Conference and Exposition*, pp. 1–5.
11. A. B. Khormizi and A. S. Nia, "*Damping of Power System Oscillations in Multi-machine Power Systems Using Coordinate Design of PSS and TCSC*," in *2011 10th International*

19 The Future

The future of renewable energy, specifically the wind and solar power and their projected growth, expressed as a percentage of the total electricity demand in the world, are discussed in this chapter. The historical data on market developments in similar basic-need industries and the Fisher–Pry market growth model indicate that PV and wind power may reach their full potential in the year 2065. The probable impact of the U.S. utility restructuring on renewable energy sources is reviewed.

19.1 WORLD ELECTRICITY TO 2050

According to the U.S. Department of Energy (DOE), electric power is expected to be the fastest-growing source of energy for end users throughout the world over the next three decades.

The world net electricity generation in 2010 was around 10 trillion kilowatts [1]. This is projected to increase to 30 and 15 trillion kilowatts for the non-OECD and OECD countries. The Organization for Economic Co-operation and Development (OECD) is an intergovernmental economic organization founded to stimulate economic progress and world trade. There are currently 37 countries as OECD members. Non-OECD country is referred to a country that is not a member of OECD.

Figure 19.1 shows a graph projecting world electricity generation up to 2050. Net electricity generation (to the grid) in non-OECD countries increases at an average of 2.3% per year from 2018 to 2050, compared with 1.0% per year in OECD countries [1]. Electricity use increases the most in the buildings sector, particularly residential, as personal incomes rise, and urban migration continues in non-OECD countries. Residential building electricity consumption more than doubles from 2018 to 2050. The share of electricity used in transportation nearly triples between 2018 and 2050 as more plug-in electric vehicles enter the fleet and electricity use for rail expands. Yet, transportation still accounts for less than 6% of total delivered electricity consumption in 2050.

Figure 19.2 shows the world net electricity generation by fuel. Renewables (including hydropower) are the fastest-growing source of electricity generation during the 2018 to 2050 period, rising by an average of 3.6% per year. Technological improvements and government incentives in many countries support such increase. By 2050, China, India, OECD Europe, and the United States will have almost 75% of the world's renewables generation. Growth in these regions results from both policy and, in the case of India and China, increasing demand for new sources of generation. Natural gas generation grows by an average of 1.5% per year from 2018 to 2050, and nuclear generation grows by 1.0% per year. The level of coal-fired generation remains relatively stable, but its share of electricity generation declines from 35% in 2018 to 22% by 2050 as total generation increases. By 2025, renewables surpass coal as the primary source for electricity generation, and by 2050, renewables account for almost half of total world electricity generation.

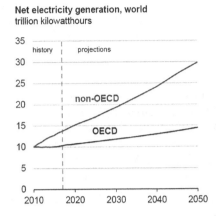 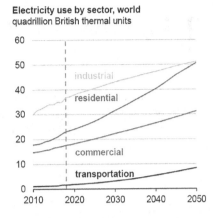

FIGURE 19.1 World electricity generation for 2010–2020. (From U.S. Energy Information Administration, Office of Energy Analysis, U.S. Department of Energy, International Energy Outlook with Projections to 2050, published in Sept. 2019.)

With modest growth, hydropower's share of renewables generation falls from 62% in 2018 to 28% in 2050 because the resource availability in OECD countries and environmental concerns in many countries limit the number of new mid- and large-scale projects. Generation from non-hydropower renewables increases an average of 5.7% per year from 2018 to 2050. By 2050, China, India, OECD Europe, and the United States are responsible for more than 80% of the world's non-hydropower renewables generation. Among renewable energy sources, electricity generation from wind and solar resources increase the most between 2018 and 2050, reaching 6.7 trillion and 8.3 trillion kilowatt hours (kWh), respectively, as these technologies become more cost competitive and are supported by government policies in many countries. By 2050, wind and solar account for over 70% of total renewables generation.

OECD and non-OECD countries have vastly different electricity demand growth profiles. OECD demand for net electricity generation (to the grid) grows at 1.0% per year between 2018 and 2050, and non-OECD demand grows at 2.3% per year on average. Although renewables are cost-competitive compared with new fossil-fired electric generating capacity additions, displacing existing non-renewable capacity requires policy incentives. In OECD countries, where more policy initiatives affect electric generation, demand growth is met primarily with renewables, which also displace some existing generation. On the other hand, growth in non-OECD electricity demand is met by a mix of renewables and nonrenewable generating technologies, generally influenced by regional resource and economic considerations.

As electricity demand continues to grow, albeit at a slower pace than in the early 2000s, renewable sources provide an increasing share of electricity generation and capacity. The electricity generation share of wind and solar grows from 12% in 2018 to 42% in 2050.

Figure 19.3 shows the world population. Worldwide, the amount of energy used per unit of economic production (energy intensity) has declined steadily for many

The Future

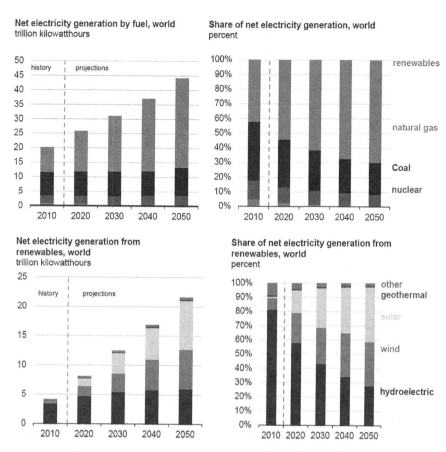

FIGURE 19.2 Net electricity generation by fuel. (From U.S. Energy Information Administration, Office of Energy Analysis, U.S. Department of Energy, International Energy Outlook with Projections to 2050, published in Sept. 2019.)

years. The amount of carbon dioxide (CO_2) emissions per unit of energy consumption (carbon intensity) declined in OECD countries since 2008 and in non-OECD countries since 2014. Energy intensity continues to decline in both OECD and non-OECD countries because of efficiency gains and gross domestic product (GDP) increases generated from low-energy-intensive services. Carbon intensity continues to decline largely because of China and other countries' move away from coal; worldwide growth in the use of non-CO2-emitting sources of energy, such as wind and solar; and improvements in process efficiencies [1].

19.2 FUTURE OF WIND POWER

The U.S. DOE reports that wind is experiencing the strongest growth among the renewable energy sources with a decreasing generation cost. Figure 19.4 shows wind vision published by the U.S. Department of Energy (DOE) [2]. As seen in 2020 there are 113.43 GW of wind generation across 36 U.S. states, both in off-shore and

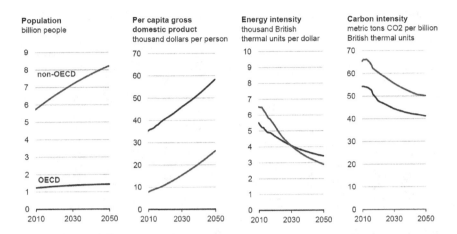

FIGURE 19.3 World population, GDP, and CO_2. (From U.S. Energy Information Administration, Office of Energy Analysis, U.S. Department of Energy, International Energy Outlook with Projections to 2050, published in Sept. 2019.)

on-shore generation. This is projected to increase to 404.25 GW by the year 2050 across 48 U.S. states, i.e. an increase of 256%.

The international renewable energy agency (IRENA) reported an average cost of onshore and offshore wind power installation at 1,497 USD/kW and 4,353 USD/kW for the year 2018 as listed in Table 19.1 [3]. The cost of installation for both onshore and offshore wind is projected to decrease by around 35% by the year 2050 while the investment per year increases my more than three-fold. By the year 2050, the cost of electricity from the wind power is projected to decrease to 3 cents/kWh and 7 cents/kWh for onshore and offshore wind, respectively. The levelized cost of electricity (LCOE) for onshore wind is already competitive compared to all fossil fuel generation sources and is set to decline further as installed costs and performance continue to improve [3]. Globally, the LCOE for onshore wind will continue to fall from an average of USD 0.06 per kWh in 2018 to between USD 0.03 to 0.05/kWh by 2030 and between USD 0.02 to 0.03/kWh by 2050. The LCOE of offshore wind is already competitive in certain European markets (for example, Germany, the Netherlands with zero-subsidy projects, and lower auction prices). Offshore wind would be competitive in other markets across the world by 2030, falling in the low range of costs for fossil fuels (coal and gas). The LCOE of offshore wind would drop from an average of USD 0.13/kWh in 2018 to an average between USD 0.05–0.09/kWh by 2030 and USD 0.03–0.07/kWh by 2050 [3].

As IRENA reported [3], by the year 2050, among all low-carbon technology options, accelerated deployment of wind power when coupled with deep electrification would contribute to more than one-quarter of the total emissions reductions needed, which is nearly 6.3 gigatonnes of carbon dioxide (Gt CO_2) annually in 2050. Wind power, along with solar energy, would lead the way for the transformation of the global electricity sector. Onshore and offshore wind would generate more than one-third (35%) of total electricity needs, becoming the prominent generation source by 2050.

The Future

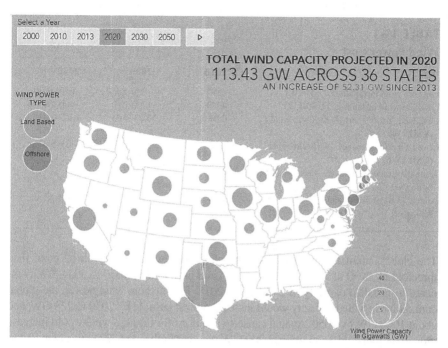

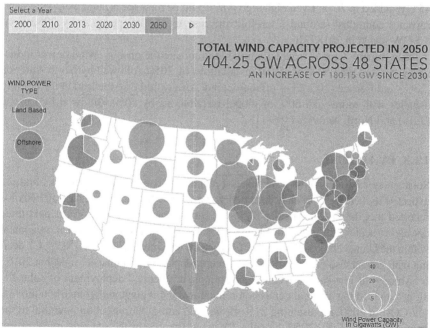

FIGURE 19.4 Wind power vision. (From U.S. Department of Energy, 2020)

TABLE 19.1
Wind Power Cost

Item	2018	2030	2050
Onshore wind installation (USD/kW)	1497	800–1350	650–1000
Offshore wind installation (USD/kW)	4353	1700–3200	1400–2800
Levelized cost of energy—Onshore wind (USD/kWh)	0.06	0.03–0.05	0.02–0.03
Levelized cost of energy—Offshore wind (USD/kWh)	0.13	0.05–0.09	0.03–0.07
Annual average investment—Onshore (USD billion/yr)	67	146	211
Annual average investment—Offshore (USD billion/yr)	19.4	61	100

The global cumulative installed capacity of onshore wind power will be more than threefold by 2030 to 1787 gigawatts (GW) and nine-fold by 2050 to 5044 GW compared to installed capacity of 542 GW in 2018. For offshore wind power, the global cumulative installed capacity would increase almost ten-fold by 2030 to 228 GW, and about 1000 GW by 2050. Annual capacity additions for onshore wind would increase more than four-fold, to more than 200 GW per year in the next 20 years, compared to 45 GW added in 2018. Even higher growth would be required in annual offshore wind capacity additions—around a ten-fold increase to 45 GW per year by 2050 from 4.5 GW added in 2018 [3].

Asia (mostly China) would continue to dominate the onshore wind power industry, with more than 50% of global installations by 2050, followed by North America (23%) and Europe (10%). For offshore wind, Asia would take the lead in the coming decades with more than 60% of global installations by 2050, followed by Europe (22%) and North America (16%) [3].

19.3 PV FUTURE

Solar energy developments have taken a significant rise in the past decade, and are expected to increase with an accelerated growth in the next three decades. IRENA [4] reported that the growth in solar PV power deployment is required in the next three decades to achieve the Paris climate goals. The Paris Climate Agreement is an agreement within the United Nations Framework Convention on Climate Change (UNFCCC), dealing with greenhouse-gas-emissions mitigation, adaptation, and finance, signed in 2016.

Among all low-carbon technology options, accelerated deployment of solar PV alone can lead to significant emission reductions of 4.9 gigatonnes of carbon dioxide (Gt CO_2) in 2050, representing 21% of the total emission mitigation potential in the energy sector [4]. By 2050 the solar PV would represent the second-largest power generation source, just behind wind power and lead the way for the transformation of the global electricity sector. Solar PV would generate a quarter (25%) of total electricity needs globally, becoming one of prominent generations source by 2050 [4].

Installed capacity of solar PV was at 480 GW in 2018, where it is projected to increase by 492% and 1675% by the year 2030 and 2050, respectively. This increase

TABLE 19.2
Solar PV Development

Item	2018	2030	2050
Installed capacity (GW)	480	2840	8519
Annual deployment (GW/yr)	94	270	372

TABLE 19.3
Solar PV Power Cost

Item	2018	2030	2050
Installation cost (US$/kW)	1210	834-340	481-165
Levelized cost of energy (US$/kWh)	0.085	0.08-0.02	0.05-0.01
Annual average investment (US$ billion/yr)	114	165	192

is due to more countries setting goals to achieve a target emission reduction as well as improved technology, and reduced cost. The total investment in solar PV is projected to be around US$ 192 billion/year by the year 2050. The installation cost is projected to reduce to 481 US$/kW by the year 2050, i.e. a 60% drop compared to year 2018. The levelized cost of solar energy is expected to continue to decline, and by the year 2030 it could be as low as 2 cents/kWh and reaches a low of 1 cents/kWh in 2050. Table 19.2 and Table 19.3 list some of the projected values for solar installed capacity, investment and energy cost up to year 2050.

19.4 DECLINING PRODUCTION COST

Economies of scale are expected to continue to contribute to declining prices. Future wind plants will undoubtedly be larger than those installed in the past, and the cost per square meter of the blade-swept area will decline with size. Figure 19.5 shows the year 2000 costs of wind turbines of various sizes in Germany. The price is in DM/m2 of the blade area, falling from 800 to 1200 DM/m^2 for small turbines to 500 to 800 DM/m^2 for large turbines of 50-m diameter. The line showing the annual energy potential per square meter rises from 630 kWh/m^2 in small turbines to 1120 kWh/m^2 in large turbines of 46-m diameter.

With the advances in technology and the economies of scale combined, the manufacturing cost of new technologies has historically shown declining patterns. The growth of a new product eventually brings with it a stream of competitors, and the learning curve decreases the cost. A learning-curve hypothesis has been commonly used to model such cost declines in new technologies. The cost is modeled as an exponentially decreasing function of the cumulative number of units produced up to that time. A standard form for a cost decline is constant doubling, in which the cost is discounted by a fraction λ when the cumulative production doubles. For renewable electricity, the production units are megawatts of capacity produced and kWh of

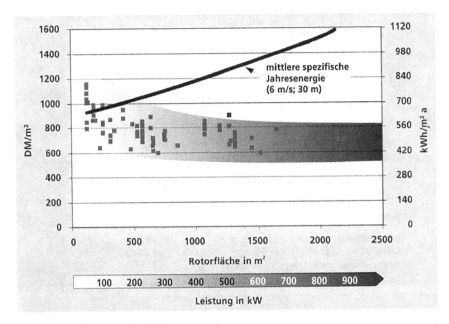

FIGURE 19.5 Economies-of-scale trends for capital cost in DM/m² and annual energy produced in kWh/m² vs. blade-swept area. (From Institute of Solar Energy and Technology, University of Kassel, Germany. With permission.)

electricity produced. If we want to monitor only one production unit, kWh is preferred because it includes both the MW capacity installed and the length of the operating experience, thus making it an inclusive unit of production.

For new technologies, the price keeps declining during the early phase of a new product at a rate that depends on the nature of the technology and the market. The pattern varies widely as seen in new technologies such as those in the computer, telephone, and airline industries. For the early wind turbines, the cost per kW capacity has declined with the number of units produced, as seen in Figure 19.6.

In general, the future cost of a new product can be expressed as follows:

$$C(t) = C_o \lambda^{\log_2(N_t/N_o)} \quad 0 < \lambda < 1 \tag{19.1}$$

where

$C(t)$ = cost at time t
C_o = cost at the reference time
N_t = cumulative production at time t
N_o = cumulative production at reference time.

Estimating the parameter λ based on historical data is complex, and the estimate itself may be open to debate. However, the expression itself has been shown to be valid for new technologies time and again in the past. It may be used to forecast future trends in the capital cost decline in any new technology, including wind and PV power.

The Future

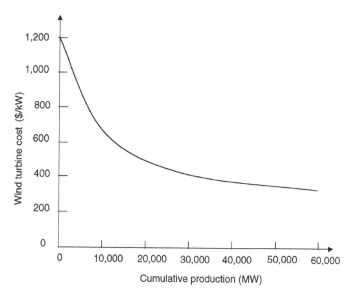

FIGURE 19.6 Learning curve of the wind turbine costs (historical data).

19.5 MARKET PENETRATION

With all market forces combined and working freely, the penetration of a new products and technologies over a period of time takes the form of an S-shaped curve, as shown in Figure 19.7. It is characterized by a slow initial rise, followed by a period of rapid growth, leveling off to a saturation plateau, and finally declining to make room for a newer technology. The S-shape hypothesis is strongly supported by empirical evidence.

One model available for predicting the market penetration of a new product is the diffusion model proposed by Bass.[3] However, it can be argued that renewable energy is not really a new product. It merely substitutes for an existing product. The Bass model, therefore, may not be appropriate for renewable power. Because electricity is a basic need of society, the penetration of renewable power technologies is better compared with similar substitutions in the past, such as in the steel-making industry, as shown in Figure 19.8. The solid lines are the actual penetration rates seen in those industries.

With ongoing developments in wind and PV power technology and with adjustments in social attitudes, the market penetration rate may follow a line parallel to the historical experience with similar products as depicted in Figure 19.8. It can be analytically represented by the Fisher–Pry substitution model.[4] In this model, the rate at which wind and PV energy may penetrate the market can be expressed as a fraction of the total kWh energy consumed every year.

If f is the fraction of the market captured by the renewables at time t, and t_o the time when f equals ½, then f can be expressed as follows:

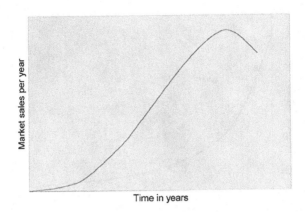

FIGURE 19.7 S-shaped growth and maturity of new products and technologies.

$$f = \frac{1}{1+e^{[-b(t-t_0)]}} \quad \text{or} \quad \frac{f}{1-f} = e^{-b(t-t_0)} \tag{19.2}$$

Here b is the growth constant, which characterizes the growth to the potential associated with a particular technology. The equation gives log-linear straight lines, as shown in Figure 19.8.

We must mention here that wind power was successfully demonstrated to be commercially viable in the 1970s and commercial production started in 1985, when a group of wind farms in California were installed and operated for profit by private investors. Based on this, and on the Fisher–Pry model just described, we can draw the dotted line in Figure 19.8, beginning in 1985 and then running parallel to the lines followed by similar substitution products in the past. That is the rate of market penetration we can expect wind power to follow in the future.

At the rate indicated by the dotted line, the full 100% potential of wind will perhaps be realized around 2065. Wind power's reaching 100% potential in the year 2065 does not mean that it will completely replace thermal and other power. It merely means that wind power will attain its full potential, whatever it may be. Experts will argue about the upper limit of this potential for a given country. However, wind and PV power, being intermittent sources of energy, cannot be the base-load provider. They can augment a base-load plant, thermal or other type, which can dispatch energy on demand. Such reasoning puts the upper limit of wind and PV power at well below 50%, perhaps around 33%. According to a 1997 study published in the U.K. by the Royal Institute of International Affairs, all renewable sources could together provide between 25 and 50% of European electricity by the year 2030. The contributions of wind and PV in meeting the total electricity demand, however, will largely depend on the operating experience gained with grid-connected plants and energy storage technologies developed during the next few decades that can remedy the nondispatchable nature of wind and PV energy. New energy storage products are being developed that incorporate battery, flywheel, fuel cell, and

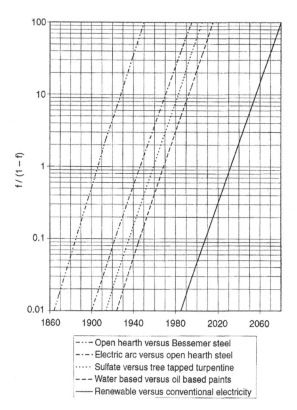

FIGURE 19.8 Market penetration of a new product substituting for existing products. Wind and PV shown in parallel with historical data. (From Fisher, J.C. and Pry, R.H., A simple substitution model of technological change, *Technology Forecasting and Social Change*, Vol. 3, 79–88, 1971.)

superconducting magnet into system solutions for the power industry, such as load leveling and power quality improvements. Electric utility restructuring and the subsequent increase in competition are creating new and higher-value markets for energy storage. Wind and PV power are the new market opportunities for large-scale energy storage products.

Sophisticated prediction models have been developed by the DOE, NREL, and others based on mathematical considerations and regression analyses of applicable data. However, there is a limit to how much such models can be confidently used to commit huge sums of capital investment. Small errors in data can cause large errors in projections, particularly decades in the future. Time-proven data on how people make investment decisions when faced with long-term uncertainties can also help. Experience indicates that investors commit funds only if the payback period is less than 5 yr; the shorter it is, the quicker the investment commitments are made. The upper limit of the payback period with significant market penetration is 4–5 yr, and large market penetration requires a payback period of less than 3 yr.

REFERENCES

1. U.S. Energy Information Administration, Office of Energy Analysis, U.S. Department of Energy, International Energy Outlook with Projections to 2050, September 2019.
2. U.S. Department of Energy, Wind Vision, 2020.
3. International Renewable Energy Agency (IRENA), FUTURE OF WIND, Deployment, Investment, Technology, Grid Integration and Socio-economic Aspects, October 2019.
4. International Renewable Energy Agency (IRENA), FUTURE OF SOLAR PHOTOVOLTAIC Deployment, Investment, Technology, Grid Integration and Socio-economic Aspects, November 2019.
5. Omid Beik, An HVDC Off-shore Wind Generation Scheme with High Voltage Hybrid Generator, Ph.D. thesis, McMaster University, 2016.
6. Omid Beik, Ahmad S. Al-Adsani, *DC Wind Generation Systems, Design, Analysis, and Multiphase Turbine Technology*, Springer, 2020.

Part D

Ancillary Power Technologies

20 Solar Thermal System

The solar thermal power system collects the thermal energy in solar radiation and uses it at high or low temperatures. Low-temperature applications include water and room heating for commercial and residential buildings.[1] High-temperature applications concentrate the sun's heat energy to produce steam for driving electrical generators. Concentrating solar power (CSP) technology has the ability to store thermal energy from sunlight and deliver electric power during dark or peak-demand periods. Its usefulness has been demonstrated on a commercial scale, with research-and-development funding primarily from the government and active participation of some electric utility companies. Therefore, CSP technology promises to deliver low-cost, high-value electricity on a large scale. This chapter covers such high-temperature solar thermal power systems.

Figure 20.1 is a schematic of a large-scale solar thermal power station developed, designed, built, tested, and operated with the Department of Energy (DOE) funding. In such a plant, solar energy is collected by thousands of sun-tracking mirrors, called *heliostats*, which reflect the sun's energy to a single receiver atop a centrally located tower. This enormous amount of energy that is focused on the receiver tower is used to melt a salt at high temperature. The hot molten salt is stored in a storage tank and used when needed to generate steam and drive a turbine generator. After generating steam, the used molten salt, now at low temperature, is returned to the cold-salt storage tank. From here, the salt is pumped to the receiver tower to be heated again for the next thermal cycle. The usable energy extracted during such a thermal cycle depends on the working temperatures. The maximum thermodynamic conversion efficiency that can be theoretically achieved with the hot-side temperature T_{hot} and the cold-side temperature T_{cold} is given by the Carnot cycle efficiency, which is as follows:

$$\eta_{carnot} = \frac{T_{hot} - T_{cold}}{T_{hot}} \quad (20.1)$$

where the temperatures are in degrees Kelvin. A higher hot-side working temperature and a lower cold-side exhaust temperature give higher plant efficiency for converting the captured solar energy into electricity. The hot-side temperature, however, is limited by the properties of the working medium. The cold-side temperature is largely determined by the cooling method and the environment available to dissipate the exhaust heat.

A major benefit of this scheme is that it incorporates thermal energy storage for several hours with no degradation in performance or for longer with some degradation. This feature makes this technology capable of producing high-value electricity for meeting peak demands. Moreover, compared with the solar photovoltaic (PV) system, the solar thermal system is economical and more efficient because it

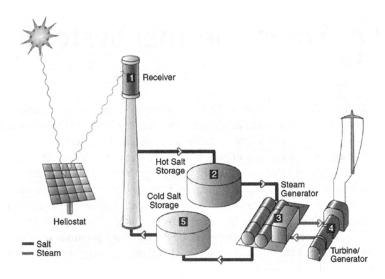

FIGURE 20.1 Solar thermal power plant schematic for generating electricity.

eliminates use of costly PV cells and alternating current (AC) inverters. It is, however, limited to large-scale applications.

20.1 ENERGY COLLECTION

CSP research and development focuses on three types of concentrators, which use different kinds of concentrating mirrors to convert the sun's energy into high-temperature heat energy. The three alternative configurations of the concentrators are shown in Figure 20.2. Their main features and applications are as described in the following subsections:

20.1.1 PARABOLIC TROUGH

The parabolic trough system is by far the most commercially matured of the three technologies. It focuses sunlight on a glass-encapsulated tube running along the focal line of a collector. The tube carries a heat-absorbing liquid, usually oil, that heats water to generate steam. More than 350 MW of parabolic trough capacity has been in operation in the California Mojave Desert since the early 1990s. It is connected to Southern California Edison's utility grid.

20.1.2 CENTRAL RECEIVER

In the central receiver system, an array of field mirrors focus sunlight on a central receiver mounted on a tower. To focus sunlight on the central receiver at all times, each heliostat is mounted on a dual-axis sun tracker to seek a position in the sky that is midway between the receiver and the sun. Compared with the parabolic trough, this technology produces a much higher concentration and hence a higher temperature of

Solar Thermal System

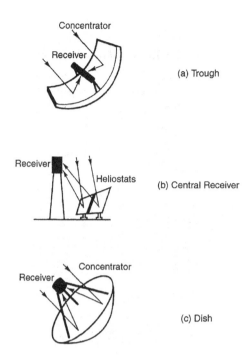

FIGURE 20.2 Alternative thermal-energy collection technologies.

the working medium, usually a salt. Consequently, it yields higher Carnot efficiency and is well suited for utility-scale power plants of tens of hundreds of megawatt capacity.

20.1.3 PARABOLIC DISH

A parabolic dish tracks the sun to focus heat, which drives a Stirling heat engine that is connected to an electrical generator. This technology has applications in relatively small capacity (tens of kilowatts) due to available engine size and wind load on the collector dishes. Because of their small size, they are more modular than other solar thermal power systems and can be assembled in capacities ranging from a few hundred kilowatts to a few megawatts. This technology is particularly attractive for small stand-alone remote applications. The three alternative solar thermal technologies are compared in Table 20.1.

20.2 SOLAR-II POWER PLANT

Central receiver power tower technology with a solar power tower, having a higher potential of generating low-cost electricity on a large scale, is getting a new developmental thrust in the U.S. An experimental 10-MW_e power plant using this technology was built and commissioned in 1996 by the DOE in partnership with the Solar-II Consortium of private investors led by Southern California Edison, the

TABLE 20.1
Comparison of Alternative Solar Thermal Power System Technologies

Technology	Solar Concentration (x Suns)	Operating Temperature (Hot Side)	Thermodynamic Cycle Efficiency
Parabolic trough receiver	100	300–500°C	Low
Central receiver power tower	1000	500–1000°C	Moderate
Dish receiver with engine	3000	800–1200°C	High

FIGURE 20.3 Solar-II plant site view in Barstow, CA. (From: U.S. Department of Energy.)

second-largest electric utility company in the U.S. It is connected to the grid and has enough capacity to power 10,000 homes. The plant is designed to operate commercially for 25–30 yr. Figure 20.3 is a photograph of this plant site located east of Barstow, CA. It uses some components of the Solar-I plant, which was built and operated at this site using central receiver power tower technology. The Solar-I plant, however, generated steam directly to drive the generator without the thermal storage feature of the Solar-II plant.

The Solar-II central receiver (Figure 20.4) was developed by the Sandia National Laboratory. It raises the salt temperature to 1050°F. The most important feature of the Solar-II design is its innovative energy collection and storage system. It uses a salt that has excellent heat-retention and heat-transfer properties. The heated salt can be used immediately to generate steam and electric power. Or it can be stored for use during cloudy periods or to meet the evening load demand on the utility grid after the sun has set. Because of this unique energy storage feature, power generation is decoupled from energy collection. For an electric utility, this storage capability is crucial, in that the energy is collected when available and used to generate high-value

Solar Thermal System

FIGURE 20.4 Experimental 1050°F thermal receiver tower for Solar-II power plant. (From: DOE/Sandia National Laboratory.)

electricity when most needed to meet peak demands. The salts selected by the Sandia laboratory for this plant are sodium nitrate and potassium nitrate, which work as a single-phase liquid that is colorless and odorless. It has the needed thermal properties up to the operating temperature of 1050°F. Moreover, it is inexpensive and safe.

Table 20.2 and Table 20.3 give the technical design features of the experimental Solar-II power plant. The operating experience to date indicates an overall plant capacity factor of 20%, and an overall thermal-to-electrical conversion efficiency of 16%. It is estimated that 25% overall efficiency can be achieved in a commercial plant design using this technology.

20.3 SYNCHRONOUS GENERATOR

The electromechanical energy conversion in the solar thermal power system is accomplished by the synchronous machine, which runs at a constant speed to produce 60-Hz electricity. This power is then directly used to meet the local loads and/or to feed the utility grid lines.

The electromagnetic features of the synchronous machine are shown in Figure 20.5. The stator is made of conductors, consisting of three interconnected phase coils, placed in slots of magnetic iron laminations. The rotor consists of magnetic poles created by the field coils carrying direct current (DC). The rotor is driven by a steam turbine to create a rotating magnetic field. Because of this rotation, the conventional rotor field coils use slip rings and carbon brushes to supply DC power from a

TABLE 20.2
Solar-II Design Features

Site
Mojave desert in California
1,949 ft above sea level
7.5 kWh/m^2 annual average daily insolation
95 acres of land

Tower
Reused from Solar-I plant
277 ft to top of the receiver
211 ft to top of BCS deck

Heliostats
1,818 Solar-I heliostats, 317.1 m^2, 91% reflectivity
108 new Lug heliostats, 95.1 m^2, 93% reflectivity
81,000 m^2 total reflective surface
Can operate in winds up to 35 mph

Receiver
New for Solar-II plant
Supplier: Rockwell
42.2-MW thermal power rating
Average flux 429 suns (429 kW/m^2)
Peak flux 800 suns
24 panels, 32 tubes per panel
20-ft tall and 16.6-ft diameter
0.8125-in. tube OD
0.049-in. tube wall thickness
Tubes 316H stainless steel

Thermal Storage System
Supplier: Pitt Des Moines
Two new 231,000-gal storage tanks, 38-ft ID
Cold tank carbon steel, 25.8-ft high, 9-in. insulation
Hot tank 304 stainless steel, 27.5-ft high, 18-in. insulation
3 h of storage at rated turbine output

Nitrate Salt–Chilean Nitrate
60% NaNO$_3$, 40% KNO$_3$
Melting temperature 430°F
Decomposing temperature 1,100°F
Energy storage density two thirds that of water
Density two times that of water
Salt inventory 3.3 million lb

Steam Generator
Supplier: ABB Lummus
New salt-in-shell superheater
New slat-in-tube kettle boiler
New salt-in-shell preheater

Turbine Generator
Supplier: General Electric Company
Refurbished from Solar-I plant
10 MW$_e$ net
12 MW$_e$ gross

Source: From U.S. Department of Energy and Southern California Edison Company.

Solar Thermal System

TABLE 20.3
Solar-II Operating Features

Thermodynamic Cycle

Hot-salt temperature: 1,050°F
Cold-salt temperature: 550°F
Steam temperature: 1,000°F
Steam pressure: 1,450 psi
Receiver salt flow rate: 800,000 lb/h
Steam generator flow rate: 660,000 lb/h

Electric Power Generator

Capacity: 10 MW$_e$
Capacity factor: 20%
Overall solar electric efficiency: 16%
Cost of conversion from Solar-I: $40 million

Source: From U.S. Department of Energy and Southern California Edison Company.

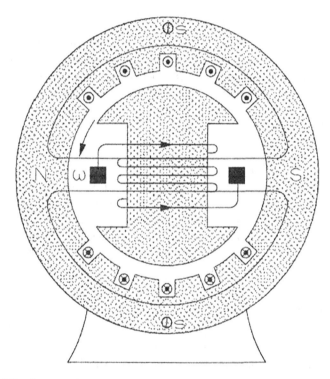

FIGURE 20.5 Cross-sectional view of the synchronous generator.

stationary source. Modern rotors, however, are made brushless by using selfexcitation with rotating diodes on the same shaft.

The stator conductors are wound in three groups connected in a three-phase configuration. Under the rotating magnetic field of the rotor, the three phase coils generate AC voltages that are 120° out of phase with each other. If the electromagnetic structure of the machine has p pole pairs and it is required to generate electricity at frequency f, then the rotor must rotate at N revolutions per minute given by the following:

$$N = 60\frac{f}{p} \qquad (20.2)$$

The synchronous machine must operate at this constant speed to generate power at the specified frequency. In a stand-alone solar thermal system, small speed variations can be tolerated within the frequency tolerance band. If the generator is connected to the grid, it must be synchronous with the grid frequency and must operate exactly at this grid frequency at all times. Once synchronized, such a machine has an inherent tendency to remain synchronous. However, a large, sudden disturbance such as a step load can force the machine out of synchronism, as discussed in Subsection 20.3.4.

20.3.1 Equivalent Electrical Circuit

The equivalent electrical circuit of the synchronous machine is represented by a source of alternating voltage E and an internal series resistance R_s and reactance X_s representing the stator winding. The resistance, being much smaller than the reactance, can be ignored to reduce the equivalent circuit to a simple form, as shown in Figure 20.6. If the machine supplies a load current I that lags behind the terminal voltage V by phase angle Φ, it must internally generate the voltage E, which is the phasor sum of the terminal voltage and the internal voltage drop IX_s. The phase angle between V and E is called the power angle. At zero power output, the load current is zero and so is the IX_s phasor, making V and E in phase with zero power angle. Physically, the power angle represents the angle by which the rotor's magnetic field leads the stator-induced rotating magnetic field. The output power can be increased, up to a certain limit, by increasing the power angle. Beyond this limit, the rotor and stator fields would no longer follow each other in a lockstep and will step out of the synchronous mode of operation. In the nonsynchronous mode, steady power cannot be produced.

20.3.2 Excitation Methods

The synchronous machine's excitation system is designed to produce the required magnetic field in the rotor, which is controllable, in order to control the voltage and reactive power of the system. In modern high-power machines, X_s can be around 1.5 times the base impedance of the machine. With reactance of this order, the phasor

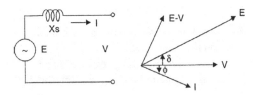

FIGURE 20.6 Equivalent electrical circuit and phasor diagram of the synchronous machine.

diagram in Figure 20.6 shows that the rotor field excitation required at the rated load (100% load at 0.8 lagging power factor) is more than twice that at no load with the same terminal voltage. The excitation system has the corresponding current and voltage ratings, with capability of varying the voltage over a wide range of 1–3, or even more, without undue saturation in the magnetic circuit. Excitation power, which is required primarily to overcome the rotor winding I^2R loss, ranges from ½ to 1% of the generator rating. Most excitation systems operate at 200–1000 V DC.

For large machines, four types of excitation systems—DC, AC, static, and brushless—are possible. In the DC system, a suitably designed DC generator supplies the main field winding excitation through conventional slip rings and brushes. Due to low reliability and high maintenance requirements, the conventional DC machine is seldom used in the synchronous machine excitation system.

Many utility-scale generators use the AC excitation system shown in Figure 20.7. A pilot exciter excites the main exciter. The AC output of a permanent-magnet pilot exciter is converted into DC by a floor-standing rectifier and supplied to the main exciter through slip rings. The main exciter's AC output is converted into DC by means of a phase-controlled rectifier whose firing angle is changed in response to the terminal-voltage variations. After filtering the ripples, this DC is fed to the synchronous generator field winding.

An alternative scheme is static excitation, as opposed to dynamic excitation described in the preceding paragraph. In static excitation, the controlled DC voltage is obtained in a rectified and filtered form from a suitable stationary AC source. The DC voltage is then fed to the main field winding through slip rings. This excitation scheme has a fast dynamic response and is more reliable because it has no rotating exciters.

Modern utility-scale generators use the brushless excitation system. The exciter is placed on the same shaft as the main generator. The AC voltage induced in the exciter is rectified and filtered to a DC voltage by rotating diodes on the shaft. The DC is then fed directly into the rotor field coil. Such a design eliminates the need for slip rings and brushes.

The excitation control system model in analytical studies must be carefully done as it forms a multiple feedback control system that can become unstable. IEEE has developed industry standards for modeling excitation systems. The model must

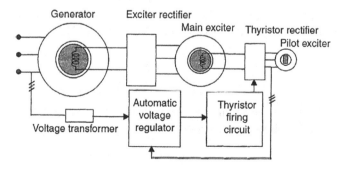

FIGURE 20.7 AC excitation system for the synchronous generator.

account for the nonlinearity due to magnetic saturation present in all practical designs. Stability can be improved by supplementing the main control signal with auxiliary signals such as for speed and power.

20.3.3 Electric Power Output

The electric power output per phase of the synchronous machine is as follows in units of W/phase:

$$P = VI \cos\phi \qquad (20.3)$$

Using the phasor diagram in Figure 20.6, the current can be expressed as follows:

$$I = \frac{E - V}{jX_s} = \frac{E < \delta - V < 0}{jX_s} = \frac{E(\cos\delta + j\sin\delta) - V}{jX_s} \qquad (20.4)$$

The real part of this current is $I_{real} = E \sin\delta / X_s$.

This part, when multiplied with the terminal voltage V, gives the output power per phase in units of W/phase:

$$P = \frac{VE}{X_s} \sin\delta \qquad (20.5)$$

Output power vs. power angle is a sine curve shown by the solid line in Figure 20.8, having the maximum value at $\delta = 90°$. The maximum power in W/phase that can be generated by the machine is therefore given by

$$P_{max} = \frac{VE}{X_s} \qquad (20.6)$$

Some synchronous machine rotors have magnetic saliency in the pole structure. The saliency produces a small reluctance power superimposed on the main power, modifying the power-angle curve as shown by the dotted line in Figure 20.8.

The electromechanical torque required at the shaft to produce this power is the power divided by the angular velocity of the rotor in units of N·m/phase:

$$T_e = \frac{VE}{\omega X_s} \sin\delta \qquad (20.7)$$

The torque also has a maximum limit corresponding to the maximum power limit and is given by the following in units of N·m/phase:

$$T_{max} = \frac{VE}{\omega X_s} \qquad (20.8)$$

20.3.4 Transient Stability Limit

The maximum power limit described earlier is called the *steady-state stability limit*. Any load beyond this value will cause the rotor to lose synchronism and hence affects the power generation capability. The steady-state limit must not be exceeded under any condition, including those that can be encountered during transients. For example, if a sudden load step is applied to a machine initially operating at a steady-state load power angle δ_1 (Figure 20.9), the rotor power angle would increase from δ_1 to δ_2, corresponding to the new load that it must supply. This takes some time, depending on the electromechanical inertia of the machine. No matter how long it takes, the rotor inertia and the electromagnetic restraining torque will set the rotor in a mass-spring type of oscillatory mode, swinging the rotor power angle beyond its new

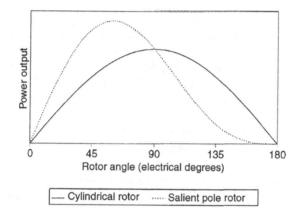

FIGURE 20.8 Power vs. power angle of cylindrical rotor and salient-pole synchronous machine.

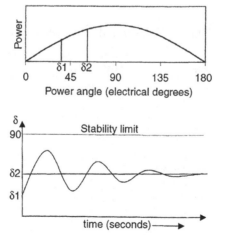

FIGURE 20.9 Load step transient and stability limit of the synchronous machine.

steady-state value. If the power angle exceeds 90° during this swing, machine stability and power generation are affected. For this reason, the machine can be loaded only to the extent that even under the worst-case load step, planned or accidental, or during all possible faults, the power angle swing will remain sufficiently below 90°. This limit on loading the machine is called the *transient stability limit*.

Equation 20.5 shows that the stability limit at given voltages can be increased by designing the machine with low synchronous reactance X_s, which is largely due to the stator armature reaction component.

20.4 COMMERCIAL POWER PLANTS

Commercial power plants using the solar thermal system are being explored in capacities of a few hundred MW_e. Based on the experience of operating Solar-II, the design studies made by the National Renewable Energy Laboratory (NREL) have estimated the performance parameters that are achievable for a 100-MW_e commercial plant. Table 20.4 summarizes these estimates and compares them with those achieved in an experimental 10-MW_e Solar-II power plant. The 100-MW_e prototype design studied showed that an overall (solar radiation to AC electricity) conversion efficiency of 23% could be achieved in a commercial plant using existing technology. For comparison, conventional coal thermal plants typically operate at 40% overall efficiency, and the PV power systems have an overall efficiency of 8–10% with amorphous silicon, 15–20% with crystalline silicon, and 30–35% with new thin-film multijunction PV cell technologies.

The major conclusions of the studies to date are the following:

1. Designing and building plants with capacities as large as 200 MW_e is possible, based on the demonstrated technology to date. Future plants could be larger. A 200-MW_e plant would require about 3 mi² of land.
2. The plant capacity factors up to 65% are possible.

TABLE 20.4
Comparison of 10-MW_e Solar-II and 100-MW_e Prototype Design

Performance Parameter	Solar-II Plant 10 MW_e (in %)	Commercial Plant 100 MW_e (in %)
Mirror reflectivity	90	94
Field efficiency	73	73
Mirror cleanliness	95	95
Receiver efficiency	87	87
Storage efficiency	99	99
Electromechanical conversion efficiency of generator and turbine	34	43
Auxiliary components efficiency	90	93
Overall solar-to-electric conversion efficiency	16	23

Source: From U.S. Department of Energy and Southern California Edison Company.

3. About 25% of the conversion efficiency of solar radiation to AC electricity is achievable annually.
4. The thermal energy storage feature of the technology can meet peak demand on utility lines.
5. The capital cost of $2000/kW_e$ for the first few commercial plants, and less for future plants, is estimated. The fuel (solar heat) is free.
6. A comparable combined-cycle gas turbine plant would initially cost $1000/kW_e$, and then the fuel cost would be added every year.

On a negative side, solar thermal power technology is less modular compared with PV and wind power. Its economical size is estimated to be in the range of 100–300 MW_e. The cost studies at NREL have shown that a commercially designed utility-scale power plant using central receiver power tower technology can produce electricity at a levelized cost of 7–10 cents/kWh, depending on the size.

20.5 RECENT TRENDS

The total installed capacity of the solar-driven steam power plants in the world today, however, is less than 1000 MW. Such solar thermal power plants are found to be economical in regions with an annual minimum direct normal solar energy of around 2000 kWh/m^2. The capital cost of a complete midsize (around 100 MW) power plant ranges between $3 and $5/W, and the energy costs ranges between 10 and 15 cents/kWh.

Solar Millennium AG of Germany's planned two 50-MW grid-connected solar thermal power plants near Granada in Spain had an estimated cost of $400 million ($8/watt) in 2010. The design uses parabolic trough concentrator technology (similar to that used in the Kramer Junction plant of U.S. Duke Solar Energy, Raleigh, NC) and plans to build and operate a 50-MW solar thermal power plant in Nevada using a nontracking parabolic concentrator, which collects almost 60% of the solar energy and reaches temperatures up to 160°C, high enough to produce steam.

Solarmundo in Belgium has developed low-cost concentrators to generate steam using flat rather than conventional parabolic mirrors, which are more expensive to manufacture. In a 2500-m^2 pilot program, the mirrors were arranged as in a Fresnel lens, i.e., fanned out in a number of different segments, each positioned at the optimum angle to the sun. The modular design is scalable up to 200 MW_e. A conceptual design study indicated that a 200-MW plant using this technology in a suitable site in North Africa could generate electricity at 4–8 cents/kWh.

Solar heat not being available on demand, solar–fossil hybrids are the next step in development of this technology. For example, an integrated combined cycle of 40-MW_e solar thermal with a 100-MW_e gas turbine power plant has been proposed at Jodhpur in Rajasthan, India.

REFERENCES

1. Mancini, T., *Solar Thermal Power Today and Tomorrow*, Mechanical Engineering, Publication of the Institution of Mechanical Engineers, London, August 1994.
2. Technology Report No. 14, Southern California Edison Company, Irwindale, CA, Fall 1995.

3. Teske, S., Solar Thermal Power 2020, *Renewable Energy World*, January–February 2004, pp. 120–124.
4. Omid Beik, Ahmad S. Al-Adsani, *DC Wind Generation Systems, Design, Analysis, and Multiphase Turbine Technology*, Springer, 2020.
5. D. Schumacher, O. Beik and A. Emadi, "Standalone Integrated Power Electronics System: Applications for Off-Grid Rural Locations," *IEEE Electrification Magazine*, vol. 6, no. 4, pp. 73–82, December 2018.
6. M. R. Aghamohammadi, and A. Beik-Khormizi, *"Small Signal Stability Constrained Rescheduling Using Sensitivities Analysis by Neural Network as a Preventive Tool,"* in *Proc. 2010 IEEE PES Transmission and Distribution Conference and Exposition*, pp. 1–5.
7. A. B. Khormizi and A. S. Nia, *"Damping of Power System Oscillations in Multi-machine Power Systems Using Coordinate Design of PSS and TCSC,"* in *2011 10th International Conference on Environment and Electrical Engineering (EEEIC)*, Rome. pp. 1–4

21 Ancillary Power Systems

There are various ways, other than conventional wind and photovoltaic (PV) systems, to generate electric power. The ultimate source of energy is, of course, the sun. This energy comes primarily in the form of heat. The sun's heat, in turn, produces wind in the air, marine currents in the ocean, and waves on the ocean surface. All of these carry energy derived from solar energy, which can be converted into electricity using various energy conversion technologies such as the wind and PV systems discussed in the preceding chapters. This chapter covers several other concepts that are being developed. Some have been successfully demonstrated by prototypes and have been proposed for large-scale power generation, seeking potential investors.

21.1 HEAT-INDUCED WIND POWER

The solar-heat-induced wind power schematic shown in Figure 21.1 uses a tall chimney surrounded by a large solar-heat-collecting roof made of glass or plastic sheets supported on a framework. Toward the center, the roof curves upward to join the chimney, thus creating a funnel. The sun heats the air underneath the roof. The heated air follows the upward incline of the roof until it enters the chimney, through which it flows at high speed and drives the wind turbines installed at the top exit. Electrical generators can be installed at the top or bottom of the chimney by means of a shaft and a gear. The large thermal time constant of the ground generates power even several hours after sunset. The overall efficiency of the plants is 3 to 5%, primarily depending on the chimney height. Because the capital cost is high, such a power plant is economical only in large capacities (>1 MW) in desert regions where the land is free or very cheap. However, the plant can be made more economical by using the land under the roof as a greenhouse for agricultural purposes.

In the arid flat land of southeastern Australia, EnviroMission of Melbourne has experimented with a solar wind tower (Figure 21.2). In their baseline design, the sun-capture area is an 11-km^2 glass-roof enclosure. The concrete chimney is 140 m in diameter and 1000 m tall. At the top, 32 wind turbines add to a total capacity of 200 MW of electric power. The estimated cost of $380 million was twice the cost of a conventional wind farm of similar capacity at the time. EnviroMission also evaluated sunny sites in the southwestern U.S. for similar power plants.

21.2 MARINE CURRENT POWER

Among the many forms of ocean energy, tidal or marine currents and waves offer two prominent short- to medium-term prospects for renewable power. The U.K. Department of Trade and Industry back in 2004 funded £50 million for marine research on harnessing wave and tidal stream power. Scotland is leading the way in this research, with the first such plant in operation on the island of Islay. The European

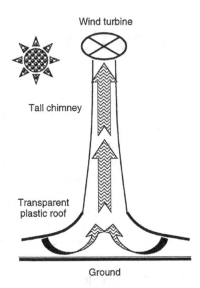

FIGURE 21.1 Solar-heat-induced wind chimney power plant schematic.

FIGURE 21.2 Proposed solar-heat-induced wind chimney power plant. (From EnviroMission Ltd, Melbourne, Australia.)

Marine Energy Center in Orkney, Scotland, offers development, testing, and monitoring of grid-connected wave energy conversion systems.

Three options available for power generation from the ocean are as follows:

- Wave energy
- Tidal energy
- Marine currents

Wave energy is ultimately a form of solar energy, which generates pressure differences in air and, hence, wind. The wind, in turn, generates waves on the ocean surface. Marine currents also result from solar heat. Tidal energy, on the other hand,

Ancillary Power Systems

results from the enormous movement of water due to gravitational forces of the sun and the moon on the seas.

Interest in marine current power has emerged only since 1990. Successful prototypes have been demonstrated using the conventional horizontal-axis turbine, although a Darrieus turbine can also be used. After a few successful demonstrations, two commercial grid-connected power generators were installed: one rated 300 kW with a horizontal-axis turbine in the U.K., and another rated 250 kW with a vertical-axis turbine in Canada. A 300-kW tidal current turbine has been installed 1 km off the shore of Devon, U.K., funded by the U.K., Germany, and the European Commission (Figure 21.3). The machine is built by the Marine Current Turbine Company and partly funded by London Electricity. It is mounted on steel pipes set into the seabed. Its 11-m-diameter blades generate 300 kW of power under tidal currents of 2.7 m/sec. The top end of the whole installation stands a few meters above the water surface. Another design was also developed that does not require anchorage to the seabed.

FIGURE 21.3 The 300-kW marine current power generator installed near Devon, U.K. (From Marine Current Turbines Ltd., Bristol, U.K. With permission.)

Marine current speeds tend to be maximum near the surface, so the turbine rotor must intercept as much of the depth as possible near the surface. The rotor can be secured in position in several ways similar to those used in the offshore oil-drilling and oil-mining platforms (Figure 21.4).

A study[1] funded by the European Commission estimated 48 TWh/yr of exploitable tidal current energy potential at more than 100 locations around Europe. Similar studies around the world have shown such energy potentials of 70 TWh/yr in the Sibulu Passage in the Philippines and 37 TWh/yr near the Chinese coasts.

The principle of converting this kinetic energy into electricity is similar to that in the wind turbine. The analytical methods covered in Chapter 3 are valid, except for the density and speed difference between the tidal current and the wind. No new theory is needed to design a tidal turbine. Density of water is 800 times that of air, but the current speed is about half that of air, so the theoretical tidal current power density per square meter of the blade-swept area is approximately 100 times higher than that in wind.

FIGURE 21.4 A proposed marine current tidal power system. (From Frankel, P., Marine Currents Turbine Ltd., Bristol, U.K. With permission.)

Tidal current turbines have two advantages over wind turbines:

- Higher power density per unit blade area.
- Tidal currents are more predictable.

For example, a marine current of 3 m/sec (6 kn) represents an impinging mechanical power of 14 kW/m^2 of the blade-swept area. A minimum current of about 1.25 m/sec, bringing 1 kW/m^2 of mechanical power, is practically required for a power plant. Thus, a potential site must have a sustained current exceeding 1 m/sec at sufficient depth for installation of a turbine of a required diameter for the desired electrical capacity.

21.3 OCEAN WAVE POWER

Ocean waves bring enormous amounts of energy to the shore, day and night, all year round. Tapping a small percentage of this energy can provide all the energy required worldwide.

The wind acting on the ocean surface generates waves. Hence, wave power is essentially wind power. The energy extracted from waves is replenished by the wind. Wave energy is thus a renewable source, more so than other renewable energy sources.

Large waves are generated by huge windstorms. Wave heights of 70 to 100 ft are possible in 70 to 100 mph storms. A wave generation mechanism, depicted in Figure 21.5, is as follows:[2]

1. Waves are generated due to the friction between the wind and the surface of the ocean. The stronger and longer the wind blows over wide, deep water, the higher the waves.
2. As the waves spread from the storm area, they become rolling swells. The water appears to be moving some distance, but it really just goes in circles. The energy of the waves, however, moves on in a domino-effect pattern.
3. When the wave hits an underwater obstruction, such as a reef or seamount, the shear friction due to the obstruction distorts the circular motion of the water, and the wave breaks up (trips over itself).

An estimated 2 to 3 million MW of power are contained in the waves breaking up on shores worldwide. At favorable coastlines, the power density could be 30 to 50 MW/km. A number of companies in the U.S., Europe, Canada, and around the world are engaged in extracting marine power to generate electricity. A dedicated facility for testing electricity generation from wave power has been established in Blyth, U.K. A 70-m × 18-m testing tank with wave-generating machines has been made from a dry dock.

A few alternative technologies can be considered for converting wave energy into electricity. One scheme uses the circulating water particles in the waves to create local currents to drive a turbine. A prototype built and connected to the Danish grid in 2002 is working well, even with irregular waves as small as 20 cm. A combination

(a) Storm generates waves by friction of wind against the water surface.

(b) As waves move farther, they become rolling swells. Water appears to be moving forward, but it only goes around in circles.

(c) An underwater reef or seamount breaks the wave, distorts the circular motion, and the wave basically trips over itself.

FIGURE 21.5 Ocean wave generation mechanism.

of Wells turbine and Darrieus rotor is used to capture the wave energy from both up-and-down and back-and-forward currents.[3] It works on a complex principle by which lift in a circulating flow is created using hydrofoil blades.

Another scheme, developed by Ocean Power Technologies (Princeton, NJ) uses the newly developed "power buoy" with a permanent magnet and a reciprocating coil tied to a floating buoy. The heaving motion of the buoy under the waves generates electrical power in the coil by Faraday's law of electromagnetic induction.

21.4 JET-ASSISTED WIND TURBINE

Stand-alone wind turbines depend on the wind blowing favorably at all times. The availability of electric power during low-wind conditions can be assured only if an extremely large energy storage battery is incorporated in the system, or if a diesel generator is available on the ground in addition to the wind generator on the tower. However, a jet-assisted wind turbine with fuel jets located at the tips of the rotor blades can assist the turbine in low wind or drive it when there is no wind. On remote farms, diesel is always available for use in farm equipment. In the absence of adequate wind, the diesel is burned and ejected in jets at the rotor tip to assist the wind turbine in maintaining the desired speed (Figure 21.6). Thus, fuel drums are sturdy,

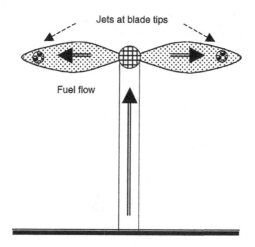

FIGURE 21.6 Jet-assisted wind turbine generator.

low-cost, and extremely reliable means of energy storage at power plants. The main savings come not from having an independent diesel generator but instead from using the same electrical generator in the wind turbine tower. Research funding has been granted by the California Energy Commission for such a scheme being developed by Appa Technology Initiatives (Lake Forest, CA).

21.5 BLADELESS WIND TURBINES

A bladeless turbine, as its name suggests, has no blades. It operates based on vibration and resonant principals created from the wind velocity. A bladeless turbine, usually, consists of a vertical cylinder that is fixed using an elastic rod. The bladeless turbine is structured in two parts, a mast and a base as shown in Fig. 21.7. It also contains an alternator, which generates AC power, a rectifier that rectifies the alternator's output and other components as seen in Fig. 21.7. As the wind velocity blows, the mast part of the bladeless wind turbine oscillates, which causes the magnets to create electrical induction in the coils of the alternator. Unlike the turbines that rotate, the movement of bladeless turbine is swinging as shown in Fig. 21.8. The alternator is therefore a non-rotational electromagnetic device. As the wind velocity varies, the strength of the induced voltage and its frequency varies. The generated voltage and currents are then rectified to DC, which are converted back to AC before connecting to the grid.

21.6 SOLAR THERMAL MICROTURBINE

In this scheme, the sun's energy is collected in the form of heat using a concentrator. The heat, in turn, is used to generate steam and drive a rotating microturbine generator (Figure 21.9). The scheme basically uses a thermodynamic cycle. The usable energy extracted during a thermal cycle depends on the working temperatures. The maximum thermodynamic conversion efficiency that can be theoretically achieved

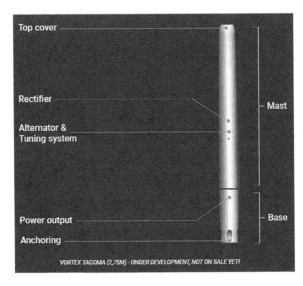

FIGURE 21.7 Bladeless wind turbine. (From Vortex Bladeless, Madrid, Spain)

with a hot-side temperature T_{hot} and a cold-side temperature T_{cold} is given by the Carnot cycle efficiency as:

$$\eta_{carnot} = \frac{T_{hot} - T_{cold}}{T_{hot}} \quad (21.1)$$

where the temperatures are in degrees Kelvin. A high hot-side working temperature and a low cold-side exhaust temperature give a high efficiency of converting the captured solar energy into electricity.

Compared to the solar PV system, the solar thermal microturbine system is economical, as it eliminates the costly PV cells and battery. The solar dynamic concentrator with a turbo alternator also offers significant advantage in efficiency and weight and hence in the overall cost over the solar PV technology.

Research funding was granted by the California Energy Commission for developing such a turbine proposed by Appa Technology Initiatives. The age-old Hero's steam engine concept has been found effective and affordable in designing a small 10 to 25 kW system suitable for residential or small-farm use. The engine consists of a single rotating composite disk having many steam jet nozzles embedded tangentially into its rim. The spinning disk drives a lightweight, high-speed electrical generator, the output of which is converted into utility-quality power by power electronics. Steam pressure in the range of 500 to 2,000 kPa yields thermodynamic efficiencies of 20 to 30% at disk speeds of 3,000 to 10,000 rpm. Simplicity and fewer components are key features of this scheme. The sun-to-electricity energy conversion efficiency is 15 to 20%, which is in the range of normal PV cell efficiency. On a sunny day, with solar radiation of 1,000 W/m², a parabolic dish of 5-m diameter can generate 3 kW of electric power using this scheme with an estimated cost of $5,000.

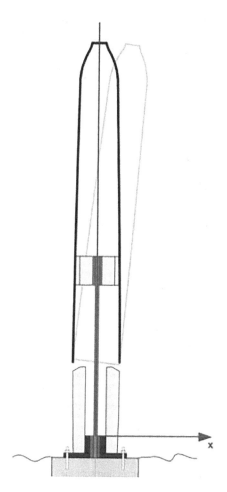

FIGURE 21.8 Operation of a bladeless wind turbine. (From Vortex Bladeless, Madrid, Spain)

21.7 THERMOPHOTOVOLTAIC SYSTEM

In the thermophotovoltaic (TPV) system, the solar heat is directed onto PV cells. The system can have a cylindrical or flat configuration, as shown in Figure 21.10. A heated surface radiates infrared (heat) onto an array of PV cells sensitive in the infrared range. A part of the energy is converted into DC, and some is reflected back and dissipated as heat. The energy conversion process is different from that in the conventional PV cell. The efficiency varies with the radiator temperature.

Most current TPVs use low-bandgap PV cells (0.55-eV InGaAs or 0.73-eV GaSb) to optimize the cell response to energy sources in the 1 to 2 μm range. The low bandgap of these cells leads to low open-circuit voltage (0.25 to 0.45 V) and poor fill factor caused by high intrinsic carrier concentration. The cell is operated at low temperatures, generally below 60°C, to have adequate output voltage.

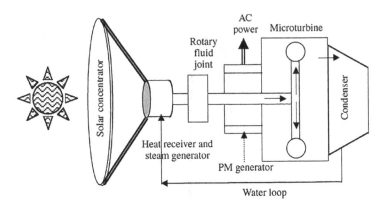

FIGURE 21.9 Solar thermal microturbine generator.

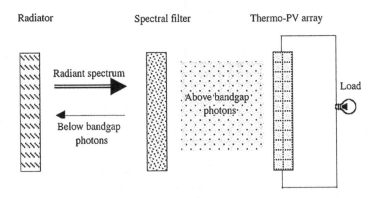

FIGURE 21.10 Thermophotovoltaic conversion scheme.

The TPV concept has recently seen renewed interest with new developments in semiconductor technology. Recent advances have produced low-bandgap (0.50 to 0.55 eV) material that is lattice-matched to GaSb substrates. At relatively low radiator temperatures, there are many viable options for the heat source, and a number of applications become attractive. Using the waste heat of a cogeneration power plant is one example. TPV ensures that power plants can generate electricity day and night, rain or shine. The extra energy collected during a sunny day is stored in a hot liquid, which is then used in TPV systems at night or on a cloudy day. The energy storage, thus, is inherent to the system design.

REFERENCES

1. Frankel, P., Energy from the Oceans, Preparing To Go On-Stream, *Renewable Energy World*, James & James, London, August 2002, pp. 223–227.
2. Myers, J.J., Holm, C.H., and McAllister, R.F., *Handbook of Ocean and Underwater Engineering*, McGraw-Hill, New York, 1968.

Ancillary Power Systems

3. Temeev, A.A. et al., *Natural-Artificial Power-Industrial System Based on Wave Energy Conversion, Proceedings of the 34th Intersociety Energy Conversion Engineering Conference*, Vancouver, BC, Canada, SAE, 2000, Paper No. 1–2556.
4. Mario, C., Ocean Power Envisions a Piezoelectric Energy Farm, U.S. 1 Business, Princeton, NJ, January 18, 1995, pp. 223–227.
5. Taylor, G.W. et al., *Piezoelectricity*, Gordon and Breach Science Publishers, New York, 1985.
6. Gross, N., Jolt of Electrical Juice from Choppy Seas, *Business Week*, November 28, 1994, p. 149.
7. Patel, M.R., *Piezoelectric Conversion of Ocean Wave Energy, Proceedings of the 1st International Energy Conversion Engineering Conference*, Portsmouth, VA, AIAA, 2003, Paper No. 6073.
8. Patel, M.R., *Optimized Design of Piezoelectric Conversion of Ocean Wave Energy, Proceedings of the 2nd International Energy Conversion Engineering Conference*, Providence, RI, AIAA, 2003, Paper No. 5703.
9. David Jesús Yáñez Villarreal, VIV Resonant Wind Generators, *Vortex Bladeless*, June 2018, pp. 1–6.
10. Omid Beik, An HVDC Off-shore Wind Generation Scheme with High Voltage Hybrid Generator, Ph.D. thesis, McMaster University, 2016.
11. Omid Beik, Ahmad S. Al-Adsani, *DC Wind Generation Systems, Design, Analysis, and Multiphase Turbine Technology*, Springer, 2020.
12. O. Beik and A. S. Al-Adsani, "Parallel Nine-Phase Generator Control in a Medium-Voltage DC Wind System," in *IEEE Transactions on Industrial Electronics*, vol. 67, no. 10, pp. 8112–8122, October 2020.
13. O. Beik and N. Schofield, "High-Voltage Hybrid Generator and Conversion System for Wind Turbine Applications," *IEEE Transactions on Industrial Electronics*, vol. 65, no. 4, pp. 3220–3229, April 2018.
14. Beik, O., Schofield, N., "*High Voltage Generator for Wind Turbines*", *8th IET International Conference on Power Electronics, Machines and Drives (PEMD 2016)*, 2016.
15. O. Beik and N. Schofield, "An Offshore Wind Generation Scheme With a High-Voltage Hybrid Generator, HVDC Interconnections, and Transmission," *IEEE Transactions on Power Delivery*, vol. 31, no. 2, pp. 867–877, April 2016.
16. O. Beik and N. Schofield, "*Hybrid Generator for Wind Generation Systems*," *2014 IEEE Energy Conversion Congress and Exposition (ECCE)*, Pittsburgh, PA, 2014, pp. 3886–3893.
17. M. R. Aghamohammadi, and A. Beik-Khormizi, "*Small Signal Stability Constrained Rescheduling Using Sensitivities Analysis by Neural Network as a Preventive Tool*," in *Proc. 2010 IEEE PES Transmission and Distribution Conference and Exposition*, pp. 1–5.
18. A. B. Khormizi and A. S. Nia, "*Damping of Power System Oscillations in Multi-machine Power Systems Using Coordinate Design of PSS and TCSC*," in *20 11 10th International Conference on Environment and Electrical Engineering (EEEIC), Rome*. pp. 1–4.

Index

Pages in *Italics* refers figures; **bold** refers table

A

AC collector grid, 115, 123
Active power filters (APFs), 236, 309–311, *311*
AC transmission, 115, 116, **116**, 125, 289
AC Wind Systems, 115–123, 128
Adiabatic temperature rise, batteries, 201, *202*
Air density, wind speed and energy, 15, 17, 20, 31
Alkaline fuel cells, 260, 262, **262**
All-DC system, 125, 126, *126*, 128–136, **134**
Alternating current (AC)
    AC-DC converter, electronics, 65, *65*, 66, 69, 83, 123, 125, 177
    AC-DC rectifier, electronics, 232–233, *233*, 244, 256
    cable, wind farm transmission, 115, 120, **121**, 122, 123, 134–136, **136**
    cycloconverter, 244
    excitation current, self-excitation capacitors, 78
Amorphous silicon PV cells, 149–150, 354
Ancillary power systems, 357–366
    heat-induced wind power, 357, *358*
    jet-assisted wind turbine, 362–363, *363*
    marine current power, 357–361, *359*, *360*
    ocean wave power, 361–362, *362*
    solar thermal, *see* Solar thermal system
    solar thermal microturbine, 363–365
    thermophotovoltaic system, 365–366, *366*
Anemometers, 33, 37, 326
    control systems, 37
Annual frequency distributions, wind energy, *28*
Array diode, PV system components, 172
Arrays, photovoltaic systems, *see* Photovoltaic power systems
Automobiles
    hybrid systems with fuel cell, 260
    solar car, 255
Average community load, system sizing, 268, 286
Axial gap induction machine, 84

B

Balanced three-phase system, 295
Battery charge/discharge converters, electronics, 246–249
Battery storage systems, 189–226
    charge regulators, 208–209
        multiple charge rates, 209
        single-charge rate, 209
    charging, 207–208
    comparison with flywheel, 212–217
    controls, 210
    design, 206–207
    equivalent circuit, 193–195
    grid-connected systems
        load scheduling, 286
        low-voltage ridethrough, 288–289
    hybrid systems
        controllers, 265
        with diesel, 257–258
        load sharing, 265–267, *267*
    lead-acid, 204–206
    management of, 209–212
        monitoring and controls, 210
        safety, 210–212
    monitoring and controls, 210
    performance characteristics
        C/D ratio, 195–196, *197*
        C/D voltages, 195, *196*
        charge efficiency, 198
        energy efficiency, 196–197
        failure modes, 61, 201–203
        internal loss and temperature rise, 200–201, *201*
        internal resistance, 197, *197*
        memory effect, 199
        random failure, 201–203
        self-discharge and trickle charge, 198
        temperature effects, 199–200
        wear-out failure, 203
    PV system components, 172–174
    resistance, internal, 197
    safety, 210–212
    solar car, 254, 255
    stand-alone systems, 253–254
    system sizing, 267–271
    temperature, internal loss and, 200
    types of batteries
        lead-acid, 192
        lithium ion, 193
        lithium polymer, 193
        nickel-cadmium, 192
        nickel-metal hydride, 192–193
        zinc-air, 193
    wind systems, components of systems, 37–45
Bearings, flywheel, 212, 214–216
Bipolar junction transistor (BJT), 230, 231, 234, 246
Birds, wind system impacts, 59, 102, 103

Bladeless wind turbines, 363, *364*, 365
Blade number and rotor efficiency, 18, *18*, 19, *19*, 27, 46, 48, *49*, 54, 58
Blades, wind systems
  design trade-offs, 52–55
  offshore wind farms, 41, 58, 59, 101, 103
  system components, 37–45
Brake systems
  fire hazards, 61
  magnetic, 80
Buck-boost converter, 179, 249, *249*, *250*
Buck converters, 172, 179, 181, 246, *248*, 249, 270
Bucket-type foundations, 110
Building-integrated photovoltaic systems, 145–147, 275
Bulk charge, 208

C

Cable size, **121**, 300
Cable systems, offshore wind farms, 58, 115
Capacity, battery
  lead-acid battery, 192
  temperature effects, *197*
Capital costs
  economies-of-scale trends, *336*
  initial, 100, 320, 323, 326, 355
  wind farm sizing, 271–273; see also Economics
Carnot cycle efficiency, 343, 364
Cascades, converter (buck-boost), 249, *249*
Cells, photovoltaic
  power system, 155–174, 189, 209, *211*, 229, 230, 232, 233, 246, 251, 254, 275, 354
  technologies, 147–153
    amorphous silicon, 149, **150**
    concentrator cell, 150
    IMM cell, 153
    multijunction cell, 150–152
    polycrystalline and semicrystalline silicon, 149
    single-crystalline silicon, 147–148
    spheral cells, 150
    thin-film cells, 149
Central receiver, solar thermal system, 344–347, 355
Certificates, renewable energy, 288
Charge converter, battery electronics, 217, 246–248
Charge/discharge (C/D) cycle
  battery design, 206–207
  battery endothermic and exothermic periods, 200, *201*
  comparison of battery types, 204
  flywheel, 212–217
  random failure, 203
  wear-out failure, 203

Charge/discharge (C/D) ratio, battery, 195–196, *197*
Charge/discharge (C/D) voltages, battery, 195
Charge efficiency
  battery operating temperature range, 199, **200**, **202**
  battery performance characteristics, 195
  temperature effects, *197*
Charge regulators, battery storage systems, 208–209
  multiple charge rates, 209
  single-charge rate, 209
Charging, battery storage systems, 207–208
Chatter, system, 265, *266*
Clamping diodes, 240, 241
Climate effects, PV array design, 166–167
Closed-loop control systems, tap-changing transformer, 245
Commercial plants, solar thermal system, 347, 354–355
Commutation, voltage control, 230, 242, 245
Compressed air storage systems, 220–223, 286, 289, **290**
Computer and Business Equipment Manufacturers Associations (CBEMA) curve, 304
Concentrating solar power (CSP) technology, 343, 344
Concentrator cell, PV systems, 150
Conductor loss, 77, 297
Constant flux linkage theorem, 82
Constant-power load, PV array, 167, *168*
Constant-TSR scheme, 50–51, 56, 91
Consumer choice, wind power, 128, 229
Controls
  battery storage systems, 210
  flywheel, 212
  grid-connected systems, interface requirements, 276–278
  stand-alone systems
    hybrid systems with diesel, 257–258
    load sharing, 265–267
    mode controller, 265
  variable-speed drive, 117
  wind systems
    components of systems, 37–45
    rate, 56–57
    speed, 55–57
    variable-speed, 49
Control signal, 119, 229, 237, 238, 352
Conversion efficiency, PV cells, 147–153, 158, 162, *163*, 255, 275
Conversion of measurement units, 147
Converters, harmonics, 301–302, **302**
Cooling, induction generator, 73, 77–78
Corrosion, offshore wind farm facilities, 108
Cosine law, 163, **164**, *164*, 168
Coupling coefficient, piezoelectric, 76, 82

Index

Current source converters, 236, *236*, 237
Current transformers (CTs), 310, 311
Current-voltage (I-V) characteristics, PV
systems array, temperature effects,
165–166, *166*
power systems, 156, 160, *161*
PV module, 160, *161*, *163*, 165–167
triple-junction PV cells, 150–152
Cutout speed selection, generator drives, 99–100
Cycloconverter electronics, 244
Czochralski process, *148*

D

Danish wind turbine, 15, 54, 55, 110, 361
Dark current, 159
Darrieus rotor, 15, 19, 362
Data processing, wind speed and energy, 29–31
Data reliability, wind speed and energy, 22, 32–33
DC wind generation schemes, 123, 125
Deadbands, mode controller, 265, *266*
Declining production cost, future prospects and trends, 335–337
Demand factors, system sizing, 267–269, **269**
Demand for energy
  future prospects and trends, 338
  green power, consumer attitudes, 3, 5, 7
  industry overview, 3–4
  modularity for growth, 7
Depth of discharge (DoD), battery, 200, 203, *203*, 205–207, 210, 217, 224, 225, 270
Design
  battery storage systems, 206–207
  electronics, DC-AC inverter, 233–235
  PV array, 270–271
  wind systems blade, 42–44
  certification process, 288
  doubly fed induction generators, 82–83
  trends in, 61–62
  turbine, horizontal *versus* vertical axis, 55
  wind systems, environmental factors in offshore wind farms
    corrosion, 108
    ocean water composition, 103–104
    wave energy and power, 105–106
  wind systems, trade-offs in, 52–55
    blade number, 54
    horizontal *versus* vertical axis, 55
    rotor upwind or downwind, 54–55
    turbine towers and spacing, 52–53
Diesel, hybrid systems with, 257–267
  load sharing, 265–267
  mode controller, 265
Digital data processing, wind speed and energy, 29–31
Diode, array, PV system components, 172

Direct current (DC)
  DC-AC inverter
    electronics, 233–235, 256
    PV system components, 172–174
  DC-DC buck converter, PV system components, 172
Direct current (DC) cable, offshore wind farms, 123
Direct current (DC) generators
  solar thermal power system, 347, 351
  wind systems, 63–66
Direct-driven generators, 83–85
Direct drive, variable speed, 98–99
Discharge converter, battery electronics, 173, 217, 246–249
Discharge efficiency, battery operating temperature range, 199, 200, **200**, **224**
Distributed power generation, 146, 262, 290–292
Doubly-fed induction (Scherbius) generators, 50, 65, *65*, 67, 82–83, 290
Downwind, terminology, 54–55
Dump heater, PV system components, 172, 173
Dump loads, 173, 249
Dynamic bus impedance and ripples, electrical performance, 300–301
Dynamic stability limit, 285, *285*

E

Earthquake, wind system hazards, 60, 61, 105
Economics, 319–327
  availability and maintenance, 320–323
  battery comparisons, 204, *205*
  battery design, 195, 206–207, 224, 270
  capital costs, initial, 320
  costs of power generation technologies, 147
  electronics
    DC-AC inverter design, 233–235
    doubly fed induction generator and, 50, 65, 67, 82–83, 290
  energy cost estimates, 323
  energy delivery factor, 319–320
  financing of project, 327
  future prospects and trends
    declining production cost, 335–337
    demand, global, 332; *see also* Future prospects and trends
  grid-connected systems
    energy storage and load scheduling, 286
    planning tools, 286–287
  hybrid systems, 257–267
  incentives for renewables, 327
  photovoltaic energy
    stand-alone PV *versus* grid line, 253–254
    trends in, 61–62
  sensitivity analysis, 323–326
    tower height effect, 324–326
    wind speed effect, 324

solar thermal microturbine, 363–365
thermophotovoltaic system, 365–366
wind energy, 9, 12, 20, 34, 69, 85, 96, 117, 223, 287
   blade design considerations, 42–44, 54
   generation costs, 9, 331–334
   incentives for renewables, 5, 12, 327
   offshore wind farms, 101–113
   Scherbius variable-speed drive, 98
   utility perspective, 7
   wind farm sizing, 271–273
Economies of scale, 9, 46, 335, *336*
Eddy loss, 296–297
Efficiency, induction generators, wind systems, 70–82
Electrical load matching, PV array design, 167–168
Electrical performance, 295–317
   component design for maximum efficiency, 296–298
   dynamic bus impedance and ripples, 300–301
   harmonics, 301–302
   lightning protection, 316–317, *317*
   model electrical system, 298–299
   National Electrical Code, 304
   quality of power, 303–311
      harmonic distortion factor, 303–304
      voltage flickers, 305–307
      voltage transients and sags, 304–305
   renewable capacity limit, 312–316
      interfacing standards, 314–316
      system stiffness, 312–314
   static bus impedance and voltage regulation, 299–300
   unit cost of energy (UCE) parameter, 323–324
   voltage current and power relationships, 295–296
Electric vehicle, 254–256
Electrochemical battery, *see* Battery storage systems
Electromagnetic features, solar thermal power synchronous generator, 347
Electromagnetic interference (EMI)
   DC-AC inverter design standards, 233–235
   wind systems, environmental aspects, 57–60
Electromechanical energy conversion, solar thermal power system, 63, 216, 347
Electromechanical torque
   inrush current and, 81, 82, 279–280
   transients, 80–82
Electronics, power, 9, 37, 44, 45, 49, 50, 55, 56, 65, 66, 85, 87, 89, 99, 120, 123, 125, 126, 128, 130, 133, 166, 173, 175–182, 215, 217, 229–251, 289, 313
   AC-DC rectifier, 232–233

   battery charge/discharge converters, 246–249
      charge converter, 246–248
      discharge converter, 248–249
   DC-AC inverter, 233–235
   flywheel, 212–217
   grid interface controls, 244–246
   hybrid system mode controller, 265
   power shunts, 249–251, *250*
   PV system components, 172–174
   switching devices, 229–232
   thermophotovoltaic system, 365–366
   wind systems
      control systems, 50
      doubly fed induction generator and, 82–83, *83*
      variable speed, 29, 44, 45, 49, **50**, 98
Emission benefits, wind power, 3, **5**
Energy balance analysis, 256, *269*, *271*
Energy capture, maximum, 49–50
Energy conversion efficiency, PV cells, 147
Energy delivery factor (EDF), 319–320, 323
Energy density, comparison of battery types, 191–193
Energy distribution, wind speed and energy, 28–29, *99*, *107*
Energy efficiency, battery performance characteristic, 195–204
Energy estimates, stand-alone system sizing, 268–269
Energy Index of Reliability (EIR), 272–273
Energy inputs, photovoltaic cell production, 196
Energy relations, flywheel, 212–214
Energy storage, *see* Battery storage systems, *see* Storage, energy
Energy, wind, *see* Wind speed and energy
Energy yield, offshore wind farms, 101, *102*
Environmental factors in wind farm design
   corrosion, 108
   ocean water composition, 103–104
   wave energy and power, 105–106
Environmental impact, wind systems
   birds, 59
   electromagnetic interference (EMI), 58
   noise, 57–58
   offshore wind farms, 102–103
Equivalent circuit model
   grid-connected systems, 283, *283*
   power quality
      system stiffness, 312–314
      voltage flickers, 305–307
   PV systems, 157–159
   solar thermal system, 343–355
   wind systems, induction generator, 75–77, 118
Europe
   energy policy, 12, 329, 330
   marine current power, 357–361
   photovoltaic energy, 143–153

Index

power quality, 303–311
wind energy
  future prospects, 331–334
  international agencies and associations, 315
  manufacturers and developers, 5, 46
  manufacturers and suppliers, 42, **44**, 50
  offshore wind farms, 101–113
  turbine wind power systems, 52–53
  wind power use in, 9–13
Excitation methods, solar thermal system, 350–352
Expected Energy Not Supplied (EENS), 272

F

Failure
  battery performance characteristics, 209
  blade design and, 43, 44
  economics, 61
  flywheel fatigue, 217
  induction generator, 82
Fatigue failure
  blade design and, 43
  induction generator, 82
Fatigue life, flywheel, 46, 217
Faults
  desynchronizing effect, 280, 281, 285
  fire hazards, 61
Feedback voltage control system, deadbands in, 299, *300*
Fermi level, 155
Filters
  DC-AC inverter, 234
  power quality, system stiffness, 312–314
  PV system components, 173
Financing of projects, 323, 327
  wind farms, offshore, 101, 102, 108
Fire hazards, wind systems, 61
Fixed-speed generator drives, 91–100
  comparison with variable-speed systems, 98
  one fixed-speed, 95–96, *96*
  two fixed-speed, 97–98
Flexible AC transmission systems (FACTS), 289
Flickers, voltage, electrical performance, 305–307
Flutter, 299, *300*
Flux linkage, induction generator, 67, 82, 85
Flywheel, 212–217
  benefits, comparison with battery, 217
  energy relations, 212–214
  system components, 214–217
Forced-commutated inverter, 234
Forces on ocean structures, 107–108
Foundation, offshore wind farms, 108–111
  gravitation, 110
  monopile, 109
  tripod, 110–111
Frequency control
  cycloconverter, 244

electronics, 246
grid-connected systems interface controls, 244–246
interface requirements, 276–278
synchronizing with grid, 278–282
synchronous operation, 281
Frequency converter, doubly-fed induction generator, 82–83
Frequency modulation index, 238, 239
Friction coefficient, terrain, 31
Fuel cells, hybrid systems with, 258–265
  load sharing, 265–267
Fuel cell stack, 264
Future prospects and trends, 329–339
  declining production cost, 335–337
  market penetration, 337–339, *339*
  photovoltaic power, 334–335
  wind power, 331–334
  world electricity demand to 2050, 329–331

G

Gamma function, wind speed and energy, 25
Gassing, battery charging, 198, 208
Gas turbine, 145, 262, 263, **320**, 355
Gate-assisted turn-off thyristor (GATT), 230, **231**
Gate current, MOSFET control, 230
Gate turnoff thyristor (GTO), 229, 230, **231**, 235, 244
Gearbox, terminology, 38, 46, 55, 64, 69, 83, 96, 115, 117, 135, 136, 305, 321, 323
Gear drives, *see* Generator drives
Gearless direct-driven generator, 83
Generator drives, 91–100
  cutout speed selection, 99–100
  selection of, 246
  speed control regions, 92–95
  types of
    one fixed-speed, 95–96
    two fixed-speed, 97–98
    variable-speed controls, 49
    variable-speed direct drive, 84
    variable-speed drive, 117
    variable-speed gear drive, 50
    variable-speed power electronics, 55
Generators, wind systems
  components of systems, 37–39, 56, 57, 82, 87, 95, 111, 115, 117, 118, 120, 128, 130, 215
  DC generator, 62, 256, 351
  direct-driven, 83–85
  doubly-fed, 82–83
  induction generator, 70–82
    construction, 71–72
    efficiency and cooling, 77–78
    equivalent circuit, 75–77

rotor speed and slip, 73–75
self-excitation capacitors, 78–79
torque-speed characteristic, 79–80
transients, 80–82
working principle, 72–73
synchronous generator, 78, 83, 125, 131, 256, 267, 277, 281, 347–354
synchronous operation, 66, 279, 280, 285
system components, 37–45
terminology, 37, 104
turbine rating, 45–47
Geographic distribution
global perspective, 9, 11
India, 11
U.S., 11–13
industry overview, modularity for growth, 3–4, 7
photovoltaic energy, 343, 357
wind power, 9–13
Europe and U.K., 11
global perspective, 9, 11
Global wind patterns, 31–32
Government agencies, wind farms in US, 41, 101
Grauer's design, induction generator, 84
Gravitation foundation, offshore wind farms, 110
Green certificate trading program, 5, 7
Green power, consumer demand, 3, 5, 7
Green pricing, 5, 6
Grid-connected systems
distributed power generation, 146, 251, 262, 290–292
electrical performance, 295–317
energy storage and load scheduling, 286
grid stability issues, 288–290
interface requirements, 276–278
operating limit, 282–286
stability limit, 284–286
voltage regulation, 282–284
photovoltaic systems, 143–153
system components, 172–174
power quality, interfacing, 288, 289
self-excitation capacitors, 78–79
synchronizing with grid, 278–282
inrush current, 279–281
load transient, 282
safety, 282
synchronous operation, 279, 280
utility resource planning tools, 286–287
wind farms, offshore cost, 101–113
integration with grid, 287–288
wind energy, 9
wind systems
components of systems, 37–45
doubly fed induction generator and, 50, 65, 67, 82, 83, 290
power electronics, variable speed, 45, 49, 55, 99, 229–251, 289, 313

Grid interface controls, electronics, 244–246
Growth, modularity of, 7

H

Half-life at double DoD, 205
Harmonic distortion factor, electrical performance, 303–304
Harmonics, 301–302
DC-AC inverter, 233–235
electrical performance, 295–317
load sharing, 265–267
power electronics-based variable speed system, 50, 99, 100
power quality, system stiffness, 312–314
Hazards, wind systems
birds, 59
earthquake, 60, 61, 105
electromagnetic interference (EMI), 58, 235
fire, 60–61
noise, 45, 47, 57–58
Heat, electrical performance, 200–202, 207, 209, 221, 250, 260, 262, 264, 267
Heaters, 167, 172, 173, 249
dump load, 173, 249
PV system components, 172–174
Heat-induced wind power, 357
Height effects, 31–32
Heliostats, 343, **348**
High-speed shaft, terminology, 38
High-switching frequency, 237
High-temperature fuel cells, 263
High voltage AC (HVAC), 115, 125
High voltage DC (HVDC), 115, 125, 126, 133, **135**, 136, 137, *137*, 138, 234, 241, *242*, 243, 289
Horizontal-axis turbine, 19, 55, 359
Horizontal axis, wind system design tradeoffs, 15–16, 19, 23, 55, 359
HVDC converters, 241–243
HVDC systems, 115, 125, 241–243
h-v-k plots, 24, *25*
Hybrid generator, 86, 87, 127, 130, 131, 133–138
Hybrid systems, 143, 253, 257–267
with diesel, 257–258
economics, 275, 287, 290
with fuel cell, 258–265
grid-connected systems, energy storage and load scheduling, 286
load sharing, 265–267
mode controller, 265
Hydrogen fuel cell, 253, 258–260, *261*, 262, 263, 265, 338
Hydro power, costs of power generation technologies, 4
Hysteresis loss, 297

Index

I

IGBT/MOSFET based converters, 235–237
Impedance, internal, series linked generator, 118, 194, 195, 266, 267, 285, 292, 299, 312
Impedance, series linked generator, 118, 195, 266, 267, 285, 292, 299, 312
Incentives for renewables, 5, 9, 12
Induction generator (IG), 38, 50, 64, 66, 67, 70–82, 115–118, 120, **122**, 123, 133, 135, 136, 256, 257, 277, 281, 290, 295, 305, 306
Induction generators, wind systems, 70–82
    construction, 71–72
    efficiency and cooling, 77–78
    equivalent circuit, 75–77
    rotor speed and slip, 73–75
    self-excitation capacitors, 78–79
    torque-speed characteristic, 79–80
    transients, 80–82
    working principle, 72–73
Induction motors, 167
Industry overview, 3–4
Inrush current, grid-connected system, synchronization, 81, 82, 279–280
Insulated gate bipolar transistor (IGBT), **117**, 118, 119, 128, 229, 230, **231**, 234–237, 239, 242, 246, 309
Insurance, 60, 323, **325**
Interface requirements, grid-connected systems, 276–278
Interfacing standards, electrical performance, 300, 302–305, 307, 314–316
Internal impedance, series linked generator, 194, 195, 267, 285
Internal loss and temperature rise, battery performance characteristics, 46, 195–204, 253
Internal resistance
    battery performance characteristics, 46, 195–204, 253
    battery storage systems, temperature effects, 175, 177, 207, 212, 214, 218–222, 226, 346
Inverters
    DC-AC
        electronics, 233–235
        line commutated, 230, 234, 235, 241, 281
        PV system components, 128–132, 172–174
    Islanding, DC-AC inverter, 235

J

Japan, photovoltaic energy, 60, 143, 220, 226, 261, 327
Jet-assisted wind turbine, 362–363
Jets, nocturnal, 31

K

Kelly cosine curve, *164*
K ratings, transformer, 304

L

Lagrangian relaxation methods, 286
Laser optical sensor, wind speed, 147
Lattice towers, 39, 59
Lead-acid batteries, 211, 290
    grid support, 50, 288, **290**
    thermal design, 201, **202**
Learning-curve hypothesis, 335
Legal aspects of offshore wind farms, U.S., 41, 60, 101–113
Legislation
    Net Metering Law, 327
    Public Utility Regulatory Policies Act (PURPA), 4
    wind farms in US, 101–113
Lift-to-drag ratio, 43
Light-activated silicon controlled rectifier (LASCR), 230
Lightning, 60, 288, 316, *317*
Line-commutated converters (LCC), 241, 242, *242*
Line commutated inverter, 234, 281
Lithium ion batteries, 191, **191**, 193, **202**, **204**, 213, 223, *224*, 226
Lithium polymer batteries, 191, **191**, 193, **202**, **204**
Load matching, PV array design, 162, 167, *168*
Loads
    distributed power generation, 146, 262, 290–292
    grid-connected systems, scheduling, 286
    system sizing, 267, 268
    victim, 303
Load sharing, hybrid systems, 203, 265–267
Load transient, grid-connected system synchronization, 282
Loss components, electrical performance, *297*
Low-speed shaft, terminology, 37, 38
Low-temperature fuel cells, 261
Low-voltage ride-through, grid stability issues, 288–289

M

Magnetic bearing, flywheel, 212, 215–216
Magnetic field, solar thermal power system, 71–74, 78, 130, 218, 347, 349, 350
Magnetic parts, losses in, 297
Magnetic saturation, harmonics generation, 301, 352

Magnet, superconducting, 217–220
Maintenance
  economics, 320–323
  wind farms, offshore, 112–113
Management, battery storage systems, 209–212
  monitoring and controls, 210
  safety, 210–212
Manufacturers and suppliers
  solar cell and modules, in U.S., 147
  wind energy, 57
    in Europe, 45
    in U.S., 41–42
Marine current power, 357–361
Marine environment, 58, 111, 112
Market penetration, future prospects and trends, 337–339
Materials, offshore wind farms, 108, 111–112
Matrix converters, 125, 243–244
Maximum energy capture, wind systems, 49–50
Maximum power operation, wind systems, 50–52
  constant-TSR scheme, 51–52
  peak-power-tracking system, 51–52
Maximum power point tracking (MPPT), 66, 117, 127, 136–138
Mean speed, defined, 24–26
Measure, correlate, and predict (mcp) technique, 21
Measurement unit conversion, 331
Mechanical energy, marine environment
  ocean wave power, 361–362
  piezoelectric generator, 357–361
Mechanical energy, marine environment, marine current power, 357–361
Medium voltage DC (MVDC), 125–127
Memory effect, battery, 192, 199
Metal oxide semiconductor field transistor (MOSFET), 178–180, 229, 230, **231**, 234–237, 239, 246
Mixed loads, PV array, 167
Mode and mean speeds, wind speed and energy, 24–26
Mode controller, hybrid systems, 258, 265, 266, *266*
Model electrical system, electrical performance, 298–299
Models, wind farms, 118–120, 133, 288
Mode, mean, and RMC speeds, wind speed and energy, 27–29
Mode speed, defined, 24–26
Modulating signal, 237
Modulation index, 238, 239
Molten salt fuel cell, 343
Monitoring
  battery storage systems, 207–208, 210
  fire hazards, 61
  synchronizing with grid, 278–282
  wind systems, components of systems, 37–45

Monopile foundation, offshore wind farms, 109
MOS-controlled thyristor (MCT), 230
Multijunction cell, xy, PV systems, 151–153
Multilevel converters, 239–241
Multi-level modular converters (MMC), 242, 243
Multilevel topologies, 239–241
Multiple charge rates, battery storage system
  charge regulators, 208–209

N

Nacelle, 38, 39, *39*, 40, *40*, 42, 52, 60, 78, 110, 133, 138
National Wind Technology Center (NWTC), 31, 34, 61, 287, 288
Net Metering Law, 184, 327
Neutral point clamped (NPC), 240, *240*, *241*
Nickel-cadmium batteries, 168, 191, **191**, 192, **202**, **204**
Nickel-metal hydride batteries, 191, **191**, 192–193, **204**
Nocturnal jets, 31
Noise, audible, 57–58
Noise, electrical, *see* Harmonics

O

Ocean environment, *see* Mechanical energy, marine environment
Ocean structure design, offshore wind farms, 106–108
Ocean water composition, offshore wind farms, 103–105
Ocean wave power, 361–362
Ocean wave technology generation cost forecast, 361–362
Off-grid electrifications, 182
Off-shore substation, 115, **116**, 120, **121**, 122, **122**, 133, **135**, 136, 138
Offshore wind farms, *see* Wind farms, offshore
One fixed-speed generator drives, 95–96
One-line diagram, 295
Open-circuit voltage, 158–160, 162, 165, 171, 172, 194, 201, 298, 365
Operating limit, grid-connected systems, 282–283
  stability limit, 284–286
  voltage regulation, 283–284
Operations, grid-connected system synchronization, 278–279
Optical wind speed sensor, 37
Overcharging, battery, 192, 193, 198, 208, 250

P

Parabolic dish, 345, 364
Parabolic trough, 344, **346**, 355

Index

Peak electrical capacity, turbine rating, 45–47
Peak power
    PV system components, 145–147
    turbine rating, 45–47
    wind system tracking systems, 51–52
Peak-power operation, PV power system, 171–172
Permanent magnet generators, 66, 85
    variable-speed direct drive, 84
Permanent magnets, flywheel bearings, 214–217
Phase control, grid-connected system interface requirements, 128, 244, 245, 351
Phase-controlled rectifier, 245, 351
Phasor diagram, voltage regulation, 283–284
Phosphoric acid fuel cells, 262
Photoconversion efficiency, PV cell, 162, 163
Photovoltaic cells, solar car, 254–256
Photovoltaic power
    cell technologies
        amorphous silicon, 149–150, 354
        concentrator cell, 150
        multijunction cell, 150–152
        polycrystalline and semicrystalline silicon, 149
        single-crystalline silicon, 147–148
        spheral cells, 150
        thin-film cells, 149
    future prospects and trends, 183–184
    projects, 184
Photovoltaic power systems, 155–174
    ancillary
        array design, 162–170
    climate effects, 166–167
        electrical load matching, *168*
        shadow effect, 163–165
        solar thermal microturbine, 363–365
        sun angle, 163
        sun intensity, 162–163
        sun tracking, 168–170
        temperature effects, 165–166
        thermophotovoltaic system, 365–366
    array sizing, stand-alone systems, 270–271
    building-integrated, 145–147
    cells, 255, 264, 344, 365
    components, 172–174
    costs of power generation technologies, 5, *6*
    economics
        comparison of costs, 175, *176*
        maintenance, 320–323
        stand-alone PV *versus* grid line, 172–174; *see also* Economics
    electronics
        battery charge/discharge converters, 246–249
        DC-AC inverter, 233–235
        DC-DC buck-boost converter circuits, 246, *246*, 249
        power shunts, 249–251
        switching devices, 229–232
    equivalent circuit, *158*
    future of, generation costs, 334–335
    grid-connected systems, 275–292
        distributed power generation, 290–292
        interface requirements, 276–278
    hybrid systems with diesel, 143
        load sharing, 265–267
    I-V and P-V curves, 160–162
    modularity for growth, 7
    module and array, 156–157
    open-circuit voltage and short-circuit current, 159
    peak-power operation, 171–172
    solar thermal system comparisons, 343–344, *346*
    stand-alone, 172–174, 253–254, 256–257
Pitch control, 44, 45, 61, 67, 69, 93, 117, 138
Pitch, terminology, 38
Planning models, wind grid-connected systems, 288
Plant capacity, system sizing, 268–269
Pole-changing stator winding, 97, *98*
Policy, energy
    Public Utility Regulatory Policies Act (PURPA), 4–5
    United States, 12
    wind energy, 12–13
Polycrystalline and semicrystalline silicon PV cells, 149
Polymers, lithium, equivalent circuit, 193–195, *194*
Power
    stand-alone system sizing, 267–271
    wind systems, power *versus* speed and TSR, 47–49
Power angle, 69, *70*, 282, 285, *285*, 350, 352–354
Power coefficient, 18, 49, 54, 69, 91, 93, 96, 97, 117
Power connection and control system (PCCU), diesel hybrid, 257–258
Power density, wind energy, 30, **33**
Power electronics, *see* Electronics, power
Power extracted from wind, 17–19
Power factor, 75, 120, 123, 233–235, 245, 256, 257, 282–284, 300, 308, 309, 311
Power level, turbine rating, 45–47
Power-limited region, rotor speed control, 92
Power output controls, 93–95
    solar thermal system, 343–355
    turbine rating, 45–47
    wind systems, lift-to-drag ratio and, 43
Power shunts, electronics, 249–251
Power-shut-off region, rotor speed control, 92

Power systems, *see* Photovoltaic power systems, *see* Wind power system
Power *versus* voltage (P-V) curves and characteristics, PV systems, *161*
Prediction of wind speed, 34
Pricing, Green, *6*
Primary battery, 189
Probability distributions
  cutout speed selection, 99–100
  wind speed and energy, 21–24
Production costs, future prospects and trends, 335–337
Proton exchange membrane, 261
Public policy, *see* Policy, energy
Pulse width converter, Scherbius variable-speed drive, 117
Pulse Width Modulation (PWM), 118, 237
PV array design, 210, 232, 250, 253, 256, 270–271

Q

Qualified facilities (QFs), 4
Quality of power
  DC-AC inverter, 233–235
  electrical performance, 295–317
    harmonic distortion factor, 303–304
    voltage flickers, 305–307
    voltage transients and sags, 304–305

R

Radial-flux permanent magnet generator, 84
  rectified DC voltage, 88, 89, 132, *132*, 133, 137
Random failure, battery, 203
Rayleigh distribution, wind speed and energy, 22, 23, *23*, 25, 26, *29*
Recent trends, solar thermal system, 355
Rechargeable battery, 189, 192
Rectifier, AC-DC, 232–233, 244, 256
Regulatory issues
  wind farms in US, 101
  wind system certification, 288
Renewable capacity limit
  interfacing standards, 314–316
  system stiffness, 312–314
Repair costs, 320–322
Residual turbulent scintillation effect, 31
Resistance, battery, 193–195
Reswitching transient, induction generator, 82
Reverse conducting thyristor (RCT), 230
Reverse diode-saturation current, 159
Ripples, 173, 300–301, 307, 351
Root mean cube (RMC) speed, 26–30

Root mean cube (RMC) voltage, thyristor control, 244
Rotor efficiency, 18, *18*, 19, *19*, 27, **27**, 46, 48, 49, *49*, 92
Rotor power coefficient, 91, *92*, 96, 97
Rotors, components of systems, 37–45
Rotor speed and slip, induction generators, wind systems, 73–75
Rotor speed, turbine power *versus*, *48*, *92*
Rotor-swept area, wind speed and energy, 17, 19–20
Rotor upwind or downwind, wind system design tradeoffs, 54–55
Rotor, variable-speed controls, 44–45, 49, 92–95

S

Safety
  battery storage systems, 191–193, 207–208, 210, 211
  grid-connected system synchronization, 282
Sags, voltage, 50, 303–305
Salt storage systems, 343, 346
Scale parameter, wind energy, 21–24, 26–29
Secondary battery, 189, 212
Self-discharge, **191**, 192, 198, 199, **200**, 200, 204, **204**, 205, **224**
  battery operating temperature range, 199, 200, **200**, **202**, 205, 206, 211, **224**, **346**, 347
  battery storage systems, lead-acid battery, 204–206, 213
  comparison of battery types, 204
  temperature effects, *197*, 199–200
Self-excitation capacitors, induction generators, wind system, 78–79, 257, 282
Semicrystalline silicon PV cells, 147–151
Sensitivity analysis, 319, 323–324
  tower height effect, 324–326
  wind speed effect, 324
Sensors
  wind systems, components of systems, 37–38
Shading, sun tracking, 168–170
Shadow effect, PV array design, 163–165
Shaft
  components of systems, 37
  wind power terminology, 37–38
Shape parameter, wind energy, 21–24, *22–24*, 27, 29
Short-circuit current, PV power systems, 82, 159, 160, 162, 165, 250
Short-circuits
  fire hazards, 61
  induction generator, 82
  power shunts, 249–251
Shorting, power shunts, 249–251

Index

Shunts, electronics, 251
Shunts, power, 249–251
Shut-off region, rotor speed control, 92, 93
Silicon cells, photovoltaic, *see* Cells, photovoltaic
Silicon controlled rectifier (SCR, thyristor), 229–231
  AC-DC rectifier, 232–233
  cycloconverter, 244
  gate turnoff (GTO), 229, 230, 244
Single-charge rate, battery storage system charge regulators, 209
Single-crystalline silicon PV cell, 147–148
Sinusoidal PWM (SPWM), 237–239
Site selection, wind farm, 312
Six-pulse converters, 242
Size of plants, modularity for growth, 7
Slip
  frequency power, 50, 82
  rotor, 73–75, 82, 96, 305, 347
Solar cars, 254–255
Solar conversion systems, 177, 180, 182
Solar energy
  organizations, societies, and associations, *4*, 33, 38, 60, 110, 207, 224, 226, 254, 256, 287, 304, 315, 327, 329, 337
  solar cell and module manufacturers in U.S., 143, 146, 147, 149
Solar II power plant, 345–347, 354
Solar power systems, *see* Photovoltaic power, *see* Solar thermal systems
Solar radiation
  array design
    shadow effect, 163–165
    sun angle, 163
    sun intensity, 162–163
  array, sun tracking, 168–170
  energy balance equation, 256, 270
Solar thermal microturbine, 363–365, *366*
Solar thermal power system, 343–356
Solar thermal systems, 343–367
  commercial plants, 347, 354, 355
  costs of power generation technologies, 357
  economics, comparison of costs, 355
  energy collection, 344–345
    central receiver, 344–345
    parabolic dish, 345
    parabolic trough, 344
  recent trends, 355
  Solar II power plant, 345–348, 354
  synchronous generator, 347–350
    equivalent circuit, 350
    excitation methods, 350–352
    power output, 352
    transient stability limit, 353–354
Solidity ratio, 20, 54
Solid oxide fuel cell, 261–263, **262**, *263*
Space vector PWM (SVM), 237

Specific energy, comparison of battery types, 204, **204**
Specific rated capacity (SRC), turbine, 46
Specific wind power, 17
Speed and power relations, 15–16
Speed control regions, generator drives, 95
Speed control, system components, 56, 95
Speed selection, generator drives, 95
Speed, wind
  control requirements, 55–57
  power *versus* speed and TSR, 91–92; *see also* Wind speed and energy speed, wind systems
Spheral PV cells, 150
Stability
  grid-connected systems, 80–82, 84, 275–292
    energy storage for, 286
    low-voltage ride-through, 288–289
    operating limit, 282–286
    synchronous operation, 281
    transients, 282
  solar thermal system, 353–354
Stack, fuel cell, 260, 264
Stall control, 37, 39, 44, 45, 95
Stand-alone systems, 253–273
  electric vehicle, 254–256
  electronics, battery charge/discharge converters, 246–249
  hybrid systems, 257–267
    with diesel, 257–258
    with fuel cell, 258–265
    load sharing, 265–267
    mode controller, 265
    sizing, 260
  hybrid systems with fuel cell, 258–265
  photovoltaic, 253–254
  self-excitation capacitors, 256–257
  system components, 172–174
  system sizing, 267–271
    battery sizing, 269–270
    power and energy estimates, 268–269
    PV array sizing, 270–271
  wind, 256–257
  wind farm sizing, 271–273
Standards
  DC-AC inverter design, 235
  distributed power generation, 290
  electrical performance, interfacing standards, 304
  harmonic spectrum, 302
  power quality, 303–311
    flicker, 305–307
    harmonics, 303–304
    interfacing, 304, 311
    voltage transients and sags, 304–305
    wind grid-connected systems, 303, 304, 307

Starting transient, induction generator, 81
State government agencies, wind farms in US, 4, 5, 12–13, 101, 184
Static bus impedance and voltage regulation, electrical performance, 171, 299–300
Stator conductors, solar thermal power system, 349
Stator winding, pole-changing, 75, 97–98
Steady-state stability limit, 246, 263, 280–282, 285
Step-down transformers, distributed power generation, 290
Step loading/unloading, power quality, 314, 315, 350
Stiffness, electrical, 312–314
Storage, energy, 189–227
  battery, *see* Battery storage systems
  comparison of technologies, 222–223
  compressed air, 220–222
  flywheel, 212–217
    benefits, comparison with battery, 217
    energy relations, 212–214, **213**
    system components, 214–217
  grid-connected systems, 286–292
    grid stability issues, 288–290
    scheduling, 286
  Solar II power plant, 345–347, 354
  stand-alone systems, battery sizing, 269–270
  superconducting magnet, 217–220, 339
  thermal power system, 189, 286
  thermophotovoltaic system, 365–366
Stress analysis, blade design, 43, 44, 49
String inverter, 235
String power loss, shadow effect, 165
Successive dynamic programming, 286
Sun angle, PV array design, 163
Sun intensity, PV array design, 162–163
Sun tracking, PV array design, 168–170
Superconducting magnet, 217–220, 339
Supersynchronous operation, induction machine, 72
Switching devices
  battery charge/discharge converters, 246–249
  electronics, **231**, 231–232
  harmonics, 233–235
Switch-mode inverters, 235
Synchronizing, grid-connected systems, 278–282
  inrush current, 279–280
  load transient, 282
  safety, 282
  synchronous operation, 281
Synchronous generator, 347–350
  equivalent circuit, 350
  excitation methods, 350–352
  power output, 352
  transient stability limit, 353–354
Synchroscope/synchronizing lamps, 279, 280, *280*

System faults, desynchronizing effect, 280, 281, 285
System integration, storage systems, *see* Battery storage system, *see* Storage, energy
Systems, *see* Photovoltaic power systems, *see* Wind power systems
System sizing, stand-alone systems, 267–271
System stiffness, electrical performance, 312–314

T

Tap changing transformers, 245
Taper charge, 207, 208
Tax credits, renewable energy, 4, **4**, 12
Technology
  photovoltaic power, 143–153; *see also* Cells, photovoltaic
  storage systems, 175, 177, 202, 212, 214, 218–222, 226, 346
Temperature
  battery design, 195, 206–207, 224, 227
  battery performance characteristics, 201
    effects of, 199–200
    internal loss and, 194
  battery storage systems
    C/D ratio effects, 195
    internal resistance effects, 157, 160, 194, 197, 199, 200, 205, 279
    lead-acid battery, 204–206, **213**, 255
    PV array design, 210, 232, 250, 253, 256, 270
Temperature effects, PV array design, 210, 232, 250, 253, 256, 270
Terminology, 37, 104
Terrain effect, wind energy, 33
Thermal storage system, solar thermal power system, 343, 345–347
Thermal systems
  ancillary
    heat-induced wind power, 357
    solar thermal microturbine, 363, 364, *366*
  costs of power generation technology, 320
  energy storage and load scheduling, 286
  solar thermal, *see* Solar thermal system
Thermal unit commitment, 286
Thermodynamic conversion efficiency, 343, 363
Thermodynamic cycle, solar thermal power system, **349**, 363
Thermophotovoltaic system, 365–366
Thevenin's equivalent model, 266, 298, 305, *306*, 312, *313*
  power quality, system stiffness, 312, *313*
  voltage flickers, 303, 305–307, 314
Thin-film cells, 149, 354
Three-blade configuration, 54
Three-phase system, balanced, 295
Thunderstorm frequency, *316*
Thyristor (silicon controlled rectifier), 128, *129*, 229, 230, 234, 241, 244, 245

Index

Thyristor, gate turnoff (GTO), 229, 230
Tidal current power, 360
Tip speed ratio (TSR), 19, 47, *49*, 49–51, 54, 56, 91, *92*, 97, 276
   constant-TSR scheme, 50–51
   generator drives, 91–100
   number of blades, 54
   power *versus* speed and, 47–49
Top head mass (THM), 40
Torque
   inrush current and, 81, 82, 279–281
   transients, **50**, 80–82, 281, 304
Torque-speed characteristic, induction generators, wind systems, 79–80, 83, 290
Total harmonic distortion (THD)
   DC-AC inverter, 233–235, 256
   power quality, 234–235, 303–311
Tower height, 31, 39, 40, *42*, 53, 108, 109, 324–326
Tower height effect, 324–326
Towers, wind systems
   bird kills, tubular *versus* lattice structures, 59
   components of systems, 37–45
   design trade-offs, 52–55
   system components, 37–45
Tracker, PV system components, 172–174
Tracking PV array, 253–256
   solar thermal power system, 343–355
   wind systems, peak-power-tracking system, 50, 171–172, 235, 255, 256, 261, 320
Transformers, **5**, 46, 66, 73, 75, 111, 115, 117–120, **122**, 123, 125, 127, 128, 133, 135, 179, 180, 235, 241, 242, 245, 275, 277, 290, *291*, 295, 299, 301, 302, 304, 308–310, 316
   distributed power generation, 290–292
   induction generator operation, 66, 70–82, 115, 117–120, **122**, 123, 133, 135, 136, 256, 257, 277, 281, 290, 295, *306*
   ratings, 53, 71, 95, 128, 180, 230, 235, 239, 241, 253, 287, 296, 304, 324, 351
   tap changing, 245, 292
Transients
   electrical performance, quality of power, 159, 193, 263, 295–317
   grid-connected system synchronization, 67, 81, 173, 253, 275–292, 295
   induction generators, wind systems, 38, 50, 64, 66, 67, 70–82, 97, 115, 117–120, **122**, 123, 133, 135, 136, 256, 257, 277, 281, 290, 295, *306*
   synchronous operation, 66, 279, 280, 285
Transient stability limit, solar thermal system, 353–354
Transmission
   distributed power generation, 146, 262, 290–292
   electronics, line commutated inverters, 234, 235, 241, 281
   equivalent circuit, 67–69, 75–79, 82, 83, 118–120, 133–134, *158*, 266, 283, 298, 299, 301, 305, *306*, 312, *313*, 350
   grid-connected systems, 67, 81, 173, 253, 275–292, 295
   wind energy, recent trends, 355
   wind farms, offshore, 60, 115, **116**, **120**, *121*, **122**, 123, 125, 126, 133, 136, 241, 289, 290
      AC cable, 115, 134, 135
      DC cable, 123, 134, 135
   wind systems, components of systems, 37–45
Triangular carrier signal, 237
Trickle charge, 198, 208, 209, 211
Triple-junction PV cell, 150–152
Tripod foundation, offshore wind farms, 110
TSR, *see* Tip speed ratio (TSR)
Tubular towers, 59
Turbine power *versus* rotor speed, 47–49
Turbines
   control requirements, 55–57
   design configurations, 15
   modularity for growth, 7
   rating, 45–47
   rotor-swept area, 17, 19–20
   step-up, gear drives for, 38, 66, 72
   system components, 37–45
   wind farm sizing, 271–273
   wind systems, components of systems, 37–45
Turbine speed
   controls, 44–45
   turbine rating, 45–47
Turbine towers and spacing, design trade-offs, 52–55
Turbulence
   nocturnal jets, 31
   offshore wind farms, 11, 41, 58, 60, 101–113, 323
Twelve-pulse converters, 242, 302, 308, 309
Twelve-pulse inverter circuit, 234, 302
Twelve-pulse line commutated bridge topology, 235
Two-blade design, 54
Two fixed-speed generator drives, 97–98

U

Ultrasound anemometers, 33, 37, 326
Uninterruptible power supply (UPS), 259, 289
Unit cost of energy (UCE), 323, 324
United Kingdom
   future prospects, 183–184
   marine current power, 357–361
   photovoltaic energy, 3

wind energy, 9, 11, 12, 20, 28, 34, 38, 41, *43*, 69, 85, 96, 117, 223, 287
  national energy policy, 330
  wind power in, 5
United States
  photovoltaic energy, 3
  thunderstorm frequency, 316
  wind energy, 5, 11, 12
    future prospects, 183–184, 331–334
    legal aspects of offshore wind farms, 41, 60, 101–113
    offshore wind farms, 41, 60, 101–113
  wind power use in, 5
United States National Renewable Laboratory (NREL), 33, 61, *148*, 149, 151–153, 287, 288, 339, 354, 355
Upwind, terminology, 38, 54, 55
Utility companies, photovoltaic energy projects, 7, 207, 287, 343
Utility perspective
  hybrid systems with fuel cell, 259
  wind power, 7, 9–13
    consumer choice, 147
    emission benefits, **5**
    modularity for growth, 7
Utility resource planning tools, grid-connected systems, 7, 286–287

V

Vane, terminology, 37, 38, 44
Variable-speed control
  comparison with fixed-speed systems, 49
  power *versus* speed and TSR, 47–49
Variable-speed drive, 117
Variable-speed generator drives
  direct drive, 67, 69, 83–86
  power electronics, 9, 44, 45, 49, 50, 55, 56, 85, 99, 126
  variable-speed drive, 117
  variable-speed gear drives, 50, 98
Variable-speed induction generator, synchronizing with grid, 278–282
Vertical-axis turbine, 15, *16*, 19, 55
Vertical axis, wind system design tradeoffs, 15, *16*, 19, 23, 55
Victim load, 303
Voltage control
  deadbands in, 265, *266*, 299, *300*
  distributed power generation, 146, 262, 290–292
  grid-connected systems
    energy storage for stability, 289–290
    interface requirements, 276–278
    low-voltage ridethrough, 288–289
    synchronizing with grid, 278–282
    grid interface controls, 244–246
Voltage current and power relationships, electrical performance, 295–296

Voltage flickers, 303, 305–307, 314
Voltage, ripple, 87–89, 132, 173, 178, 180, 181, 233, 300–301, 351
Voltage source converter (VSC) technology, 66, 117–120, **117**, *119*, **122**, 125–128, 135, **136**, 236, *236*, 239, *240*, 241–243, *241*
Voltage transients and sags, 304–305

W

Walney wind farm, 115, *116*, **120**, 133–135, **136**
Water-pumped storage, grid support, 222, **290**
Wave energy and power, offshore wind farms, 105–106
Wear-out failure, battery performance characteristics, 203
Weibull probability distribution, wind speed and energy, 21–23
Wind energy
  international associations, 332, 338
  manufacturers and developers in Europe, 46, 146
  organizations, societies, and associations, 110
  periodicals/publications, *277*, 320
  suppliers in U.S., **44**, 50
  university programs in U.S., 5, 6, 9, 103, 146, 155, 254, 256
Wind farms
  grid-connected systems, integration with grid, 287–288
  spacing of towers, 39–40, 52–53, 316
  stand-alone system sizing, 172–174, 253–273
Wind farms, offshore, 101–113
  corrosion, 108
  costs, offshore, 101, 107, *108*, 110, 111
  environmental impact, 102–103
  forces on ocean structures, 107–108
  foundation, 108–109
  future of, 331–334
  gravitation, 110
  maintenance, 112–113
  materials, 111–112
  monopile, 109
  ocean structure design, 106–108
  ocean water composition, 103–104
  projects, 101, 102, 108
  transmission of power to shore, 107
    AC cable, 115, 134, 135
    DC cable, 123, 134
  tripod, 110
  U.S. legal aspects in, 101
  wave energy and power, 105–106
Wind power
  future prospects and trends, 55, 60–62
  geographic distribution
    Europe and U.K., 5, 7, 11, 12, 16, 23, 38, 41, 43, 58, 59

Index

global perspective, 3, 9, 11, 45
India, 5, 11, 12
U.S., 9, 12–13
incentives for renewables, 9, 12
industry overview, 3–4
utility perspective, 7
 consumer choice, 7
 emission benefits, **5**, 7
 modularity for growth, 7
Wind power equation, 17–20, 22, 26–28, 30, 40, 57, 72, 76, 77, 79, 88, 89, 91, 133, 134
Wind power systems, 37–62
 ancillary
  assisted wind turbine, 362–363
  heat-induced wind power, 357
 components, 37–45
  blades, 42–44
  speed control, 44–45
  tower, 39–41
  turbine, 41–42
 control requirements, 55–57
  rate, 56–57
  speed, 55–56
 design trade-offs, 52–55
  blade number, 54
  horizontal *versus* vertical axis, 55
  rotor upwind or downwind, 54–55
  turbine towers and spacing, 52–53
 design trends, 61–62
 economics
  sensitivity analysis, 323–326
   tower height effect, 324–326
   wind speed effect, 324, *see* Economics
 electrical generator, *see* Generators, wind systems
 electronics, 9, 37, 44, 49, 50, 55, 56, 63, 65, 66, 83, 85, 87, 89, 99, 120, 123, 125, 126, 128, 130, 133, 166, 173, 175, 177, 215, 217, 223, 229–251
  DC-AC inverter, 233–235, 256
 energy storage for stability, 189–226
 environmental aspects
  birds, 59
  electromagnetic interference (EMI), 57–60
  noise, 57–58
 grid-connected systems, 275–292
 hybrid systems, 143, 253, 257–267
  with diesel, 143, 257–267
  lightning protection, 316–317
  load sharing, 203, 265–267
 interface requirements, 276–278
 maximum energy capture, 49–50
 maximum power operation, 50–52

constant-TSR scheme, 50–51
peak-power-tracking system, 51–52
 potential catastrophes, 60–61
  earthquake, 61
  fire, 60–61
 power quality, system stiffness, 312–314
 power *versus* speed and TSR, 47–49
 stand-alone, 73, 78, 143, 172–174, 189, 246, 249, 253–273, 327, 345, 350, 362
 switching devices, 180, 229–232, 244, 309
 terminology, 37, 104
 turbine rating, 45–47
Wind speed and energy, 15–34
 air density, 15, 17, 20, 31
 global wind patterns, 31, 32
 power extracted from wind, 17–19, 48, *66*, 167
 rotor-swept area, 17, 19–20, 46, 324
 speed and power relations, 15–17
 turbine power *versus* rotor speed characteristics, 47, *48*, 49, *92*
 turbine rating, 45–47
 wind speed distribution, 21–34
  data reliability, 22, 33
  digital data processing, 29–31
  energy distribution, 28–29
  hub height, effect of, 31–32
  mode and mean speeds, 24–26
  mode, mean, and RMC speeds, 27–28
  root mean cube speed, 26
  Weibull probability distribution, 21–24
 wind speed prediction, 34
Wind speed effect, economics, sensitivity analysis, 324
Wind turbines, 12, 15, 31, 39, *42*, 44, 50, 52–54, 58–60, 63, 65, 69, 82, 83, 101, 103, 109, 110, 112, 115, 125, 126, 135, 235, 272, 273, 276, 288, 289, 313, 321, 335, 336, 357, 361–363
World electricity demand to 2050, 329–331
Wound field (WF), 130, 136
Wound rotor synchronous machines, 84

Y

Yaw and tilt control, 44, 45
Yaw mechanism
 components of systems, 37–38
 terminology, 37

Z

Zinc-air batteries, storage systems, 191, **191**, 193, **204**, 256

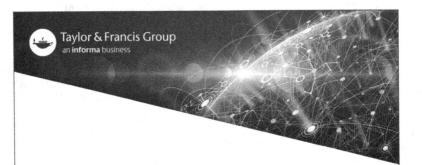

Printed in the United States
By Bookmasters